CONTENTS

One   Introduction   1
Geographical setting   1
Geological history   1
History of research   3

Two   Devonian and Carboniferous rocks   5
Upper Devonian and Lower Carboniferous rocks   7
  Hyner Shale and Trusham Shale (undifferentiated)   7
  Combe Shale   7
  Teign Chert   8
Upper Carboniferous rocks   10
  Crackington Formation   10
    Ashton Shale Member   11
    Crackington Formation (undivided part)   12
    Biostratigraphy   13
  Bude Formation   22

Three   Lower Carboniferous igneous rocks and associated metamorphism   24
Dolerites   24
  Petrography   25
  Geochemistry   25
  Contact metamorphism and metasomatism   26
Spilitic basalt   27
  Petrography   27
  Geochemistry   28
Tuffs   28
  Geochemistry   28

Four   Permian and Triassic rocks (New Red Sandstone): introduction   29
Biostratigraphy   31
Correlation   32
Summary of depositional environments   34

Five   Permian rocks: Exeter Group   35
Introduction   35
Stratigraphy   40
  Exeter area   40
    Knowle Sandstone   40
    Whipton Formation   40
    Alphington Breccia   45
    Heavitree Breccia   47
    Monkerton Formation   50
    Dawlish Sandstone   51
  Crediton Trough   59
    Upton Formation   59
    Cadbury Breccia   59
    Higher Comberoy Formation   62
    Bow Breccia   62
    Knowle Sandstone   63
    Thorverton Sandstone   65
    Creedy Park Sandstone   67
    Crediton Breccia   68
    Yellowford Formation   70

    Newton St Cyres Breccia   72
    Shute Sandstone   74
Summary of the sedimentary history of the Exeter Group   75

Six   Triassic rocks: Aylesbeare Mudstone Group and Sherwood Sandstone Group   77
Aylesbeare Mudstone Group   77
  Lithology and sedimentary features   77
    Depositional environments   79
    Nodules and reduction features   79
    Geochemistry of mudstones   83
  Southern area   83
    Exmouth Mudstone and Sandstone   83
    Littleham Mudstone   84
  Northern area   85
    Clyst St Lawrence Formation   85
    Aylesbeare Mudstone Group (where undivided)   86
Sherwood Sandstone Group   87
  Budleigh Salterton Pebble Beds   87
    Depositional environments   91
  Otter Sandstone   92
    Depositional environments   93

Seven   Permian igneous rocks   96
Exeter Volcanic Rocks   96
  Geochemistry and petrogenesis   96
  Magnetic anomalies   97
  Lamprophyres   102
  Basalts   104
  Agglomerate   111
  Rhyolite   111
Igneous clasts in the Permian breccias   111
Dartmoor Granite   113
  Lithology and petrology   113
  Igneous enclaves   114
  Megacrystic layering   115
  Geochemistry   116
Metamorphic aureole of the Dartmoor Granite   116
  Argillaceous hornfels   117
  Meta-cherts   117
  Metadolerites   117
Bridford Lamprophyre   118

Eight   Cretaceous and Palaeogene rocks   119
Cretaceous: Upper Greensand   119
Palaeogene: Tower Wood Gravel   119

Nine   Structure   120
Deep basement   120
Magnetic anomalies associated with Carboniferous rocks   120
Form of the Crediton Trough   120
Form of the Permo-Triassic rocks outside the Crediton Trough   124

Structure of the Devonian and Carboniferous rocks   125
    Folds in Lower Carboniferous strata   125
    Folds in Upper Carboniferous strata   125
    Faults in Lower Carboniferous strata   126
    Faults in Upper Carboniferous strata   128
    Faults in the Dartmoor Granite   128
Structure of the Permo-Triassic rocks   128
    Faults   128
    Folds   129

Ten   Quaternary   130
Introduction   130
Head   132
    Blanket head and regolith   132
    Older head   133
    Valley head   133
Periglacial structures   134
River terrace deposits   134
Buried channel deposits   137
Marine deposits   137
Alluvium   137
Alluvial fan deposits   138
Peat   138
Made ground and worked ground   138
Landslip   139

Eleven   Economic geology   141
Soils and land use   141
    Soils on Carboniferous and igneous rocks   141
    Soils on the Permo-Triassic rocks   141
    Soils on drift deposits   141
    Land use   142
Construction materials   142
    Sand and gravel   142
    Roadstone   143
    Building stone   143
    Brick clay   146
    Marl and lime   146
Metalliferous minerals   146
    Baryte   146
    Lead, zinc and copper   147
        Newton St Cyres area   147
        Teign Valley   147
    Manganese   147
        Crediton Trough   147
        Teign Valley   149
    Vanadium and uranium   149
Engineering geology   150
    Engineering geology classification   150
        Strong rocks   150
        Interbedded strong rocks and mudstones   151
        Mudstones, shales and clays   151
        Granular rocks and deposits   154
Hydrogeology and water supply   157
    Carboniferous rocks   158
    Exeter Group: below the Dawlish Sandstone   158
    Igneous rocks   159
    Exeter Group: Dawlish Sandstone   159
    Aylesbeare Mudstone Group   159
    Budleigh Salterton Pebble Beds and Otter Sandstone   159

Quaternary (river terrace deposits)   160
    Waste disposal at landfill sites   160

References   161

Appendices   169
1   Major- and trace-element analyses of the Exeter Volcanic Rocks of the district   169
2   Geological Survey photographs   171

Index of fossils   173

General index   174

FIGURES

1   Sketch map illustrating the physical features of the district   xii
2   Sketch map of the simplified geology of the district   2
3   Distribution of Devonian and Carboniferous rocks in the district   6
4   (a) Histogram of bed thickness of sandstones in the Crackington Formation   13
    (b) Current rose diagram for the Crackington Formation   13
5   Distribution of igneous rocks in the Lower Carboniferous of the Exeter district and part of the Newton Abbot district   24
6   Distribution of the Exeter Group formations of the district   36
7   Correlation of Exeter Group sequences in the district   37
8   Diagram of stratigraphical relationships of the Exeter Group in the Crediton Trough   38
9   Plots of ratios of selected elements in the fine-grained fractions of samples from the Exeter Group of the district   39
10   Cross-section through boreholes in the central area of Exeter city to illustrate lithological variation in the Whipton Formation   41
11   Cross-section through Whipton Formation and Quaternary deposits in the Shilhay area of Exeter   42
12   Representative graphical sedimentary log and interpretation of the Heavitree Breccia in the M5 Motorway cuttings near Exminster   47
13   Representative graphical sedimentary log and interpretation of the Dawlish Sandstone at Bishop's Court Quarry   53
14   Geometry of aeolian dune and interdune facies in the Dawlish Sandstone, Bishop's Court Quarry   54
15   Graphic logs of the Dawlish Sandstone in boreholes in the Brampford Speke area   55
16   Graphic logs of water boreholes in the Dawlish Sandstone of the Bussell's Farm (Rewe) and Columbjohn areas   57
17   Representative graphical sedimentary log and interpretation of the Dawlish Sandstone at Brampford Speke   58
18   Representative graphical sedimentary logs and interpretation of the Bow Breccia and the Knowle Sandstone   64

Geology of the country around Exeter

This memoir incorporates general and detailed accounts of all the geological formations in the Exeter district, based on 1:10 000 scale mapping. Used in conjunction with the 1: 50 000 series Sheet 325 Exeter, it will prove of value and interest to many users, including planners and engineers, and those in the bulk mineral and water supply industries, as well as professional and amateur geologists and those with a general interest in the countryside.

The mainly rural district is centred on the cathedral city of Exeter, and encompasses a wide variety of landscapes reflecting the underlying geology. The Dartmoor Granite forms a small area of moorland in the south-west, and is flanked by attractive country underlain by the oldest sedimentary rocks in the district, folded cherts and shales with dolerite sills, of Upper Devonian and Lower Carboniferous age. Folded shales and sandstones of Upper Carboniferous age give rise to hilly ground mainly west of Exeter, but also in the Ashclyst Forest and north of Crediton.

The Carboniferous rocks are overlain by the relatively undeformed New Red Sandstone, of Permian and Triassic age. The Permian rocks fill a fault-controlled basin — the Crediton Trough — in the north of the district, and also form mainly low-lying ground around Exeter. Mudstones of presumed Triassic age (Aylesbeare Group) form lowland in the eastern third of the district, bounded to the east by the prominent scarp of the Triassic Budleigh Salterton Pebble Beds.

Microscopic plant fossils of Late Permian age (about 260 million years old) discovered during the survey, have given the first palaeontological confirmation of the presence of Permian rocks in Devon. Older New Red Sandstone rocks contain lavas dated radiometrically at about 291–282 million years, and there is an unconformity within the New Red Sandstone between this Early Permian sequence and the Late Permian sequence, representing about 22 million years. The gap in deposition was probably caused by uplift and regional erosion associated with the intrusion of the Dartmoor Granite about 280 million years ago. There is a full account of the interesting volcanic rocks associated with the New Red Sandstone.

The memoir also describes the structure of the district, the small outcrops of Cretaceous and Palaeogene rocks near its southern boundary, and the Quaternary (drift) deposits. The economic geology chapter includes a description of the engineering geology, together with an account of the hydrogeology and water supply, and details of the former metalliferous mining activity.

Cover photograph

St Peter's Cathedral, Exeter, and part of the Bishop's Palace, viewed from the Bishop's Palace gardens. The cathedral is built mainly of Salcombe Stone, a sandstone from the Upper Greensand (Cretaceous) of east Devon. The Bishop's Palace is constructed of reddish brown Heavitree Breccia (Late Permian) and purple lava (Early Permian). (Photographer: P Thomas)

Heavitree Breccia in cuttings [SX 932 881] for the M5 Motorway near Exminster.
(Photographer: Dr J C Davey)

BRITISH GEOLOGICAL SURVEY

R A EDWARDS and
R C SCRIVENER

Geology of the country around Exeter

Memoir for 1:50 000 Geological Sheet 325
(England and Wales)

CONTRIBUTORS

Stratigraphy
C M Barton
C R Bristow
B J Williams
A J J Goode

Geophysics
J D Cornwell
C P Royles
S J Self

Biostratigraphy
G Warrington
N J Riley
N Turner

Sedimentology
N S Jones
S A Smith

Hydrogeology
R J Marks

Petrology and igneous geochemistry
N J Fortey
R D Morton
G E Strong
S J Kemp

Engineering geology
A Forster

Sedimentary geochemistry
L Ault
H W Haslam

Geochronology
J Chesley

London: The Stationery Office   1999

The grid used on the figures is the National Grid taken from the Ordnance Survey map. Figure 11 is based on material from Ordnance Survey 1:50 000 scale maps numbers 191 and 192.
© Crown copyright reserved
Ordnance Survey licence No. GD272191/1999.

ISBN 0 11 884527 6

Bibliographical reference

EDWARDS, R A, and SCRIVENER, R C. 1999. Geology of the country around Exeter. *Memoir of the British Geological Survey*, Sheet 325 (England and Wales).

Authors

R A Edwards, BSc, PhD
R C Scrivener, BSc, PhD
British Geological Survey, Exeter

Contributors

C M Barton, BSc, PhD
C R Bristow, BSc, PhD
A J J Goode, BSc
British Geological Survey, Exeter

J D Cornwell, BSc, PhD
C P Royles
S J Self
G Warrington, DSc
N J Riley, BSc, PhD
N Turner, BSc, PhD
N S Jones, BSc, PhD
N J Fortey, BSc, PhD
G E Strong, BSc
S J Kemp, BSc
A Forster, BSc
L Ault, BSc
H W Haslam, BSc, PhD
British Geological Survey, Keyworth

R J Marks, BSc   *British Geological Survey, Wallingford*

B J Williams, BSc
S A Smith, BSc, PhD
formerly British Geological Survey

R D Morton, BSc, PhD   *University of Alberta*

J Chesley, BSc, PhD   *University of Arizona*

Printed in the UK for the Stationery Office
TJ000849   C6   12/99

Other publications of the Survey dealing with this district and adjoining districts

BOOKS
British Regional Geology
South-west England, 4th edition 1975
Memoirs
Okehampton (324), 1968
Chulmleigh (309), 1979
Wellington and Chard (311), 1906*
Sidmouth and Lyme Regis (326), 1911*
Newton Abbot (339), 1984
Dartmoor (338), 1912*

MAPS
1:584 000
Tectonic map of Great Britain and Northern Ireland
1:625 000
South sheet (Geological)
South sheet (Quaternary)
South sheet (Aeromagnetic)
1:250 000
Portland (Solid geology)
Portland (Sea bed sediments)
Portland (Gravity)
Portland (Aeromagnetic)
1:100 000
Hydrogeological map of the Permo-Trias and other minor aquifers of South West England
1:50 000 and 1:63 360
Sheet 324 (Okehampton) (1969)
Sheet 309 (Chulmleigh) (1980)
Sheet 310 (Tiverton) (1969) (provisional)
Sheet 311 (Wellington) (1976)
Sheet 326/340 (Sidmouth) (1974)
Sheet 339 (Newton Abbot) (1976)
Sheet 338 (Dartmoor Forest) (1995) (provisional)

* Out of print

19 Representative graphical sedimentary logs and interpretation of the Crediton Breccia at Chapel Downs and near Raddon Cross 69

20 Graphic logs of boreholes in the Yellowford Formation 71

21 Representative graphical sedimentary logs and interpretation of the Newton St Cyres Breccia at 'Cromwell's Cutting', Crediton, and Newton St Cyres 73

22 Distribution of the Aylesbeare Mudstone Group and the Sherwood Sandstone Group in the district 78

23 Graphic sedimentary log of the Venn Ottery Borehole 80–83

24 (a) Graphic sedimentary log of the Budleigh Salterton Pebble Beds and the Otter Sandstone at Beggars Roost Quarry (b) Detail of an exposure of the contact between the Budleigh Salterton Pebble Beds and the Otter Sandstone at Beggars Roost Quarry 90

25 Distribution of the Exeter Volcanic Rocks and the Dartmoor Granite in the district 97

26 Residual aeromagnetic anomaly map of the Crediton Trough area 99

27 Residual aeromagnetic anomaly map of the area around Exeter and the eastern Crediton Trough 100

28 Magnetic data for the Killerton Park area 101

29 Magnetic data for the Burrow Farm area, Broadclyst 102

30 Aeromagnetic map of the district and adjacent areas 121

31 (a) Bouguer gravity anomaly map of the district and adjacent areas (b) and (c) Residual gravity anomaly profiles 122

32 Alternative models for the form of the Crediton Trough, based on three-dimensional interpretation of residual Bouguer gravity anomalies 123

33 Simplified geological map of the Dunsford–Bridford–Doddiscombsleigh area showing locations and names of the major folds and faults 124

34 (a) and (b): stereograms of poles to bedding planes in the Crackington Formation 125
    (c) fold types in the Crackington Formation 125
    (d) variation in fold type and dip of axial planes 125

35 Section at Bonhay Road, Exeter, showing structure of the Crackington Formation 126

36 Distribution of faults in the district 127

37 Distribution of peat, alluvium, river terraces and marine deposits in the district 131

38 Classification of landslip types on the Crackington Formation of the district 139

39 Engineering geology map of the district 152–153

TABLES

1 Classification of the Devonian and Carboniferous rocks of the district 5

2 Ammonoid zones in the Crackington Formation of the district 17

3 Carboniferous fossils from the district 18

4 Analyses of the Scanniclift Copse Spilite and associated tuffs, and the Ryecroft Dolerite 25

5 Analyses of the Ryecroft Dolerite, Scanniclift Copse Spilite, Christendown Clump Tuff and the Dartmoor Granite 25

6 Evolution of the nomenclature of the New Red Sandstone of the Crediton Trough 29

7 Evolution of the nomenclature of the New Red Sandstone of the district outside the Crediton Trough 30

8 New Red Sandstone stratigraphy of the district 31

9 Palynological study of the New Red Sandstone 32

10 Variscan and post-Variscan events in south-west England in relation to the Exeter Group of the district and contiguous areas 33

11 Succession of clast types in the Exeter Group of the district 35

12 Composition of clasts in the Budleigh Salterton Pebble Beds of the Exeter and adjacent districts 89

13 Radiometric ages for igneous rocks of the district 98

14 Major- and minor-element analyses of acid igneous rocks from the district and the adjacent Dartmoor Granite 112

15 Samarium/neodymium (Sm/Nd) isotope data for acid igneous rocks, including clasts from the Exeter Group of the Crediton Trough 113

16 Summary of the feldspar megacryst data from the Dartmoor Granite 114

17 Modal analyses of the Dartmoor Granite 114

18 Stages of the British Quaternary 130

19 Landfill waste disposal sites in the district 139

20 Licensed abstraction data of the Environment Agency 158

PLATES

Frontispiece Heavitree Breccia in M5 Motorway cuttings near Exminster

1 (a) Teign Chert at Mistleigh Copse, Doddiscombsleigh 14
   (b) Crackington Formation at Pinhoe Brickpit, Exeter 14
   (c) Sole markings on Crackington Formation sandstone at Pinhoe Brickpit, Exeter 15
   (d) Bude Formation at East Henstill Quarry 15

2 Fossils from the Crackington Formation 16

3 Late Permian fossils, Exeter Group 43

4 (a) Scarp feature of the Heavitree Breccia near Shillingford Abbot 48
   (b) Thickly bedded Heavitree Breccia in M5 Motorway cuttings near Exminster 48
   (c) Detail of bedding in Heavitree Breccia in A30 Link Road cuttings near Exminster 49
   (d) Sand-filled pipes in Heavitree Breccia at Heavitree Quarry, Exeter 49

5 (a) Coarse Alphington Breccia at The Quay, Exeter 51
   (b) Dawlish Sandstone aeolian dune and interdune deposits at Lake Bridge, Brampford Speke 51
   (c) Aeolian cross-bedding in the Dawlish Sandstone at Clyst St Mary 52

6 (a) Cadbury Breccia at Raddon Hill Farm   60
   (b) Crediton Breccia at Forches Cross, Crediton   60
7 (a) Horizontally bedded gravel in the Budleigh
        Salterton Pebble Beds at Beggars Roost Quarry   88
   (b) Ventifact-bearing bed at the top of the Budleigh
        Salterton Pebble Beds, Beggars Roost Quarry   88
   (c) Otter Sandstone overlying Budleigh Salterton
        Pebble Beds at Beggars Roost Quarry   89
8 (a) Feature of the Budleigh Salterton Pebble Beds on
        Littleham Mudstone at Bicton Common   93
   (b) Otter Sandstone at Stoneyford, Hawkerland   93
   (c) Basal Otter Sandstone at Beggars Roost
        Quarry   94
9 Photomicrographs of Exeter Volcanic Rocks   103
10 (a) Basalt at School Wood Quarry, Dunchideock   105
   (b) Basalt at Pocombe Quarry, Exeter   106
   (c) Base of lava flow at School Wood Quarry,
        Dunchideock   106

11 (a) Heltor Rock, Dartmoor   115
   (b) Megacrysts in the Dartmoor Granite at
        Burnicombe Down   115
12 Blanket head and regolith at Springdale,
   Whitestone   132
13 Typical features of active landslips on the
   Crackington Formation west of Exeter   140
14 (a) Heavitree Breccia and lava building stone in
        Exeter city wall, Western Way, Exeter   143
   (b) 'Pocombe Stone' in a wall at The Quay,
        Exeter   144
   (c) 'Killerton Stone' in Broadclyst Church   144
   (d) Pebbles from the Budleigh Salterton Pebble Beds
        in a wall at Woodbury   145
   (e) Manganiferous Teign Chert at Bridford Barytes
        Mine Quarry   145

PREFACE

This memoir provides a detailed account of the geology of the district covered by the 1:50 000 series Sheet 325 Exeter, published in 1995, and is best read in conjunction with the map. The district is centred on the cathedral city of Exeter, the administrative capital of Devon; outside the city and its surrounding suburbs, trading and industrial estates, the district is predominantly agricultural.

Until the recent remapping there had been no detailed survey since that on the one-inch scale, carried out before 1887. The survey of Exeter city and the surrounding areas was supported by the Department of the Environment in recognition of the particular need there for accurate up to date geological maps to aid land use planning and to identify bulk mineral resources. A full set of 1:10 000-scale geological maps is now available for use in conjunction with this memoir and with the Technical Reports produced during the course of the survey. These publications provide basic geological information which is of value to a variety of users. For example, they can act as a guide to developers in their initial assessment of the likely ground conditions at particular sites and, with this use in mind, the memoir contains an account of the engineering geology of the formations within the district, together with an engineering geology map. Planning authorities will also find the maps and memoir of value in arriving at decisions about land use planning. The results of the survey are also of use in resource assessment by the bulk minerals and the water supply industries. I am confident that both professional and amateur geologists, and those interested in the countryside, will find much of scientific interest in the pages of the memoir.

The district is underlain mainly by two major divisions of sedimentary rocks, which give rise to contrasting areas of landscape and land use. The first comprises strongly folded rocks of mainly Carboniferous age, the 'Culm Measures', laid down in an ancient sea; and the second consists of gently dipping undeformed red rocks of Permian and Triassic age, the 'New Red Sandstone', formed in desert conditions.

The survey of the district has combined classical geological mapping with the latest specialist techniques. This fusion of disciplines has enabled a substantial revision and fresh understanding of the stratigraphy, age, structure and sedimentology of the New Red Sandstone and older formations. The Exeter area has a particular significance for me, as Director, in that it was one of the first areas mapped by the founding Director of the Survey, Sir Henry De la Beche. This memoir shows how much we have learned since that early work. At the same time, it also graphically illustrates the continuing value of the observations made by De la Beche and his colleagues as long as 160 years ago. The fact that some of the data originally obtained by them have been incorporated into this memoir and the accompanying map is testimony to the quality of these early observations.

David A Falvey, PhD
Director

British Geological Survey
Kingsley Dunham Centre
Keyworth, Nottingham
NG12 5GG

HISTORY OF THE SURVEY OF THE EXETER SHEET

The original geological survey of the Exeter district was carried out on the one-inch scale by Sir Henry T De la Beche, and published as part of his geological map of Devon in 1835. Subsequently, after the formation of the Geological Survey, this mapping was incorporated in the old series one-inch geological Sheets 21, 22, 25 and 26, before 1845. His 'Report on the Geology of Cornwall, Devon and West Somerset' (1839) was the first memoir of the Geological Survey. A one-inch resurvey of the area was carried out by W A E Ussher between 1873 and 1887, and the results were published as the new series one-inch geological Sheet 325 (Exeter) in 1899, with an explanatory memoir by Ussher published in 1902.

Small areas along the western side of the district were mapped in 1962–64 by Dr M Williams and Mr J E Wright as part of the overlap from the survey of the Okehampton Sheet (324). Dr E C Freshney mapped a corridor approximately 1 km wide along the route of the proposed A30 Trunk Road in the Tedburn St Mary area in 1969.

The survey on the 1:10 000 scale began in 1982–83 with the mapping of a block of nine sheets, totalling 225 km², around and including the city of Exeter, as part of a programme commissioned by the Department of the Environment (DoE); the surveyors were Dr C R Bristow, Dr R A Edwards, Dr R C Scrivener, and Mr B J Williams. An area of 20 km² around Thorverton was mapped in 1984 by Dr Edwards as an extension to the main DoE contract. From 1985, the survey of the sheet continued using Science Budget funds until its completion in 1991. This phase of the survey was carried out mainly by Dr Edwards and Dr Scrivener, with parts surveyed by Dr C M Barton and Mr A J J Goode. The survey was carried out under the direction of Dr R W Gallois and Dr P M Allen, Regional Geologists, and Mr B J Williams, Acting Regional Geologist.

The component 1:10 000 scale National Grid sheets of Geological Sheet 325 are shown on the diagram below, together with the dates of survey and the initials of the surveyors. The surveying officers were C R Bristow, R A Edwards, E C Freshney, A J J Goode, R C Scrivener, B J Williams, M Williams and J E Wright. Uncoloured dye-line copies of the maps are available for purchase from the British Geological Survey, Keyworth, Nottingham NG12 5GG.

Technical reports, which are available or in preparation for most of the Exeter district, are shown on the diagram below, together with the dates of publication and the authors' names. They contain more detail than appears in the memoir, but may also contain interpretations and stratigraphical terminology which have been superseded by later work. In addition to the reports indicated on the diagram, a report entitled 'Geology of Exeter and its environs', summarising the geology of nine 1:10 000 sheets around Exeter (sheets SX 99 NW, NE, SW, SE; SX 98 NW, NE; SY 09 NW, SW; SY 08 NW), is available. The engineering geology of the district is described in a separate report. Copies of these reports are available for purchase from the British Geological Survey, Keyworth, Nottingham NG12 5GG and from 30 Pennsylvania Road, Exeter, Devon EX4 6BX.

NE MW & JEW 1962 & 1964	NW RCS 1990	NE RCS 1990	NW RAE 1990	NE RAE 1990	NW RAE 1990	NE RAE 1991
SS 70 / SS 80 / SS 90 / ST 00						
SE RCS and MW 1962 and 1991	SW RCS 1988 and 1989	SE RCS and RAE 1987 and 1984	SW RAE 1984 and 1985	SE RAE 1985 and 1986	SW RAE 1986 and 1987	SE RAE 1987 and 1988
NE MW, JEW, RCS and RAE 1962, 1963 & 1991	NW RCS 1989 and 1990	NE RCS 1986	NW RCS 1982	NE CRB 1982	NW CRB 1983	NE RAE and CRB 1987, 1988 & 1983
SX 79 / SX 89 / SX 99 / SY 09						
SE RAE, RCS, ECF, JEW 1991, 1969 & 1962	SW RAE, RCS and ECF 1991 and 1969	SE RAE and AJJG 1987 and 1991	SW RCS 1983	SE CRB and BJW 1982 and 1983	SW RAE 1983	SE RAE 1988 & 1989
NE CMB and JEW 1962 and 1990	NW CMB 1990 and 1991	NE RAE 1990	NW CRB 1983	NE BJW 1982	NW RAE 1982 and 1983	NE RAE 1989 and 1991
SX 78 / SX 88 / SX 98 / SY 08						

Component 1:10 000 sheets for the district with initials of the surveyors.

SS 80 SE (SCRIVENER and EDWARDS, 1991)	SS 90 SW (EDWARDS, 1987a)	SS 90 SE (EDWARDS, 1987b)	ST 00 SW (EDWARDS, 1988)	ST 00 SE (EDWARDS, 1989a)
SX 89 NE (SCRIVENER 1988)	SX 99 NW (SCRIVENER, 1983)	SX 99 NE (BRISTOW, 1983)	SY 09 NW (BRISTOW, 1984b)	SY 09 NE (EDWARDS, 1989b)
parts of SX 89 SE and SW, and part of SX 79 SE (EDWARDS, 1997)	SX 99 SW (SCRIVENER, 1984)	SX 99 SE (BRISTOW and, WILLIAMS, 1984)	SY 09 SW (EDWARDS, 1984b)	SY 09 SE (EDWARDS, 1990)
SX 88 NW and part of SX 78 NE (BARTON, 1992)	SX 88 NE (EDWARDS, 1991)	SX 98 NW (BRISTOW, 1984a)	SX 98 NE (WILLIAMS, 1983)	SY 08 NW (EDWARDS, 1984c) SY 08 NE (EDWARDS, 1996)

Technical reports of the district.

ACKNOWLEDGEMENTS

This memoir has been largely written and compiled from the Technical Reports and notes of the surveyors — Dr R A Edwards, Dr R C Scrivener, Dr C R Bristow, Dr C M Barton, Mr B J Williams and Mr A J J Goode. In addition, it includes contributions, identified below, from many specialists. Dr N J Riley identified the Carboniferous macrofossils. Dr N Turner examined a suite of samples (collected by Dr Riley) from the Crackington and Bude formations for palynomorphs. Dr H W Haslam examined the geochemistry of the Crackington and Bude formation sandstones. Dr P Grainger of the Earth Resources Centre, Exeter University, provided information on the geology and distribution of landslips in an area west of Exeter. He has also kindly given permission to incorporate figures and plates from his PhD thesis into the memoir (Plate 13; Figures 35 and 38).

The chapter on the Lower Carboniferous igneous rocks and associated metamorphism and the Upper Carboniferous chapter include petrographic data from an unpublished PhD thesis by Professor R D Morton of the University of Alberta.

Dr G Warrington identified miospores in the New Red Sandstone and contributed an account of the biostratigraphy of the New Red Sandstone. Sedimentological work on the Exeter Group and Aylesbeare Group by Dr N S Jones, and on the Sherwood Sandstone Group by Dr S A Smith (formerly with the British Geological Survey) have been incorporated in the relevant chapters. Miss L Ault and Dr Haslam contributed information on the geochemistry of the New Red Sandstone rocks. Dr J Chesley of the University of Arizona carried out $^{40}Ar/^{39}Ar$ analyses on samples from the Exeter Volcanic Rocks and clasts from the Permian breccias. Mrs D P F Darbyshire, NERC Isotope Geology Laboratory, has provided Sm/Nd and Rb/Sr isotope data on the acid igneous rocks and the Exeter Volcanic Rocks. Information from an unpublished PhD thesis by Dr J C Davey is included, with his kind permission, in the account of the Dawlish Sandstone in Chapter 5.

The chapter on the Permian igneous rocks includes an account of the petrology, mineralogy and geochemistry of the Exeter Volcanic Rocks extracted from Technical Reports by Dr N J Fortey. The account of the Dartmoor Granite and its metamorphic aureole, and the description of the Bridford Lamprophyre, were largely contributed by Dr Barton, with the addition of some data supplied by Dr J R Hawkes and Mr J Dangerfield, formerly of the British Geological Survey. Mr S J Kemp carried out X-ray diffraction analysis of samples from the Exeter Volcanic Rocks. The description of the magnetic anomalies associated with the Exeter Volcanic Rocks was extracted from a Technical Report by Dr J D Cornwell, Mr C P Royles, and Mrs S J Self. Mr G E Strong and Dr N J Fortey provided petrographic descriptions of thin sections from various formations in the district.

The account of the structure and geophysics includes extracts from a Technical Report on the form and thickness of the New Red Sandstone rocks as deduced from Bouguer gravity data, by Dr Cornwell, Mr Royles, and Mrs Self. Dr R K Westhead and Dr E C Freshney advised on the interpretation of structural data.

The economic geology chapter includes a contribution on the engineering geology of the district by Mr A Forster, and Mr R J Marks contributed an account of the hydrogeology and water supply. Dr T J Shepherd has provided microthermometric determinations on fluid inclusions, and Dr R Clayton, University of Exeter, carried out chemical and mineralogical analyses of manganese oxide and carbonates.

The survey of 245 km^2 of the district, mainly around the city of Exeter, was supported by the Department of the Environment.

The collaboration and assistance of various landowners throughout the district is gratefully acknowledged. Dr C Nicholas of ECC Quarries Ltd is thanked for arranging access to their quarries, and for valuable discussions. Dr C Tubb and his staff of South West Water were similarly most helpful.

The photographs were taken mainly by Mr T P Cullen, with some earlier photographs by Mr J Rhodes, Mr J M Pulsford, Mr C J Jeffery and Mr H J Evans; a complete list of available photographs is given in Appendix 2.

The memoir was compiled by Dr Edwards and edited by Dr R W Gallois and Mr J I Chisholm.

NOTES

The word 'district' used in this memoir means the area included in 1:50 000 scale Geological Sheet 325 (Exeter).

Figures in square brackets are National Grid references; places within the Exeter district lie within the 100 km squares SS, ST, SX and SY. The grid letters precede the grid numbers.

Dips are given in the form '50° to 287°'. The first number is the angle of dip in degrees and the second is its full-circle bearing measured clockwise from True North.

The authorship of fossil species is given in the index of fossils.

Numbers preceded by A refer to photographs in the Geological Survey collections.

Numbers preceded by E refer to specimens in the English Sliced Rock collection of the British Geological Survey.

The grain-size scale used in this memoir is the Wentworth (1922) grade scale, modified to the phi (ø) scale by Krumbein (1934). The classification of sorting and skewness used is that of Folk and Ward (1957).

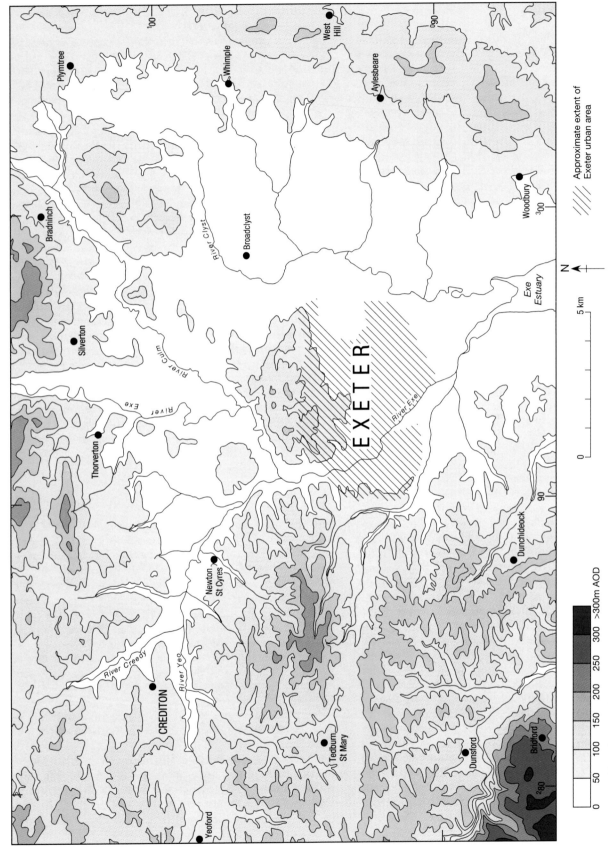

Figure 1 Sketch map illustrating the physical features of the district.

ONE

Introduction

GEOGRAPHICAL SETTING

The district lies in south Devon (Figure 1). The River Exe and its tributaries, the rivers Culm, Creedy, Yeo and Clyst, drain into the Exe estuary; small areas in the south-west and south-east of the district lie within the catchments of the Rivers Teign and Otter respectively. The main settlement is the cathedral city of Exeter (population c. 100 000) which is centred on the former Roman town of Isca; it is the administrative capital of Devon. Outside the old city and its surrounding suburbs and light industrial estates, the district is predominantly agricultural.

The older sedimentary rocks can be divided into two groups which give rise to contrasting areas of landscape and land use. The folded 'Culm Measures' are mainly Late Carboniferous in age, with small areas of Late Devonian and Early Carboniferous rocks in the south-west of the district. Most of the Late Carboniferous sequence consists of shales with thin sandstones; they give rise to dissected hills, which at Whitestone reach 245 m above Ordnance Datum (OD) but which commonly form ridges with crests mainly between 150 and 180 m above OD. These rocks give rise to brown or grey-brown (locally red-stained) soils for which the term 'Dunland' is used. The Culm rocks are unconformably overlain by undeformed rocks of Permian and Triassic age, the 'New Red Sandstone'. The Permo-Triassic rocks form 'Redland' soils.

The Permo-Triassic rocks in the district underlie much of the ground below 60 m above OD (Figure 1). They occur mostly in the eastern half of the district, but also occupy an east–west belt up to 6 km wide extending through the town of Crediton (Figure 2). In this belt, the Permian rocks fill a partly fault-bounded trough, possibly up to 1 km deep, termed the Crediton Trough. The Permo-Triassic formations consist of red breccias, sandstones, mudstones and pebble beds, each of which gives rise to a distinct landscape type. The breccias form hills which are locally prominent, for example along the northern side of the Crediton Trough where the Raddon Hills rise to 235 m above OD, and also south-west of Exeter where they form dissected hills rising to 112 m above OD.

The sandstones give rise to mainly gently undulating lowlands with some steeper slopes and are present both in the Crediton Trough and in a 3 km-wide north–south belt extending from Broadclyst to Topsham; the contrast with the higher, steeper breccia topography is usually well defined.

The eastern part of the district is a lowland floored by red mudstones, with a few intercalated sandstones which form discontinuous ridges. This landscape unit is well defined between the sandstone lowlands on the west and the Budleigh Salterton Pebble Beds on the east. The Pebble Beds outcrop as a dissected north–south ridge with a pronounced west-facing scarp; they give rise to heathland south of Aylesbeare, and form the watershed between the drainage basins of the Rivers Exe and Otter. The youngest Triassic formation of the district consists of sandstones, which overlie Pebble Beds in the south-east of the district.

Scattered outcrops of lava (the Exeter Volcanic Rocks) occur at or near the base of the New Red Sandstone sequence around and south-west of Exeter, and also are present within the New Red Sandstone of the Crediton Trough. These igneous rocks locally give rise to prominent ridges and hills, as for example at Killerton Park, near Broadclyst.

The Dartmoor Granite crops out over a small area in the south-west, where it forms the highest ground in the district (over 300 m above OD); typical granite tors, such as that of Heltor Rock (Plate 11a) are developed.

Near Dunchideock, the youngest solid formations in the district — the Cretaceous Upper Greensand and overlying Palaeogene flint gravels — form a small outlier on which stands the prominent local landmark of Lawrence Castle. This outlier marks the northernmost limit of the flat-topped Haldon Hills, which are a distinctive feature of the Newton Abbot district to the south.

Along the main river valleys, the flat floodplains are underlain by Quaternary alluvial silt, clay and gravel. Small patches of peat occur on the Dartmoor Granite and in the valley of the River Clyst. On the sides of the main river valleys, the remnants of river terraces at eight different levels represent the deposits of former floodplains. Other Quaternary deposits include head, formed largely in periglacial conditions when ice sheets lay not far to the north of the south-west England peninsula. The marine deposits of the Exe estuary were formed during the Flandrian Transgression, when sea level rose following the melting of the ice sheets after the most recent Pleistocene glaciation. The rock succession proved in the Exeter district is summarised on the inside front cover of this memoir.

GEOLOGICAL HISTORY

The oldest exposed rocks in the district (Hyner Shale and Trusham Shale) crop out in the cores of anticlines adjacent to the Dartmoor Granite. The Hyner Shale spans the Devonian and Carboniferous boundary; the Trusham Shale is Early Carboniferous in age. Their fauna and fine-grained lithologies indicate deposition in a marine basin.

The Hyner and Trusham shales are overlain by black mudstones with white silt laminae (Combe Shale) into

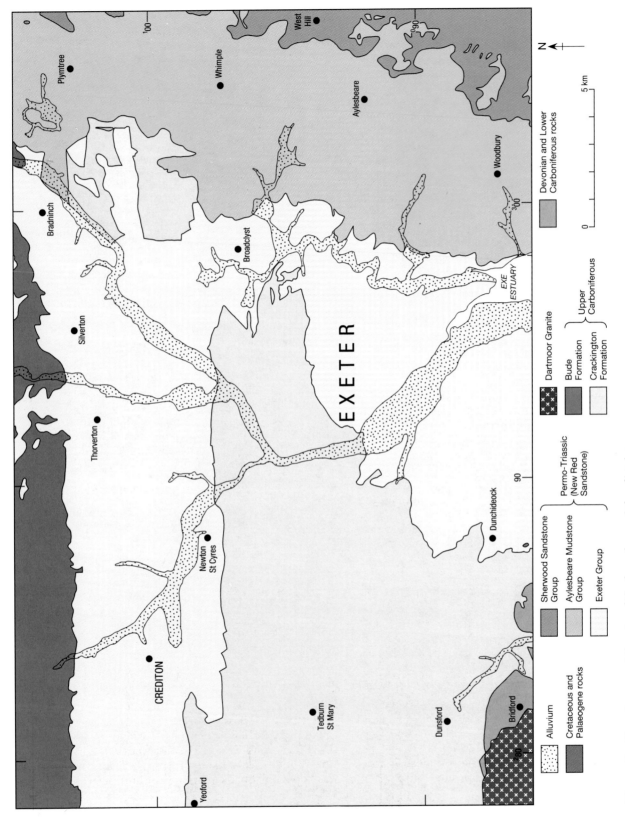

Figure 2   Sketch map of the simplified geology of the district.

which were intruded albite-dolerite sills; the lithologies and lack of fauna indicate deposition in euxinic marine environments. The Combe Shale is succeeded by bedded radiolarian chert (Teign Chert). The sporadic eruption of volcanoes is indicated by the presence of tuff horizons within the cherts. Towards the end of the Early Carboniferous, chert deposition gave way to siliceous mudstones with a few limestones and cherts.

The formation of deep-sea muds with a few thin sandstones (Ashton Shale) at the beginning of the Namurian (Late Carboniferous) heralds a long period when the district was the site of a deep-sea trough which accumulated a thick sequence of shales with turbidite sandstones. This sequence, the Crackington Formation, occupies most of the outcrop of Carboniferous rocks in the district (Figure 2), and is largely of Namurian age.

During the Westphalian, thick-bedded sandstones, siltstones and shales of the Bude Formation, which crop out north of the Crediton Trough, were laid down either in a fresh- to brackish-water lake or on a broad, gently sloping, submarine fan fed directly by delta systems.

In the late Westphalian, the Devonian and Carboniferous sediments were affected by the last compressive mountain-building phase of the Variscan Orogeny. The Variscan Mountains were rapidly eroded in the Early Permian, and possibly from the Late Carboniferous (Stephanian) onwards. Huge volumes of debris from the disintegration of the highlands were deposited as alluvial fans adjacent to the mountain range. Deposition was mainly by debris flow and sheet-flooding during occasional heavy rainstorms in a semi-arid climate. These coarse-grained deposits are now preserved as the breccias of the Exeter Group. A marked break in sedimentation is present between the Early Permian and the Late Permian parts of the Group. During deposition of the earlier sequence, volcanic eruptions produced lava flows about 291 to 282 million years ago. The younger sequence is 250 to 260 million years old. The unconformity between the two was caused by uplift and denudation associated with the emplacement of the Dartmoor Granite about 280 million years ago. The breccias were succeeded by sands formed as desert dunes, with some water-laid interdune deposits (Dawlish Sandstone).

In the Early Triassic, mudstones (Aylesbeare Mudstone) were deposited in an extensive continental sabkha-playa lake, and overlie Permian and Carboniferous strata unconformably. The mudstones were succeeded by coarse gravels (Budleigh Salterton Pebble Beds) which were deposited by braided rivers flowing from the south. A thin bed containing wind-faceted pebbles caps the Pebble Beds, and indicates a sedimentary break of uncertain length in the Early Triassic. The Otter Sandstone, the youngest Triassic deposit in the district, was largely deposited on a sandy braidplain, with some aeolian dunes.

There is a marked break in the sedimentary record of the district above the Triassic rocks; Jurassic strata probably once covered the district, but have been removed by erosion. The break was followed by deposition of glauconitic sands (Upper Greensand) of Cretaceous (Upper Albian–Cenomanian) age, preserved in very small outcrops in the south of the district; they are the deposits of

a shallow shelf sea, probably laid down within 5 km of a shoreline to the west. The former existence of the Chalk sea over the district is indicated by residual flint gravels, overlying the Upper Greensand, derived by in-situ solution of chalk. They are of Palaeogene age.

During the Quaternary glaciations, the district lay to the south of the ice sheets, but the presence of soliflucted deposits (head) and frozen-ground features show that it experienced a periglacial climate. Stages in the Quaternary evolution of the river systems are indicated by the presence of river terraces at eight levels. Present-day deposition along the river valleys is producing alluvium — silt, clay, gravel and minor peat. The melting of the last ice sheets resulted in a rise in sea level, which flooded the Exe estuary and produced marine deposits. Buried channels beneath the estuary were probably cut at times of low sea level during the glacial period.

HISTORY OF RESEARCH

Early accounts include that of Polwhele (1797), who recognised the volcanic nature of the igneous rocks (Exeter Volcanic Rocks) associated with the New Red Sandstone, and De la Beche's *Report on the geology of Cornwall, Devon, and West Somerset* (1839). Sedgwick and Murchison (1840) introduced the name 'Culm Measures' and recognised their Carboniferous age. Shapter's (1842) account contains a geological map of the Exeter district, and many of Ussher's papers on the Permian and Triassic rocks of south-west England (e.g. 1875, 1876, 1877) refer to the district. The first detailed account of all aspects of the geology of the district is the Geological Survey Memoir (Ussher, 1902).

Various aspects of the Culm rocks of the district have been reported on by Collins (1911), Butcher and Hodson (1960), Chesher (1968) and Selwood et al. (1984).

Early descriptions of the Permo-Triassic rocks included those of De la Beche (1839), Ussher (1902), Conybeare and Buckland (quoted in Murchison, 1867), Ussher (1892 a), Hull (1892), and Irving (1892). Irving (1888) divided the south Devon sequence into Permian and Triassic, taking the boundary at the base of the Budleigh Salterton Pebble Beds. Hutchins (1958, 1963) studied the Permian sequence of the Crediton Trough. Laming (1966, 1968) described the sedimentology and stratigraphy of the Permo-Triassic rocks of Devon. Henson (1970) and Selwood et al. (1984) refined the stratigraphy of the south Devon Permo-Triassic. Bristow and Scrivener (1984) and Bristow et al. (1985) described Permian stratigraphy and structure around Exeter. The discovery in the basal Permo-Triassic sequence at Exeter of Late Permian palynomorphs was reported on by Warrington and Scrivener (1988, 1990).

De la Beche (1835, 1839), Vicary (1865), Hobson (1892), Ussher (1902), Tidmarsh (1932) and Knill (1969) described the Exeter Volcanic Rocks, and their geochemistry and petrogenesis were examined by Cosgrove (1972), Exley et al. (1983), Thorpe et al. (1986), Grimmer and Floyd (1986), Thorpe (1987) and Leat et al. (1987). Miller et al. (1962) and Miller and Mohr (1964) gave radiometric dates for the Exeter Volcanic Rocks. Their palaeomagnetism and magnetic properties were studied by Cornwell (1967) and Cornwell et al. (1990).

Aspects of the structure of the district were examined by Selwood and McCourt (1973) and Durrance (1985). Davey (1981 b) carried out gravity surveys across the Crediton Trough. Kidson (1962) studied the river terraces and drainage evolution of the River Exe. Durrance (1969, 1974) identified buried channels and terraces in the Exe estuary. Sherrell (1970), Davey (1982) and Cradock-Hartopp et al. (1982) described aspects of the hydrogeology of the district.

TWO

Devonian and Carboniferous rocks

A small area of Upper Devonian to Lower Carboniferous rocks is present in the south-west of the district (Figure 3). Upper Carboniferous rocks crop out over an area of about 200 km², mainly in the west of the district, and are probably concealed beneath the New Red Sandstone elsewhere. The Carboniferous rocks are traditionally termed Culm Measures in south-west England. The stratigraphical classification of the Devonian and Carboniferous rocks is given in Table 1.

During the latest Devonian and Carboniferous, the district lay towards the northern margin of the Rheno-Hercynian tectonic zone (Leeder, 1982, 1987; Cope et al., 1992). This is now preserved as a narrow, elongate, structurally complex region of deformed deep-water marine sedimentary rocks that were deposited in a series of late Devonian to Dinantian back-arc basins stretching from south-west Ireland through south-west England (but excluding southern Cornwall) into Germany. The palaeogeography of this region is difficult to reconstruct because the crust has been severely deformed by thrusting and nappe emplacement, which migrated progressively northward to reach the present district in late Westphalian

Table 1 Classification of the Devonian and Carboniferous rocks of the district.

Subsystem	Series	Formation	Member or informal unit
Upper Carboniferous (Silesian)	Westphalian	Bude Formation	
	Namurian	Crackington Formation	Ashton Shale Member
Lower Carboniferous (Dinantian)	Viséan	Teign Chert	Posidonia beds
	Tournaisian	Combe Shale	
		Trusham Shale	
Upper Devonian	Famennian	Hyner Shale	

times. The terminal position of the thrusting, and the northern limit of Rheno-Hercynia, is now marked by the Variscan Front, which lies along the Bristol Channel. The associated crustal shortening was a response to plate subduction and collision along a suture to the south, extending from the Carboniferous position of Iberia (in the Bay of Biscay), through Armorica, the Massif Central and the Vosges Mountains. Closure began in the Mid-Devonian and culminated during the Variscan Orogeny with emplacement of Late Carboniferous granites (such as the Dartmoor Granite) along the axis of the south-west England peninsula. Sedimentation in Rheno-Hercynia was predominantly hemipelagic, with localised submarine volcanogenic input. The sea was colonised by ammonoid, bivalve, conodont, trilobite and ostracod faunas tolerant both of low oxygen levels (dysaerobic) and of aphotic (dark) conditions. As a consequence, standard coral/brachiopod and foraminiferal biozonation (Riley, 1993) appropriate to correlation in the carbonate platforms and ramps common in the British Dinantian north of the Variscan Front, are not applicable to the Exeter district.

At times, the hemipelagic depositional environment was remarkably uniform across Rheno-Hercynia (Jackson, 1984, 1985, 1992), black phosphatic/siliceous shales being widespread in the mid-Courceyan (main part of the Combe Shale), and radiolarian cherts being typical in the late Chadian (upper part of the Combe Shale), Asbian and Brigantian (Teign Chert). These intervals are thought to indicate widespread chemical stratification events in the water column, due to eustatic high-stand in the case of the black shales, and to submarine volcanism in the case of the cherts.

As northward closure of Rheno-Hercynia continued into the Namurian, nappe emplacement encroached nearer to south-west England and there was a resultant increase in clastic debris, represented by the deposition of sandstone turbidites of the Crackington Formation and a change in the clay mineralogy of the shales. The marine fauna, though still hemipelagic, was denied continuous access to the basins owing to a combination of eustatic and tectonic processes, and discrete marine faunal units (marine bands) were formed, some of which coincide with temporary interruption of sandstone turbidite supply. North of the Variscan Front, strata between these same marine bands are repeatedly associated with deltaic progradation and palaeosol formation; however, the Exeter district remained in relatively deep water. Such cyclicity is thought to have been driven by glacio-eustatic changes of sea level, brought about by the glaciation of the Gondwanan continent which lay at that time across the South Pole.

The Namurian and early Westphalian also saw the transformation of Rheno-Hercynia from a series of inter-

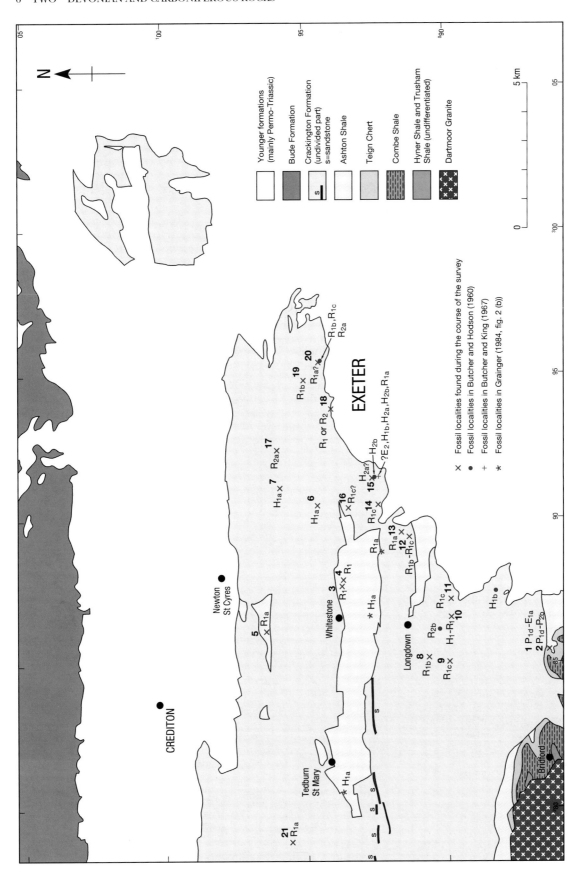

Figure 3   Distribution of Devonian and Carboniferous rocks in the district, showing also fossil localities and ammonoid ages. Locality numbers refer to Table 3.

connected back-arc basins into more isolated foreland basins lying in front of the area of nappe emplacement. The stratigraphy became more parochial as marine influence waned, and the Bude Formation, the youngest Carboniferous sedimentary unit known in the district, was deposited in lacustrine and deltaic environments.

UPPER DEVONIAN AND LOWER CARBONIFEROUS ROCKS

Hyner Shale and Trusham Shale (undifferentiated)

In the district, the Hyner Shale and overlying Trusham Shale have everywhere been metamorphosed by the Dartmoor Granite, and the two formations cannot be readily distinguished. Hornfelsed beds attributed to the Trusham Shale crop out within anticlinal fold cores adjacent to the granite at two localities [SX 809 872 and SX 818 863] near Bridford (Figure 3); the uppermost part of the Hyner Shale may be present within the axial zones of these folds. Both formations were first described from the adjacent Newton Abbot district (Selwood et al., 1984), where the Hyner Shale comprises greenish grey or bluish grey silty shale with small calcareous nodules, and passes up conformably into the Trusham Shale. In the Newton Abbot district, the Hyner Shale spans the Devonian/Carboniferous boundary; the lower part yields ostracods indicating the *hemisphaerica-dichotoma* Zone (Late Famennian), and the upper part contains ostracods of the *latior* Zone (Tournaisian) (Selwood et al., 1984, p.34).

The Trusham Shale of the Newton Abbot district consists of up to 60 m of greenish grey to olive-green, micaceous, sparsely laminated silty shale with partings of medium grey silty shale. In the Exeter district, the Trusham Shale outcrops comprise pale and medium grey banded hornfels. The Trusham Shale is Early Carboniferous in age, having yielded ostracods indicative of the *latior* Zone (Tournaisian) (Selwood et al., 1984, p.61). The bivalve *Sanguinolites? ellipticus* is locally common at higher levels in the formation, together with indeterminate juvenile brachiopods and bryozoans.

Bridford

Pale-weathering buff hornfels with a ferruginous layering or lamination occurs as brash near Lowton [SX 8098 8723]. The low-lying ground in this area occupies an anticlinal fold core adjacent to the granite (Birch Down Anticline, Figure 33). Pale or medium grey, ribbon-textured hornfelses occupy the core of the Bridford Anticline. The hornfels is composed of silt-grade quartz in a matrix of sericitic clay, and forms brash on the slopes south-east of Bridford [mid-point SX 8175 8628].

Combe Shale

The outcrop of the Combe Shale defines a series of kilometre-scale folds within the metamorphic aureole of the Dartmoor Granite (Figure 3). It is progressively cut out by the granite towards the west, and is absent along the northern margin of the granite west of the 80 easting [SX 8010 8764]. East of the granite aureole, the Combe Shale crops out in the core of the Doddiscombsleigh Anticline (Figure 33).

The Combe Shale consists of medium to dark grey or black shale, locally with white silty lamination. Within the granite aureole, the formation is typically blocky and glassy-textured hornfels, and can be difficult to distinguish from black cherts in the overlying Teign Chert. In the Bridford Syncline and Bridford Anticline (Figure 33) it consists of dark grey, fine-grained or flinty hornfels or slate. Zones of silicification and hydrothermal brecciation occur locally, in which medium grey slates or massive silicified slates are veined by a ferruginous and manganiferous stockwork. A zone of this type occurs north of the Bridford Sill [SX 8216 8672] in the upper part of the folded succession. The base of the formation is taken at an upward change from greenish grey shales (Trusham Shale) to black shales. Several thick, laterally continuous sills of albite dolerite occur within the Combe Shale and have caused thermal metamorphism of the adjacent shales (Chapter 3). The effects of tectonic thickening make it difficult to estimate the thickness of the formation. The minimum thickness in the Doddiscombsleigh Anticline is about 130 m, but the thickness may locally exceed 200 m in the Bridford Anticline.

The Combe Shale has yielded no fauna, although small fragments of wood have been reported (Selwood et al., 1984). The formation was probably deposited in a marine environment with restricted circulation and stagnant or anaerobic conditions.

Bridford

Cuttings from a borehole [SX 8067 8752] at Lowton consist of medium to dark grey and black flinty hornfels with white lamination. Extensive brash of ferruginous-weathering, dark grey or black, massive or brecciated hornfels and medium grey fine-grained hornfelsed shales is present on Higher Lowton Down [SX 813 872] south-east of Lowton; an excavation [SX 8130 8699] exposed brown or locally rusty-weathering, medium grey, fine-grained and blocky micaceous hornfels.

The north-west face of Stone Copse Quarry [SX 8237 8603], near Stone, exposes the upper part of the Combe Shale, where it occurs in a steep east- or north-east-plunging open fold. The lithology is mottled orange-brown-weathered, medium to dark grey, fine-grained, siliceous hornfels, locally with faint white silty laminations. An unidentified pale yellow, crystalline to botryoidal, secondary mineral coats fractures and some exposed surfaces. Well-defined beds, 0.2 m thick, dip at 54° to 060°. A near-vertical, east-inclined cleavage is present throughout.

Doddiscombsleigh

About 300 m west of Mistleigh Copse, south of Doddiscombsleigh, bedding in small exposures [SX 8504 8594] of black shale with white silty laminations dips 30° to 073°; fracture cleavage dips 66° to 170°. About 150 m south of the above exposure, crags [SX 8502 8579] show at least 6 m of pale olive-grey, fine-grained, locally banded rock, possibly adinole, dipping 30° to 095°. Morton (1958) noted that the rock was similar to that south of Harehill Plantation [SX 8571 8596], about 600 m south of Doddiscombsleigh, but differed in the presence of abundant pseudomorphed idiomorphic porphyroblasts. The pseudomorphs consist of a concentration of green mica, and are surrounded by a quartz/albite 'halo', free from mica and somewhat coarser in grain. Commonly, spots are visible and consist of a coarser patch of quartz and albite, free from mica. The upper 1 m of the section consists of adinoles free from pseudomorphs.

West of Harehill Plantation, fragments of black shale with white laminations are common in the soil [SX 8570 8605]. About 100 m south, crags [SX 8571 8596] show about 15 m of yellowish grey to pale olive-grey, hard, splintery, locally laminated, fine-grained hornfels between the Ryecroft Sill and the Harehill Sill (Figure 5). Bedding dips at 34° to 078° and is cut by a fracture cleavage which dips 73° to 147°. Morton (1958, p.165) noted that the rocks show very little variation and differ from the adinoles at Hill Copse [SX 840 844] near Lower Ashton (outside the present district) in that no calcite is present; the quartz/albite mosaic contains abundant yellowish, weakly pleochroic mica, with very pale green fibrous chlorite; no pseudomorphs after porphyroblasts occur. Round spots, up to 0.2 mm in diameter, are present, but are little coarser in grain than the matrix, and contain no mica.

Teign Chert

The Teign Chert occurs in a series of tight folds within the metamorphic aureole of the Dartmoor Granite, and also in a series of overturned folds east of the aureole (Figure 3). The formation is progressively cut out westwards by the granite and is absent west of the 79 easting; cherts reappear west of the 75 easting in the Okehampton district (Edmonds et al., 1968). In the Doddiscombsleigh area, the formation gives rise to prominent ridges which can be traced around the limbs of the major easterly plunging folds.

The Teign Chert consists of well-bedded or colour-banded cherts (Plate 1a), interbedded with siliceous shales. The cherts, particularly those in the basal part of the formation, are bluish grey or black and glassy in texture; white, greenish white, green, pale grey and red varieties are also present and occur throughout the succession. Individual chert beds generally range from a few millimetres to over 0.3 m thick, and are interbedded with grey to greyish green shale and mudstone a few millimetres to about 8 cm thick. Radiolaria are locally abundant in the cherts, and can be recognised by the presence of white casts and hollows. Several units of tuff, locally thick enough to map, are present within the Teign Chert, and spilitic basalt occurs within the lower part of the formation in Scanniclift Copse [SX 844 862], 700 m south-west of Woodah Farm, and at Stone Copse Quarry [SX 824 860] (see Chapter 3). Also present within the chert sequence are impersistent manganiferous beds, which were worked in the past (Chapter 11). The formation base is taken at the lowest bedded chert above the Combe Shale; locally the formation rests on spilitic basalt or albite dolerite emplaced above the Combe Shale.

The upper 30 m of the formation consist of siliceous shales with interbedded cherts and scarce thin limestones; black shale containing the bivalve *Posidonia becheri*, and overlain by a bed with spirally striate ammonoids, marks the base of this unit, which was informally named the *Posidonia* beds by Selwood et al. (1984).

The apparent thickness of the Teign Chert is in the range 140 to 240 m in the present district. The stratigraphical thickness may be less, because of the effects of folding; tectonic thickening is apparent in the axial zones of large-scale folds like the Bridford Syncline (Figure 33).

The lower part of the Teign Chert has yielded no fauna except radiolaria (Hinde and Fox, 1895). Morton (1958, p.68) reported the radiolarian *Cenellipsis* sp. from Harehill Plantation [SX 8581 8611], Doddiscombsleigh. In the Exeter district, *Posidonia becheri* has been reported from several localities around Doddiscombsleigh and Bridford (see details). *Bollandoceras micronotum*, indicating a late Asbian B_2 Zone age, has been reported (Butcher and Hodson, 1960) from near Woodah Farm [SX 8498 8664], Doddiscombsleigh. From the *Posidonia* beds at Down Lane [SX 8545 8666], Doddiscombsleigh, Riley (1991) reported cf. *Posidonia membranacea* (late Brigantian P_{1d} Zone to Pendleian E_{1c} Zone), cf. *Sudeticeras* sp. (early Brigantian P_{1c} Zone to early Pendleian E_{1a} Zone), and *Lusitanites* or *Neoglyphioceras* spp. (late Brigantian P_{1d}–P_{2c} zones). The *Posidonia* beds may therefore extend from the late Viséan (Brigantian P_{1d} Zone) into the Namurian (Pendleian E_{1a} Zone). The *Posidonia* beds in the Newton Abbot district have yielded ammonoids indicating a range from the B_1 Zone, represented by *Entogonites grimmeri* (Riley, 1991), to the P_{2a} Zone, with all the intermediate P_1 subzones present (Selwood et al., 1984, p.62). *Goniatites sphaericostriatus* (P_{1c}), recorded by Butcher and Hodson (1960) from the Crackington Formation at Bonhay Road [SX 9143 9273 to 9146 9257] in Exeter, is considered to have come from a block of Westleigh Limestone, commonly used as a building stone in the city, and not from an in-situ locality.

Most of the bedded cherts are thought to have precipitated from colloidal silica on the sea floor. The silica content of the water may have been influenced by nearby eruptions of submarine volcanoes; the presence of tuffs in the sequence indicates contemporaneous volcanic activity during chert deposition. However, some cherts may represent mudstones at different stages of silicification, and toward the top of the formation the cherts may include silicified limestones.

Within the granite aureole

Hornfelsed chert [SX 795 878] adjacent to the northern margin of the granite, about 1 km north-north-west of Heltor Rock, forms a fault-bounded segment with about 600 m strike length. Coarsely recrystallised white or greenish white chert with thin manganiferous partings forms a typical brash, while manganiferous and ferruginous wad and brecciated crumbly white chert occur locally [e.g. SX 7947 8796] near the top of the formation, about 500 m north of Heltor.

South of Bridford Wood and immediately adjacent to the faulted margin of the granite [SX 8014 8765], there is a small exposure and numerous large blocks of regularly layered and colour-banded, very fine-grained chert. The mesoscopic layering is on a 5 to 10 mm scale, and is defined primarily by the preferred planar concentration of iron oxides and chlorite within the siliceous host. Joint surfaces are coated with manganese oxides. A poorly exposed section, approximately 150 m farther east-north-east [SX 8028 8767 to 8026 8776], consists of sugary-textured and recrystallised chert with indistinct pale grey layers near the formation base; it passes up into colour-banded chert approximately 40 m above.

Banded chert near Lowton [SX 8086 8760] contains a near-vertical, north-inclined layering and a vertical, north-west-trending cleavage. Individual layers are sharply defined, medium grey, white or pale green and 2 to 20 mm thick. These cherts have unusual mineralogical compositions that include petalite-allanite-pistacite-ferrohastingsite-axinite-sphene-apatite assemblages (Morton, 1958).

Jaspilitic or hydrothermal chert breccias are present in brash between Copplestone Down [SX 8156 8763], about 500 m south-east of Woodlands, and Birch [SX 8225 8725]. They consist of layered, brick red and very pale green chert in which the former layers commonly replace the latter to give an irregular, breccia-like appearance. Some varieties [SX 8204 8746] contain clasts of silicified wallrock and brown tourmalinised mudstone entrained in a cherty matrix of quartz and tourmaline. The clasts show complex internal cross-cutting veinlets of quartz and tourmaline that indicate multiple fracturing during hydrothermal activity.

Chert in the Bridford Syncline consists of very fine-grained, layered or colour-banded greenish grey and greyish green silica, with manganese coatings along joints and fractures. Tourmalinised breccias occur north of the Bridford Sill [SX 8213 8682], and consist of 1 to 10 mm-size angular rock fragments in a silicified matrix of comminuted material together with cross-cutting, quartz-filled fractures (E 65523). The rock fragments appear to be chert or siliceous shale, and have been extensively silicified. The metasomatic quartz is charged with minute prismatic grains of dravitic tourmaline, and this mineral has also replaced micaceous layers in the original rock fragments. Angular fragments of flow-banded, fine-grained tuff occur in a similar breccia [SX 8261 8685] 150 m south-south-west of Birch Aller Mine, about 350 m south of Neadon Farm.

The south-eastern extension of Teign Chert in the Bridford Syncline is well seen in the disused Bridford Barytes Mine Quarry [mid-point SX 8296 8656] where a section, within the middle and upper parts of the formation, shows the following:

	Thickness m
Siltstone with fine-grained tuffs, bleached pale to dark grey, massive and silicified in part, with cross-cutting baryte veins 20 to 300 mm wide; *Posidonia becheri* reported	at least 5.0
Baryte, ferruginous weathering, white, coarsely crystalline, parallel with layering in underlying cherts	3.0
Chert, metalliferous and manganese-rich, well-defined layering, near-vertical and inclined north-east	2.0
Shale, pale grey, silicified, some interlayered sandstone and silicified sandstone	3.0
Chert, ferruginous and manganiferous	2.0
Section missing	c. 8.0
Chert, well-defined layering on 5 to 20 mm scale, some fine-grained internal planar structure, ferruginous and manganiferous	7.0
Chert, glassy, black, some interlayered grey silicified shale	14.0
Tuff, pale grey, with black lithic fragments	5.0

Ussher (1913) recorded *Posidonia becheri* from ploughed fields near a lower, now overgrown, Bridford Barytes Mine Quarry [SX 8299 8655], and Morton (1958) also found this fossil within a 75 mm-thick black shale bed overlain by about 6 m of thin cherts and siliceous shales in the same quarry.

The north-western limb of the Christow Common Syncline (Figure 33) is defined predominantly by a ridge of black, fine-grained to glassy chert, locally veined by white quartz. The basal part of the succession, exposed [SX 8250 8615] in forestry tracks near Stone north of Rookery Brook, and in the south-east face of Stone Copse Quarry [SX 8243 8600], is mottled medium grey and pale greenish grey, massive to layered, fine-grained chert with patchy manganese oxides. Well-bedded cherts occur as loose blocks in the quarry and consist of millimetre-scale, white, grey, black, pale green and orange-banded silica with layers rich in iron and manganese oxides. Pale green layers weather bright orange in part and contain chlorite and carbonate. A thin-section (E 65530) shows laminae of fine-grained opaque material within microcrystalline quartz and some silicified debris, possibly radiolaria. Morton (1958) recorded various assemblages that include diopside, tremolite/actinolite, epidote and ferrohastingsite within the colour-banded cherts at Stone Copse Quarry.

East of the granite aureole

Chert within the Doddiscombsleigh Anticline, south of Woodah Farm, is exposed in a small disused quarry [SX 8481 8648]. The principal lithology is a dark grey, very fine-grained, slightly micaceous, siliceous rock that contains a patchy anastomosing and vuggy alteration to white carbonate and quartz. A mesoscopic fold, plunging at about 20° east-north-east, occurs within the quarry. A thin section (E 66094) shows numerous opaque or sub-opaque inclusions in a microcrystalline quartzose matrix with numerous 0.1 mm-size spherulitic quartz aggregates, probably recrystallised radiolaria.

A quarry [SX 8498 8664], 200 m south-east of Woodah Farm, is largely overgrown and part-flooded, but exposes an inverted section of siliceous shales and cherts within the upper part of the Teign Chert. Thinly bedded (up to 80 mm thick) siliceous shales and cherts, approximately 5 m thick, and extensively veined by white quartz and in part mineralised with goethitic or limonitic veins, form the southernmost strata exposed. A sequence approximately 3 m thick has been quarried, possibly as a source of carbonate, from the overlying strata. The footwall sequence, 2 m of which is seen, is orange stained and ferruginous weathering, consisting of dark grey or black siliceous shales cut by fractures coated with manganese oxides. An inverted section in this quarry is given by Morton (1958) in correct stratigraphical order, as follows:

	Thickness m
Limestone	0.23
Shale	0.08
Limestone	0.13
Shale, with thin (50 mm) limestones	1.04
Limestone	0.10
Shale with very thin limestones	1.83
Limestone, with subordinate shales	0.42
Limestone	0.11
Shale	0.17

The strata dip south at about 65°. The limestones are pale bluish grey, consist largely of calcite with some angular quartz silt, and are strongly veined with quartz and calcite. Ussher (1902) recorded *Posidonia becheri* and *Phillipsia cliffordi*, and Collins (1911) recorded *Glyphioceras spirale* and an unnamed brachiopod. Selwood et al. (1984, p.176) reported *Posidonia becheri*, and Butcher and Hodson (1960) figured *Bollandoceras micronotum*, indicating the B_2 Zone of late Asbian age.

In Down Lane, south-west of Doddiscombsleigh, scattered exposures are present in the floor of the lane from near Christendown Clump [SX 8504 8634 to 8544 8634], and more continuous exposures in the banks of the lane from there to a place [SX 8547 8652] near the eastern end of the lane. Dips are all to the south. The section, totalling 59 m, has been recorded by Morton (1958) as follows:

	Thickness m
POSIDONIA BEDS	
Shale and mudstone, blue, hard, cleaved, siliceous; infrequent fossils in the basal 1 m: *Posidonia becheri*; *Chonetes* sp.; *C. (Semenewia)* sp.; *Aviculopecten* cf. *losseni*; poorly preserved smooth ostracods; and a fish scale. Collins (1911) recorded, probably from the same exposure: *Seminula ambigua?*; *Posidonomya becheri*; *Stroboceras sulcatum*; and *Glyphioceras*, probably the adult stage of *G. reticulatum*.	10
TEIGN CHERT BELOW THE POSIDONIA BEDS	
Chert, in beds 8–10 cm thick, interbedded with 5 cm- thick shale beds; chert becoming scarcer towards the top; at the top, predominantly cleaved, pale blue-grey mudstone with rare very thin chert beds	16
Chert, dark grey, medium-bedded (units up to 0.3 m thick), alternating with very thin grey shale	6
Gap in section	c. 12
Chert, blue-grey, bedded, cleaved	2
Gap in section	c. 8
Chert, blue-black, bedded, radiolarian	1.2
Gap in section	12
Tuff, blue-grey, hard, bedded	1

Riley (1991) reported cf. *Posidonia membranacea* and cf. *Sudeticeras* sp., and a trilobite free cheek from exposures [SX 8545 8666] on the west side of Down Lane. *P. membranacea* has a stratigraphical range from the P_{1d} to E_{1c} zones (late Brigantian to Pendleian), while *Sudeticeras* ranges from the P_{1c} to E_{1a} zones (early Brigantian to early Pendleian) (Table 2). About 5 m uphill (stratigraphically below the preceding locality), another locality in Down Lane yielded *Lusitanites* or *Neoglyphioceras* spp.. Riley (1991) noted that these samples represent the 'spirale beds' which are recorded from north Devon. *Lusitanites* and *Neoglyphioceras* are restricted to the late Brigantian (P_{1d} to P_{2b} zones).

About 100 m south-west of Shippen, Doddiscombsleigh, J A Chesher (MS fieldslip, 1966–68) recorded shale and chert with *Posidonia* in debris from a trench [SX 8571 8628]. Exposures [SX 8581 8611] in Harehill Plantation, about 600 m south of Doddiscombsleigh, next to workings for manganese, show dark grey chert and laminated shale, dipping at 28° to 219°. This locality is close to that described in Morton (1958, p.48), as follows: 'In Harehill Plantation....6 ft. of radiolarian cherts are exposed by an old manganese mine adit. The cherts are blue-grey and have vugs and small veins of quartz. Lighter bands in the chert are found to be highly fossiliferous, with better preserved radiolaria than anywhere else in this area...' The radiolaria were all *Cenellipsis* sp. (Morton, 1958, p.68).

Outcrops of massive jasper chert are present on the hilltop [SX 8591 8589] about 250 m south-south-east of Harehill Plantation. Morton (1958, pp.48–49) described this rock as follows: 'In hand specimen, this recrystallised chert is blood red with glassy colourless spots in it. In thin section it is seen to consist of large spherulites of radiate quartz with cores of radiating chalcedony. Between adjacent spherulites, there is often a coarse granoblastic mosaic of quartz. Minute dusty inclusions of haematite? are concentrated in the granoblastic quartz and in the outer peripheries of the spherulites; this obviously causes the red colour of the rock. Quartz veins also cut the rock.'

Selwood et al. (1984) recorded *Posidonia* aff. *becheri* [small form] from a lane exposure [SX 8510 8625] 400 m west of Cherry Tree Cottage, Doddiscombsleigh. They also reported (1984, p.176) a nearby section [SX 8537 8622] yielding *Posidonia* aff. *becheri* [small form], *P.* sp., crinoid stems, trilobites, and brachiopods.

A section [SX 8548 8625] behind Cherry Tree Cottage shows *Posidonia* beds, consisting predominantly of grey to dark grey shale and cherty shale, with sporadic thin beds (< 5 cm, locally 10 cm) of chert. At the south end of the exposure, the beds dip 60° to 050°; in the middle of the exposure the dip is 87° to 348°; and at the north end, 58° to 171°. Selwood et al. (1984) reported *Posidonia* cf. *becheri* from debris thrown out from excavations at Cherry Tree Cottage.

Exposures in bedded cherts in a small quarry [SX 8543 8599] on the east side of Mistleigh Copse, 600 m south of Doddiscombsleigh, consist of at least 14 m of well-bedded black chert beds ranging from a few centimetres to 0.3 m thick, separated by thin grey to grey-green shale and mudstone interbeds and partings from less than 1 cm to 8 cm thick (Plate 1a). The exposure is faulted at the south end, where the strata dip at 44° to 155°; in the middle of the exposure the dip is 38° to 147°; and at the north end, 52° to 147°.

Details of tuffs and manganiferous beds in the Teign Chert are given in Chapters 3 and 11.

UPPER CARBONIFEROUS ROCKS

Crackington Formation

The Crackington Formation of the district consists predominantly of folded grey shales with subordinate interbeds of turbiditic sandstone (Plate 1b). Shale: sandstone ratios are mainly between 2:1 and 4:1. The basal part of the formation (the Ashton Shale Member) consists of shale with relatively few thin sandstones. The Crackington Formation is commonly red-stained adjacent to the New Red Sandstone unconformity. Elsewhere, oxidation of minor pyritic and carbonaceous material has replaced original dark grey colours with pale brown and pale grey; such weathering is probably mostly of Quaternary age. Much of the outcrop is mantled by blanket head and regolith (Chapter 10) mostly less than 2 m thick. Exposures below the regolith are generally sparse.

The formation crops out extensively in the west of the district and also forms inliers around Ashclyst Forest and east of Hele at the eastern end of the Crediton Trough (Figure 3). It gives rise to moderately dissected hills with ridges commonly trending east-west. Many of the steep-sided valleys have been excavated along the strike of the rocks or along fault lines. The Ashton Shale between Exeter and Tedburn St Mary forms lower ground with less steep slopes.

Geochemical studies of turbiditic sandstones from the Crackington and Bude formations (Haslam, 1990; Haslam and Scrivener, 1991) show that they are depleted, relative to average upper continental crust, in elements that are susceptible to chemical weathering and enriched in elements that occur in minerals resistant to chemical weathering. These chemical comparisons are typical of turbidites from passive margin settings, derived from earlier sedimentary and metasedimentary rocks (Bhatia and Crook, 1986).

ASHTON SHALE MEMBER

The Ashton Shale Member succeeds the *Posidonia* beds, probably conformably (Selwood et al., 1984). The base is taken at the change to black shale from the underlying siliceous shales and cherts; there is a marked topographical change at the junction, with the ridge of the Teign Chert rising steeply above the low-lying Ashton Shale outcrop.

The member occupies the core of an anticline that extends between Exwick and Tedburn St Mary. Localities yielding H_{1a} ammonoids (Figure 3; Table 3) at Exwick Barton [SX 9050 9465] and Pynes Water Works [SX 9112 9602], near Cowley, may represent small anticlinal cores of Ashton Shale too small to represent on the map. An anticline containing Ashton Shale is present east of Longdown, and small outcrops occur south of Newton St Cyres (Figure 3). In the south-west of the district, the member is in stratigraphical contact with the Teign Chert along the northern edge of the Dartmoor Granite and around Doddiscombsleigh; its outcrop east of Doddiscombsleigh is bounded on the north by a fault along the valley of the Shippen Brook.

Outside the metamorphic aureole of the Dartmoor Granite, the Ashton Shale consists predominantly of dark grey to black shale with a few thin (7 cm or less) interbeds of fine-grained sandstone and siltstone; the ratio of shale to sandstone and siltstone is 10:1 or greater. Distinctive dark grey splintery shales and slaty shales are locally present, as are 'pencil' shales formed by the intersection of bedding and cleavage. Where the Ashton Shale forms the country rock of the Dartmoor Granite, it is extensively recrystallised and consists of hornfels. Farther east in the aureole, for example around Bridford Wood [SX 798 882], the member consists of splintery pyritous shale with sparse, fine-grained sandstone and siltstone interbeds. The thickness of the Ashton Shale is difficult to calculate owing to the effects of folding, but is estimated at between 250 and 430 m.

The sandstones of the Ashton Shale were examined by Morton (1958), who described them as fine- or medium-grained quartzose lithic greywackes, consisting of quartz, lithic fragments, a little plagioclase, and some plant debris. The lithic fragments include siltstone, claystone, chert, greywacke, quartzite, rare quartz-muscovite schist and devitrified pumiceous volcanic glass. All the fragments are set in a matrix of clay with a little silt and carbonaceous material. Volcanic glass is not recorded from the Crackington Formation above the Ashton Shale, and indicates that at the time of Ashton Shale deposition, a recently active volcanic terrain was being eroded, probably that from which the Lower Carboniferous tuffs were erupted.

The thin beds of fine-grained sandstone and siltstone in the Ashton Shale are interpreted as distal turbidites which flowed into a marine basin which was not necessarily deep, but was characterised by restricted water circulation and anoxic bottom conditions (Grainger, 1983).

Differences in clay mineralogy between shales in the Ashton Shale and in the overlying Crackington Formation have been demonstrated by Grainger and George (1978). The Ashton Shale consists of non-chloritic shale (illite-kaolinite), whereas the overlying sequence is chloritic (illite-kaolinite-chlorite). These differences in clay mineralogy influence the weathering, slope stability and types of soil developed on the Crackington Formation (Grainger, 1983, 1984). The non-chloritic Ashton Shale is weaker and more prone to slope instability, so that landslips are more widespread on its outcrop than on other parts of the formation (Chapter 11).

Venny Tedburn to Newton St Cyres

Exposures [SX 8330 9661] in the Kelland Brook about 1.5 km south-east of Venny Tedburn show possible Ashton Shale, consisting of dark grey shale with thin (less than 5 cm) beds of dark grey to medium grey siltstone; the dip is 80° to 190°.

Just south of Woodley Farm [SX 869 975], sporadic exposures [e.g. SX 8649 9721] in the Crooklake Brook show very dark grey shale dipping north at 10–25°. An exposure [SX 8680 9727] shows 4 m of very dark grey shale with sparse sandstones and siltstones, mostly less than 4 cm thick; the shale:sandstone ratio is approximately 15:1. Sporadic stream-bed exposures [SX 8529 9643 to 8559 9636] of shales with subordinate thin sandstones and flaggy siltstones are present along the southern side of Hundred Acre Copse, south-west of Northridge. Exposures [SX 8625 9655] in Northridge Copse yielded poorly preserved ammonoids, possibly indicating the R_{1a} Zone of Kinderscoutian age (Table 3, locality 5).

A disused quarry [SX 8691 9638] in Lower Western Copse, west of Coombland Wood, shows 1.6 m of weathered dark grey shale with sporadic thin beds of sandstone and siltstone; the ratio of shale: sandstone is 6:1. Vertical joints are prominent. The succession is right way up and dips 10° to 355°.

Tedburn St Mary–Whitestone–Exwick

Sporadic exposures [SX 8010 9345 to 8073 9399] in Dillybridge Brook, south-west of Tedburn St Mary, show a north-dipping sequence of shales with thin sandstones. Most beds are steeply dipping and inverted; a few shallower right-way-up beds are present. Specimens collected by Mr K Page from exposures in the Tedburn St Mary area were identified by Riley (1983). An exposure [SX 798 933] near Dilly Bridge west of Tedburn St Mary contained *Homoceras subglobosum*. A loose nodule in the stream [SX 8078 9400] near Westwater, Tedburn St Mary, showed decalcified conches of cf. *H. subglobosum*, lacking details of the ornament. The large number of small individuals in these and adjacent samples, and the lack of associated ammonoids, suggest a horizon within the Chokierian (H_{1a} Zone).

Exposures in cuttings for the A30 road between Tedburn St Mary and Pathfinder Village were recorded by Dr P Grainger (personal communication, 1992) during construction of the road in 1976. The cutting [SX 804 935] north of Hackworthy showed mainly grey brittle shales and silty shales with sporadic thin siltstones and local ironstone nodules. South of Winslakefoot, the cutting [SX 820 935] showed dark grey shales with micaceous siltstones and occasional sandstones. The shales are mainly platy, but locally have a prominent prismatic structure. The cutting [SX 829 931] near South Park Copse, south of Great Huish Farm, showed at the western end very deeply weathered, pale yellowish brown and grey, silty clay with bleached sandstone, passing down to pale grey, rotten shales and very broken, orange-brown sandstone and siltstone. Sandstones, some thick, are commoner at the eastern end of the cutting. South of Pathfinder Village, the cutting [SX 844 931] showed, at the base, black carbonaceous shale with sporadic siltstones and thin sandstones, becoming brown and pale brown near surface.

Two localities in the Nadder Brook at Halsfordwood [SX 8772 9375 and 8792 9373] showed dark grey to black shales with some sandstones. Both yielded *Vallites striolatus*, a long-ranging homoceratid within the Kinderscoutian (R₁ Zone) (Table 3, localities 3 and 4; Plate 2 (2, 4, 5, 6, 8, 10)).

Sporadic exposures are present along the valley of the Alphin Brook. At West Town [SX 8561 9296], Whitestone, vertical dark grey to black shales with sporadic thin (about 2 cm) sandstone beds strike at 093°. About 500 m east of West Town [SX 8616 9297], dark grey shale with pyrite nodules and grey, micaceous, cross-laminated siltstone beds dip 80° to 065°. Several exposures are present in the stream about 400 m west of Cutteridge Gate, near Osbornes Farm, as follows: [SX 8732 9273] black shale with nodules dips 80° to 163°; [SX 8738 9272] black shale with sporadic sandstone beds up to 0.2 m thick with load casts on the base dips 60° to 153°; [SX 8797 9271] black and dark grey shale dips vertically, striking 091°, and contains a muddy, cross-laminated, 0.05 m sandstone bed; [SX 8803 9272] dark grey shale with a 0.15 m-thick sandstone dips at 65° to 351° (inverted); [SX 8808 9270] dark grey to black shale or mudstone with siliceous or chert nodules dips vertically and strikes at 093°.

Nodules collected from the core of an anticline [SX 9050 9465] near Exwick Barton yielded fossils indicative of the early Chokierian H₁ₐ Zone (Table 3, locality 6). At Pynes Water Works, Cowley, an excavation [SX 9112 9602] showed very dark grey shales with scattered thin sandstones and slightly calcareous siltstones. Fossils indicate the early Chokierian H₁ₐ Zone (Table 3, locality 7).

Bridford

The Ashton Shale west of the 79 easting forms the country-rock of the Dartmoor Granite and is extensively recrystallized. Massive, dark grey, fine-grained hornfels, together with pale grey, siliceous hornfels, are exposed [SX 7861 8811] in a track north of Leign Farm and form a splintery brash in adjacent ground.

A quarry [SX 7985 8818] beside the B 3212 road in Bridford Wood shows brown-weathering, locally ferruginous, medium to dark grey, slightly micaceous, splintery pyritous shales with sparse pale grey, fine-grained and sparsely laminated sandstones or siltstones. The thin sandy to silty layers are not easy to distinguish from the shale; although the sequence is bedded and dips north at about 56°, the presence of a slightly steeper, north-inclined pencil cleavage has obscured most of the original sedimentary features. The suggestion of a normally graded sandstone within a single layer indicates that the strata young towards the north.

An exposure [SX 8058 8815] along a track in Bridford Wood, within the upper part of the Ashton Shale, shows very dark grey or black, recrystallised shale with numerous 20 to 70 mm-thick, medium grey sandstones. Indistinct cross-lamination in one sandstone suggests that the sequence faces north.

Doddiscombsleigh

A section [SX 8579 8651] in the track south of Town Barton, Doddiscombsleigh, shows 4.7 m of yellowish-brown-weathering, grey shale with sporadic, thin sandstones mostly less than 6 cm thick, indicating a sandstone:shale ratio of about 1:10. The beds dip at 58° to 182°. Some concretions with grey silt cores are present.

About 500 m east-south-east of St Michael's Church, Doddiscombsleigh, exposures [SX 8623 8635] show shales with thin sandstones dipping at 57° to 171°. A section [SX 8654 8608] in Tick Lane shows red-stained shales with thin sandstones. A

minor antiform plunges at 17° to 115°; the axial plane dips at 76° to 202°.

CRACKINGTON FORMATION (UNDIVIDED PART)

Above the Ashton Shale, the remainder of the Crackington Formation consists of rhythmically interbedded grey shales and turbiditic sandstones. The base is defined by the gradational increase in the proportion of sandstone beds from 10 per cent or less of the total in the Ashton Shale to 20 or 30 per cent in the overlying sequence. This change in sandstone proportions is accompanied by an upward change in clay mineralogy (p.11).

The sequence is folded and faulted and lacks distinctive marker bands, so that its thickness is difficult to calculate; at least 1000 m of strata are estimated to be present in the district. There is some variation in the proportion of shale to sandstone, but there are insufficient data to decide whether the variation has stratigraphical significance. Over most of the district, the sequence is uniform and shows little lithological variation, except in the west where sandstone units of mappable thickness, with individual thickly bedded sandstones up to 2.1 m thick, are present south and south-west of Tedburn St Mary along the southern side of an anticline cored by Ashton Shale (Figure 3). Thicker beds (up to 2.4 m) of massive sandstone are also present [e.g. ST 0137 0221] on the northern side of the Ashclyst Forest inlier. Their apparent absence elsewhere in the district suggests that they are probably lenticular. Similar sandstones were recorded from the eastern Okehampton district (Edmonds et al., 1968, p.81, pl. 2A).

Excluding these thick sandstones, the origin of which is uncertain, turbidites in the sequence range from siltstone beds only several millimetres thick to sandstones up to 0.4 m thick. A histogram of thicknesses of sandstone and siltstone beds in the formation between Exeter and Dunsford (Figure 4 a) shows that 87 per cent of the beds are thinner than 7 cm, and 60 per cent thinner than 3 cm. The maximum bed thickness recorded in the district was 40 cm at two localities. The sandstones are medium to dark grey where unweathered, well-cemented, fine- to medium-grained greywackes. Typically, they consist of subrounded to subangular quartz grains, opaque minerals and lithic fragments, in a matrix (forming up to about 30 per cent of the total volume) of recrystallised clay minerals (sericite and chlorite) and silt-grade quartz. Mica commonly occurs on bedding surfaces of siltstones and fine-grained sandstones. Carbonised plant fragments are locally concentrated on bedding surfaces, but occur throughout the Crackington Formation.

The mudrocks of the Crackington Formation range from very carbonaceous and pyritic shale to slightly carbonaceous silty shale, locally with sideritic mudstones. Claystone nodules cemented by ankerite and siderite or pyrite are present in some shale bands (Grainger, 1983). Illite dominates the clay mineral assemblages in the shales of the Crackington Formation, and is accompanied by kaolinite, chlorite, and a mixed-layer mineral (Grainger and George, 1978).

The sandstones and siltstones show most of the sedimentary features of a moderately distal turbidite suite.

a.

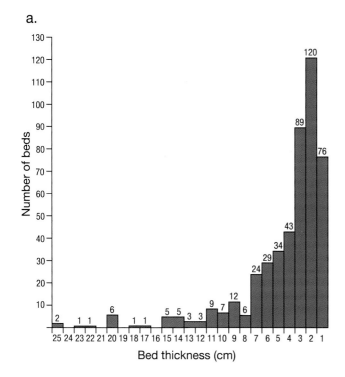

b.

Figure 4 (a) Histogram of bed thickness of sandstones in the Crackington Formation between Exeter and Dunsford (477 measurements). (b) Current rose diagram for the Crackington Formation of the Pinhoe Brickpit [SX 955 947] based on flute and groove casts (23 measurements).

The sandstones are laterally persistent sheets of even thickness with sharp bases and gradational tops. The bases of the thicker beds show a wide variety of sole markings including flute, groove, and load casts (Plate 1c). These enable the way-up of the strata to be determined, and are also of value in determining the directions of flow of the turbidity currents. Graded bedding is developed particularly in thicker sandstone beds, and material of coarse sand or granule size is present within large flute casts at the base of some beds. Cross-lamination or parallel lamination is also locally developed, especially towards the tops of thicker sandstones. The thinner beds of fine-grained sandstone and siltstone are generally parallel-laminated.

Directional data deduced from measurements of sole markings in the outcrop west of Exeter indicate that the turbidity currents flowed approximately east–west. Measurements at Pinhoe Brickpit [SX 955 947] indicate a principal derivation of sediment from the south-west, with a contributory input from the north-east (Figure 4 b).

BIOSTRATIGRAPHY

The Crackington Formation of the district is sparsely fossiliferous; faunas collected during the present survey are listed in Table 3, with localities shown in Figure 3. Photographs of selected fossils are shown in Plate 2. The fauna is typically dysaerobic (tolerant of low levels of oxygen at the substrate surface) and includes the bivalves *Caneyella*, *Dunbarella* and *Posidonia*, gastropod spat, ammonoids and orthocone nautiloids. Marine fauna is concentrated in thin shale horizons and in rare cases as decalcified impressions or clasts in the bases of sandstone turbidite beds. Typically, preservation is in small lenticular decalcified nodules. Faunal intervals correspond to the marine bands of the Pennine Basin in northern England (Ramsbottom, 1977) where they are usually separated by faunally barren strata which are considered to represent more brackish or freshwater environments associated with deltaic progradation. However, in the Exeter district, turbidite deposition was maintained throughout the Namurian.

Fragmentary plant material is present throughout the formation, but identifiable material is scarce. Morton (1958, p.111) recorded a poorly preserved 25 mm-long pith-cast of a stem from *Calamites* sp. at a locality [SX 8320 8717] near Bridfordmills. Similar stems were also recorded from the base of a sandstone at Pinhoe Brickpit [SX 9550 9458]. Arber (*in* Collins, 1911) commented on the fragmentary nature of even the best fossil plant material from the Exeter district, and the consequent difficulty of identification. She noted that *Neuropteris* appears to be common, including possible *N. schlehani*. Another fern-like plant may be *Urnatopteris tenella*. Several *Calamites* occur, but the species are uncertain.

Phillips (1841) recorded ammonoids from the Exeter district, and attributed them to *Goniatites inconstans* (R_{1a} Zone?). Other early records of fossils are given in Collins (1911). Re-examination by Butcher and Hodson (1960) of the collections of Ussher (1902) and Collins (1911) indicated the presence of the following zones: H_{1b} (*Homoceras beyrichianum*) from Idestone Hill [SX 8755 8826]; H_{2b} (*Homoceras undulatum*) from Bonhay Road [SX 9143 9273 to 9146 9257], Exeter; R_{1b} (*Reticuloceras regularum*, *R. moorei* and *R. nodosum*) from Stoke Road, Exeter and from Pinhoe Brickpit (*R. nodosum*) [SX 955 944]; R_{1c} (*Homoceras striolatum*) from Pinhoe Brickpit; R_{2a} (*Reticuloceras*

Plate 1a   Teign Chert at Mistleigh Copse [SX 8543 8599], Doddiscombsleigh, consists of well-bedded black chert in units from a few centimetres to 0.3 m thick, separated by thin grey to grey-green shale and mudstone beds and partings from less than 1 cm to 8 cm thick. The hammer is 0.3 m long (A 15267).

Plate 1b   Crackington Formation in the eastern part [SX 9551 9460] of Pinhoe Brickpit, Exeter. The section shows inverted strata, consisting of grey shale with subordinate thin interbeds of turbiditic sandstone, dipping north at 60° (A 15274).

aff. *gracile*) from Pinhoe Brickpit; and R_{2b} (*Reticuloceras bilingue*) from Perridge Tunnel [SX 862 903], Longdown (House and Selwood, 1964). Butcher and King (1967) recorded ?E_2 to R_{1a} zone ammonoids of ?Arnsbergian to Kinderscoutian age from the Bonhay Road section [SX 9143 9273 to 9146 9257] (Figure 35). Grainger and Witte (1981) noted three localities in the Ashton Shale yielding H_{1a} Zone ammonoids, and one near Pocombe Bridge indicating the Kinderscoutian (R_1 Zone). In the Newton Abbot district, the Ashton Shale has yielded E_{2b} and E_{2c} zone ammonoids of Arnsbergian age (Selwood et al., 1984, p.65). Grainger (1983) reported three ammonoid localities in the Tedburn St Mary area indicating that the Ashton Shale extends up into the H_{1a} Zone. Localities yielding H_{1a} Zone ammonoids (Table 3) at Exwick Barton [SX 9050 9465] and Pynes Water Works [SX 9112 9602], Cowley, may represent small anticlinal cores of Ashton Shale not shown separately on the map. The youngest Ashton Shale occurs at Nadder Brook, Halsfordwood [SX 8772 9375 and 8792 9373], where *Vallites striolatus* occurs, indicating a Kinderscoutian (R_1 Zone) age. Since possible *Homoceras beyrichianum* (H_{1b} Zone) and *H. undulatum* (H_{2b} Zone), representing the Chokierian and Alportian stages, are known from beds above the Ashton Shale, the top of the member is apparently diachronous.

Plate 1c Crackington Formation at Pinhoe Brickpit [SX 9550 9458]. The base of an inverted sandstone bed shows sole markings, including load casts, groove casts, prod marks and slide marks. The hammer is 0.4 m long (A 15272).

Plate 1d Bude Formation at East Henstill Quarry [SS 8158 0413]. Beds of massive sandstone up to 1.05 m thick are interbedded with blocky and locally nodular siltstone. The hammer is 0.3 m long (A 15275).

The earlier collections noted above, and those made during the course of the present survey (Table 3) indicate that the Crackington Formation sequence overlying the Ashton Shale ranges from Chokierian (H_{1b} Zone) to Marsdenian (R_{2b} Zone) (Table 2). The Chokierian and Alportian sequence is believed to be complete, but only the marine bands of Isohomoceras subglobosum ($H_{1a1–3}$), Homoceras beyrichianum (H_{1b1}) and H. undulatum (H_{2b1}) have been located. The overlying Kinderscoutian Stage appears to occupy much of the formation in the district. Faunas including the Hodsonites magistrorum (R_{1a1}), Reticuloceras nodosum (R_{1b2}), R. stubblefieldi (R_{1b3}), R. reticulatum ($R_{1c1–3}$) and R. coreticulatum (R_{1c4}) marine bands have been recognised, and it is probable that the sequence is complete, although the other Kinderscoutian faunas have not been proved in the district. The youngest Crackington Formation in the district is of Marsdenian age, and is represented by faunas from the Bilinguites gracilis (R_{2a1}) and overlying B. bilinguis ($R_{2b1–2}$) marine bands. Elsewhere in south-west England, younger Namurian faunas are known, and in north Devon the Crackington Formation extends into the Westphalian and includes the Amaliae Marine Band (Ramsbottom et al., 1978) (Table 2).

The Crackington Formation yielded only sparse to very sparse assemblages of palynomorphs (Turner, 1992), and

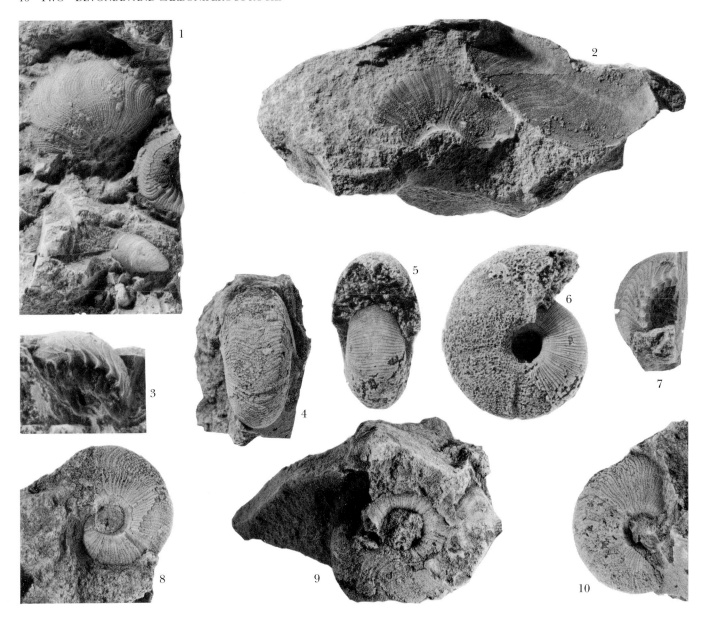

Plate 2 Fossils from the Crackington Formation. Localities are shown on Figure 3 and in Table 3.

1) Nodule with fragmentary *Reticuloceras* cf. *pulchellum* (left) and *R. nodosum* (right) with the bivalve *Caneyella* sp. (bottom); specimen BRI 1633, locality 19, × 3.

2) *Vallites striolatus*, umbilical fragment showing spiral lirae; specimen RHR 482, locality 3, × 3.

3, 7) *R. coreticulatum*, lateral view of juvenile fragments; 3–specimen Ro 7529, × 6; 7–specimen Ro 7526, × 3. Both from locality 14.

4, 5, 6, 8, 10) *V. striolatus*, ventral and lateral views; 4, 8, 10–Specimen GC 72, × 2; 5, 6–specimen GC 73, × 3. All from locality 4. Note pathological repair to transverse ornament in fig. 5.

9) *R. reticulatum*, lateral view; specimen RH 5150, locality 11, × 1.

Photographs by Mr T P Cullen

principally contains fragments of inertinite. The presence of *Lycospora pusilla* in most samples indicates a Carboniferous age. A sample [SX 8613 9134] from near Longdown yielded *Lycospora pusilla*, *Densosporites anulatus*, *Calamospora* spp., *Crassispora kosankei* and *Raistrickia saetosa*, indicative of a late Namurian or younger age.

Ford Brook

The Ford Brook south of Neopardy shows sporadic exposures from a point near Winstode [at about SX 7927 9732] to where it passes out of the district. The lithologies, recorded during the survey of the Okehampton district, are grey shaly mudstones with sandstone beds up to about 0.3 m thick. The sequence is

Table 2 Ammonoid zones in the Crackington Formation of the district.

Permian rocks					
Upper Carboniferous	Westphalian	G₂			Bude & Bideford formations — North Devon
	Namurian	Yeadonian	G₁	b / a	Amaliae Marine Band
		Marsdenian	R₂	c / b X / a X	Crackington Formation (undivided part)
		Kinderscoutian	R₁	c X / b X / a X	
		Alportian	H₂	c / b X / a X	Exeter district
		Chokierian	H₁	b X / a X	Ashton Shale
		Arnsbergian	E₂	c X / b / a	
		Pendleian	E₁	c / b / a	? — ?
Lower Carboniferous	Viséan (pars) / Brigantian (pars)				Teign Chert

X   zones recorded in the Crackington Formation of the Exeter district.

mainly north-dipping, with inverted beds common. Near East Ford, a stream section [SX 7892 9560] shows grey shaly mudstone with thin sandstone beds and R₁ₐ Zone fossils (Table 3, locality 21). Nearby, west of East Ford, sporadic stream exposures [SX 7850 9550 to 7872 9550] show tight sigmoidal folds overturned to the south in grey shaly mudstone with sandstone beds up to 0.3 m thick. Minor faults are present parallel to the folds.

Venny Tedburn

Exposures [SX 8070 9648] in the valley north-west of Harford show an 80 m section in dark grey shales (65 per cent of the total) with pale grey, fine-grained sandstone interbeds up to 30 cm thick (35 per cent of the total).

Sporadic exposures of shales with thin sandstone and siltstone interbeds are present in the Kelland Brook [SX 8290 9682 to 8375 9656], south-east of Venny Tedburn. An exposure [SX 8405 9618] in a tributary of the Kelland Brook, about 200 m north of Copperwalls Farm, shows black shale with sandstone and siltstone beds up to 5 cm thick, dipping north at 35°.

Newton St Cyres–Whitestone–Nadderwater

A section [SX 8782 9538] in Newton Wood shows up to 3.6 m of weathered grey shales interbedded with thin (mostly less than 10 cm) fine-grained sandstones, dipping at 40° to 340°. The shale:sandstone ratio is 3.7:1. A flat-lying minor fold plunges at 20° to 280°. A track section [SX 8642 9581] in Whitestone Wood shows inverted shales and sandstones dipping at 80° to 004° and 30 m south-west, a right-way-up succession dipping 43° to 010°. Another section [SX 8626 9566] in Whitestone Wood exposes near-isoclinal minor folds in near-vertical dark grey shales. The fold axes trend east-west.

A north–south section [SX 8637 9535] in a cutting north of Seven Stone Ball, south of Whitestone Wood, shows shales with subordinate thin sandstones in folds overturned to the south with the inverted limbs dipping more steeply than the right-way-up limbs. The section is cut by numerous small strike faults. The ratio of shale to sandstone is 3.5:1.

Sporadic exposures in the Nadder Brook between Church Bridge [SX 8684 9392], Whitestone, and a point [SX 8754 9396] near Halsfordwood show grey shales with sandstones mostly between 5 cm and 20 cm thick.

Roadside exposures south-east of Hackworthy Corner [SX 8757 9416 to 8806 9398] show shale with sandstone beds 2 to 5 cm thick [SX 8764 9413] in the north side of the road, dipping at 25° to 338°. Farther east [SX 8775 9405] are exposures of grey shale with sandstone beds 10 to 15 cm thick, with a minor fold plunging at 20° to 263°; the beds dip at 50° to 353° (inverted). Nearby [SX 8788 9401], grey to medium grey shale with thin sandstone beds up to 0.05 m dip at 70° to 348° (inverted).

Sporadic roadside exposures are present along Rowhorne Road [SX 8855 9410 to 8905 9375], north-west from Whitestone Cross [SX 8948 9316], Nadderwater. At one point [SX 8898 9382], weathered brown and grey shale and thin sandstone beds are folded asymmetrically; the shallow limb dips at 48° to 347° and the steeper limb is vertical and strikes at 067°. Nearby [SX 8905 9375], pale greyish brown shale with thin (less than 20 cm) sandstone beds shows a fold with the shallow limb dipping at 40° to 313° and the steeper overturned limb dipping at 84° to 327°.

Cowley–Stoke Woods

A disused quarry [SX 9013 9667] south-east of Half Moon Village shows a sequence of 5 m of grey shales with thin sandstones, dipping at 40° to 328°. The shale:sandstone ratio is about 3:1. A bullion in a mudstone bed yielded orthocone fragments and indeterminate juvenile ammonoids. Farther south, a road cutting [SX 9056 9586] at Cowley Hill shows a right-way-up sequence of shales and sandstones dipping at 40° to 342° (northern part) and 45° to 332° (southern part).

An inverted sequence of shales with subordinate thin sandstones, some of which show prominent load casts, is exposed in a 10 m cutting [SX 9247 9617] along the north side of Stoke Woods; dips are 45° north at the top, and 54° to 342° at the base where the shales show pencil cleavage. Nearby [SX 9248 9614], fossils indicated the R₂ₐ Zone (Table 3, locality 17). A second cutting [SX 9333 9670] along Stoke Road shows an overturned fold in shales and sandstones with the inverted southern limb

Table 3   Carboniferous fossils from the district.

Locality number in Figure 3	Grid reference	Stratigraphic unit	Fauna	Zone or Marine Band index (Ramsbottom et al., 1978)
1	SX 8545 8666	TCh	cf. *Posidonia membranacea*, cf. *Sudeticeras* sp., trilobite free cheek	$P_{1d}–E_{1a}$
2	SX 8544 8666	TCh	*Lusitanites* or *Neoglyphioceras* spp.	$P_{1d}–P_{2b}$
3	SX 8772 9375	AnSh	*Vallites striolatus* (Plate 2 (2))	R_1
4	SX 8792 9373	AnSh	*Vallites striolatus* (Plate 2 (4, 5, 6, 8, 10))	R_1
5	SX 8625 9655	CkF	cf. *Hodsonites magistrorum* (poor flank moulds)	R_{1a1}
6	SX 9050 9465	?AnSh	*Caneyella semisulcata*, anthracoceratid or dimorphoceratid indet., *Isohomoceras subglobosum*	H_{1a}
7	SX 9112 9602	?AnSh	*Caneyella semisulcata*, *Straparollus* sp., anthracoceratid or dimorphoceratid, *Isohomoceras subglobosum*	H_{1a}
8	SX 8527 9070	CkF	turreted gastropod, *Reticuloceras* cf. *stubblefieldi*	R_{1b3}
9	SX 8506 9001	CkF	*Caneyella* sp., *Posidonia minor*, *Posidonia obliquata*, turreted gastropods, orthocone fragments, anthracoceratid indet., *Reticuloceras* cf. *reticulatum*, *R.* cf. *coreticulatum*, *Vallites striolatus*	R_{1c4}
10	SX 8666 8988	CkF	homoceratid? indet.	$H_1–R_1$
11	SX 8730 8994	CkF	anthracoceratid indet., *Reticuloceras reticulatum*, *Vallites striolatus* (Plate 2 (9))	$R_{1c1–3}$
	SX 8892 9170	CkF	spat, *Posidonia* sp. juv.	
12	SX 8941 9136	CkF	spat, *Caneyella minor* (including elongate varieties)	$R_{1b}–R_{1c}$
13	SX 8961 9158	CkF	spat, hexactinellid sponge spicules, cf. *Hodsonites magistrorum* (juveniles), cf. *Vallites* sp.	R_{1a1}
	SX 8960 9154	CkF	spat, *Dunbarella* cf. *rhythmica*, cf. *Vallites striolatus* (juvenile)	R_1
14	SX 9055 9245	CkF	turreted gastropod spat, *Reticuloceras coreticulatum* (Plate 2 (3, 7))	R_{1c4}
15	SX 9143 9265	CkF	*Caneyella* sp. poor juv., *Homoceras* cf. *smithi* juv., conodonts (indet. moulds)	H_{2a1}?
16	SX 9040 9357	CkF	*R.* cf. *coreticulatum*, spat	R_{1c4}?
17	SX 9248 9614	CkF	*Bilinguites gracilis*	R_{2a1}
18	SX 9389 9417	CkF	anthracoceratid or dimorphoceratid, reticuloceratid indet.	R_1 or R_2
19	SX 9489 9520	CkF	*Posidonia obliquata*, turreted gastropod, anthracoceratid or dimorphoceratid, *Reticuloceras nodosum*, *R.* cf. *pulchellum*, (Plate 2 (1))	R_{1b2}
20	SX 9550 9465	CkF	hexactinellid spicules, *Dunbarella* fragment, *Homoceratoides* fragment?, *Reticuloceras pulchellum*	R_{1a}?
21	SX 7892 9560	CkF	*Dunbarella* sp., *Vallites henkei*, *Reticuloceras* sp. [*R. circumplicatile* group], *Reticuloceras* cf. *pulchellum*, mollusc spat	R_{1a}
	SX 8681 9722	CkF	orthocone nautiloid (juvenile fragment), ammonoid (juvenile pachyconic)	Namurian?
	SX 9024 9662	CkF	orthocone fragment, ammonoid juveniles indet.	Namurian?

CkF: Crackington Formation (undivided part)

AnSh: Ashton Shale

TCh: Teign Chert

dipping at 60° north and the right-way-up limb at 42° to 012°. The axial plane of the fold is faulted.

Exwick–Redhills–Exeter

A temporary exposure [SX 9040 9357] at Exwick yielded ammonoids tentatively assigned to the R. coreticulatum Marine Band (R_{1c4}) (Table 3, locality 16).

Shales and sandstones are exposed in the cutting of St Andrew's Road between Exwick Barton [SX 9052 9478] and Exwick Mill [SX 9071 9344]. At one point [SX 9057 9462], 180 m south-south-east of Exwick Barton, shales with subordinate thin sandstone interbeds are inverted and dip steeply north; 110 m south, a fault zone is indicated by disturbed shales and fractured sandstones with much brown iron staining and quartz veining. At 135 m farther south, the section continues in inverted shales with sandstone interbeds dipping at 65–85° north. Near Exwick Barton Cottage [SX 9066 9443], a fold shows a steep inverted northern limb dipping at 65° to 342°, and a shallow right-way-up limb dipping at 35° to 352°. The section continues 70 m south of Exwick Barton Cottage in inverted shales with subordinate sandstone interbeds. The shale: sandstone ratio is about 4:1 and the dip is 65–70° to the north. Throughout the interval, the shales are harder and less weathered and are, in places, slightly calcareous. A nodule bed is present 114 m south of the cottage. The section continues to the south in shales with thin flaggy sandstones, all inverted and dipping steeply to the north. At 183 m south of the cottage an 0.2 m-thick sandstone, right way up and with a coarse base, dips at 70° to 352°. Between 195 m south and 215 m south of the cottage, shales with thin sandstone interbeds are inverted and dip at 80–85° north. At 235 m south of the cottage, an anticline shows shearing along the axial plane, and at 240 m south, beds of shale with thin sandstones are vertical. The section continues to a place [SX 9072 9413] 150 m north of Exwick Mill in weathered, brown-stained shales with thin sandstone interbeds that are right way up.

An exposure [SX 9055 9245] in Redhills, above the buttress wall at the junction with the main road, yielded an R_{1c} Zone fauna (Table 3, locality 14; Plate 2 (3,7)).

East of the River Exe, a section extends northwards from the main entrance [SX 9115 9398] of Birks Halls, University of Exeter: between 20 m and 36 m north of the entrance, grey shales with scattered thin sandstone interbeds dip at 70° north. At 37 m north, the beds are vertical and strike east–west, and at 38 m north an inverted sandstone 0.32 m thick dips at 65° to 358°. The section between 40 m and 51 m north of the entrance shows a recumbent anticline in shales with sandstone interbeds.

The river cliff in Bonhay Road, Exeter, provides a north–south section [SX 9142 9272 to 9146 9257] in Crackington Formation. The sequence is shown diagrammatically in Figure 35, together with the approximate locations of ammonoids of the ?E_2, H_{1b}, H_{2a}, H_{2b} and R_{1a} zones, recorded by Butcher and King (1967). Faunas collected from one locality [SX 9143 9265] in the section during the present survey indicated the H_{2a} Zone (Table 3, locality 15). The record of *Goniatites sphaericostriatus* (Butcher and Hodson, 1960), indicative of the P_{1c} Zone, was probably from a block of Westleigh Limestone, not in situ. Steep strike faults downthrowing to the north are apparently later than shallower, mainly northerly dipping reverse faults.

A road cutting [SX 9343 9477 to 9346 9455] at Stoke Hill, east of the Mincinglake Stream, shows recumbent folds in shales with subordinate sandstone interbeds; the shale:sandstone ratio is about 2:1. The inverted limbs dip at 50–65° north, and the right-way-up limbs dip at 15–35° north. The stream bed [SX 9384 9424 to 9406 9401] of the Mincinglake Stream, Stoke Hill, exposes a succession of tight folds in cleaved shales with subordinate sandstone interbeds up to 30 cm thick. The right-way-up limbs dip 35–40° north and the inverted limbs 50–60° north. The shales are locally purplish red stained, and nodules of haematite and siderite occur in scattered bands. Small-scale fractures typically trend at 347° with dextral displacements of about 1.1 m. Nodules and shale from an exposure [SX 9389 9417] 240 m north-west of Calthorpe Road, Stoke Hill, yielded indeterminate ammonoid remains indicative of the Kinderscoutian (R_1) or Marsdenian (R_2) zones (Table 3, locality 18).

Pinhoe

A shale bed in stream-bed exposures [SX 9489 9520] in the Pin Brook yielded an R_{1b} fauna (Table 3, locality 19; Plate 2 (1)).

The northern part of the Pinhoe Brickpit shows medium grey shaly mudstones, with regular interbeds at about 0.5 to 1.5 m intervals of thin (up to 10 cm), pale grey, well-cemented sandstone (Plate 1b). An atypical sandstone in the northernmost part of the pit [SX 9558 9480] is soft, yellowish brown, fine-grained greywacke (E 58857). It consists of subangular to subrounded quartz, 0.05 to 0.15 mm in size, with some muscovite and a little chert set in a matrix of clay, mica, clay minerals, opaque iron oxide and a little authigenic quartz. Opaque iron oxide constitutes about 10 per cent of the rock and between 10 and 15 per cent of the matrix. In the central part of the pit [SX 955 947], a very coarse-grained (particles average between 1 and 2 mm) infill to a flute cast consists mainly of angular

quartz in a ferruginous matrix; other particles include angular chert, weathered and vesicular igneous rocks, fine-grained greywacke and rare flake of muscovite. On the south side of the pit [SX 9550 9458] there are many long (up to 20 m) exposures of the overturned bases of the sandstones. Almost all are covered with sole marks (Plate 1c); drifted plant remains, including *Calamites* (horsetail) stems up to 3 cm diameter and 0.5 m long, sponge spicules and ammonoids occur on the bases of the sandstone. Analysis of sole markings indicates a principal derivation of sediment from the south-west, with a contributory input from the north-east (Figure 4 b).

A fauna collected during the survey [SX 9550 9465] indicates the R_{1a} Zone (Table 3, locality 20). The bivalve *Dunbarella* sp. is locally common on shale bedding surfaces. An earlier part of the workings at Pinhoe [SX 956 944], now backfilled, yielded the ammonoids *Vallites striolatus*, *Reticuloceras nodosum* and *Bilinguites* aff. *gracilis*, indicative of the *R. nodosum* (R_{1b}) and *Bilinguites gracilis* (R_{2a}) zones (Butcher and Hodson, 1960).

Clyst Hydon–Clyst St Lawrence

Near Washbeerhayes Farm, about 2 km west-north-west of Clyst Hydon, thicker more massive sandstone beds are present in the Crackington Formation. Stream-bed exposures [ST 0139 0219] show 0.4 m of purple micaceous sandstone, dipping at 24° to 132° (way up uncertain). A thin section (E 65491) shows it to be a fine-grained quartz-wacke composed of quartz, mica, opaque minerals and secondary iron oxides, in a clay-grade matrix. Nearby [ST 0137 0221] are about 2.4 m of purple, hard, massive to flaggy fine-grained sandstone with shale interbeds, way up uncertain, dipping at 38° to 320°.

About 0.8 km west of Clyst Hydon, sections occur sporadically along the course of a stream south of Roach Copse [ST 0220 0139 to 0257 0146]. An exposure [ST 0234 0140] shows 5.15 m of shale and mudstone with interbeds up to 0.3 m thick of purple, locally micaceous, sandstones. One sandstone is a coarse silt to very fine sand-grade quartz-wacke (E 65490).

In Clyst St Lawrence, exposures [ST 0254 0003] show 2 m of purple shale with sporadic sandstones up to 0.2 m thick, mostly 5 cm or less. The sequence dips at 52° to 351° (inverted).

Ashclyst Forest

Exposures of red-stained grey shaly mudstones with sandstone beds up to a maximum of 0.4 m thick occur principally in the two eastward-draining streams [SY 0015 9955 to 0120 9898 and 0085 9970 to 0140 9935] and a track [SY 0080 9945 to 010 994] within the Ashclyst Forest, and in road ditches [SY 0142 9994 to 0162 9994] and road banks [SY 018 997 to 0194 0000] east of the forest.

Great Fulford

A quarry [SX 7970 9219], about 750 m north-east of Great Fulford, shows thickly bedded sandstones differing from the normal Crackington Formation type. The section is as follows:

	Thickness m
Shale, grey, and mudstone, with beds of hard grey sandstone 0.1 to 0.3 m thick. Dip 72° to 345°, inverted	1.5
Sandstone, thickly bedded	0.9
Shale	0.4
Sandstone	0.4
Gap in section—probably mainly thickly bedded sandstone	4 to 5

	Thickness m
Sandstone, hard, grey, slightly feldspathic and micaceous, fine-grained, in beds up to 0.3 m thick	2.1
Shale, with sandstone beds up to 0.3 m thick	c. 3.0
Sandstone	(seen) 0.6

Similar sandstone is also present in small exposures [SX 7926 9252] at Melhuish Barton.

Exposures [SX 8112 9198] at Windout show 3.6 m of yellowish-brown-weathered, grey and dark grey shale with interbedded hard, grey, fine-grained sandstones, mostly less than 7 cm thick, some with laminated tops and bases with flute casts. The beds dip at 72° to 330° (inverted). The shale:sandstone ratio is 2.3:1. About 60 m north of Windout, an exposure [SX 8112 9203] shows well-developed cleavage, dipping at 53° to 345°, which cuts indistinct bedding dipping at 70° to 070°.

Dunsford

Crags above the River Teign at Steps Bridge [mid-point SX 8038 8843] show a sequence about 10 m thick of hornfelsed shale with thin sandstones dipping at 46–56° to 350°. Sandstones 1.5 to 8 cm thick typically have undulose lower surfaces that resemble load structures, massive basal intervals and finely laminated tops. A thin-section (E 65541) of a finely parallel-laminated sandstone shows it to be very fine-grained quartz-wacke, with slight grading from fine to very fine sand-grade and a sharp boundary with mudstone. Cross-laminated sandstones occur locally, and scarce sandstone stringers or lenses, about 0.5 cm thick, contain ripple cross-lamination. Bottom structures are rare; where present, sole markings are subhorizontal and oriented east-west. An isolated face contains two sets of sole markings; one plunges at 9° to 084° and the other plunges 30° to 080°. A single small-scale intrafolial fold, with a 0.3 m wavelength and axial surface inclined at 65° toward the north, may represent a slump structure.

Exposures in the river at Steps Bridge [SX 8035 8843 to 8045 8835] show a similar north-inclined sequence with a variably developed vertical cleavage. Sandstones are up to 20 cm thick and the shale:sandstone ratio is in the range 3:1 to 5:1.

An exposure [SX 8253 8837] north-east of Weeke Barton shows highly deformed, south-inclined, dark grey shales with interbedded pale grey, 6 cm-thick sandstones. A conspicuous cleavage is vertical or inclined north at 80°; minor folds, with 0.2 m wavelength, plunge at 20° to 240° and contain axial surfaces inclined north at 65°. River exposures adjacent to the previous locality and as far south as Burnwell [SX 8282 8791] show south-dipping, right-way-up sequences with shale:sandstone ratios of 4:1 or less. A small waterfall has developed over a 0.3 m-thick sandstone at Burnwell, and forms part of a relatively sandy sequence in which 3 cm-thick cross- and parallel-laminated sandstones account for about 30 per cent of the strata.

Bridford

Crackington Formation shales and sandstones are exposed in a 10 m-long near-vertical section behind the garage at Bridfordmills [SX 8340 8708]. The sequence is mainly grey shales, with 3 to 8 cm thick sandstones that account for 20 per cent or less of the strata. Intermittent exposures along the Teign Valley road (B 3193) [SX 8386 8636 to 8396 8623] south of Bridfordmills consist of medium grey, slightly micaceous sandstones, 2 to 3 cm and up to 15 cm thick, and brown-weathering shales. The beds dip south or south-south-east at moderate or steep angles, and are vertical along the line of a major east-north-east-trending high-angle fault [SX 8396 8623]. South of this fault, a thrust slice of Crackington Formation, typically inclined east-south-

east at moderate angles, occurs repeated within the Lower Carboniferous sequences.

Along the disused railway near Leigh Cross, a 90 m-long cutting [midpoint SX 8350 8750] exposes thinly bedded, rusty-weathering, pale to medium grey, quartzose, slightly micaceous laminated sandstones with partings of fine-grained slaty mudstone. The sandstone beds are typically 2 to 3 cm thick (maximum 20 cm), and exhibit a faint planar lamination and occasional cross-lamination. The beds in the central part of the section are near-vertical and dip north; the southernmost 40 m of the section is tightly folded on east–west subhorizontal axes with steep, north-inclined axial surfaces. The northernmost 25 m of the section includes beds inclined at 35° to 020° A small section behind a house [SX 8347 8745] near Leigh Cross consists of a highly cleaved, south-inclined shale and sandstone sequence in which the bedding dips at 76° to 342° and the cleavage dips at 85–90° to 355°.

Doddiscombsleigh

In a laneside section [SX 8503 8707] north of Lake Cottage, Doddiscombsleigh, grey shale with beds of grey sandstone and siltstone 2 to 7 cm thick dip at 34° to 159°. The section is cut by a small north-west–south-east fault. A section [SX 8503 8688] at Lake Farm (Lake House), Doddiscombsleigh, shows 4.2 m of locally quartz-veined grey shales and interbedded sandstones between 1 cm and 0.2 m thick; the shale:sandstone ratio is 2.7:1. The beds dip at 54° to 171°.

A section at Easternhill Farm [SX 8480 8838 to 8485 8834] exposes a highly deformed, sandstone-rich sequence. Sandstone beds are micaceous, 1 to 2 cm and up to 10 cm thick. They are cut by numerous white quartz veins about 0.5 cm wide. The sandstones have loaded bases, laminated tops, and in the north-western part of the section are inclined north at moderate angles and are the right way up. The rest of the section is highly folded and faulted; cleavage is repeated by faulting, and has more than one orientation. Sulphides, including chalcopyrite, are conspicuous. A near-vertical fault or zone of shear deformation passes through the section, inclined at 88° to 172°. Bedding dips 24 to 30° toward the north or north-west; cleavage is inclined at 60° to 325° and 82° to 168°.

A quarry [SX 8440 8798] in Lowley Wood exposes a 15 m-deep and 20 m-wide section through a sandy facies of the Crackington Formation. Sandstones, accounting for 30 and locally 50 per cent of the sequence, are typically 4 to 10 cm thick and are brown-weathering, medium grey and slightly micaceous. The upper intervals are faintly laminated, while the basal sections are sulphide rich and scoured into metalliferous shales. The shales are dark grey, contain abundant sulphides and exhibit a manganese-rich bloom on weathered surfaces. Beds are inclined at 83° to 162° while cleavage is oriented at 46° to 325°. Current lineations and possible prod marks plunge toward 165° and confirm that the sequence is inverted.

Longdown

Exposures [SX 8305 9068] about 200 m west-south-west of Ford, Dunsford, show about 5 m of Crackington Formation shales and sandstones, dipping at 52° to 345°. Yellowish-brown-weathered pale and dark grey shales are interbedded with hard grey fine-grained sandstones up to 12 cm thick (mostly less than 8 cm), some with groove casts, load casts and flutes on the base and laminated, locally planty and micaceous tops. The shale:sandstone ratio is 3:1.

A stream-bed section [SX 8350 9009] near Langdale shows 5.7 m of grey to dark grey shale with sparse sandstones up to 8 cm thick, dipping at 80° to 342°. The shale:sandstone ratio is

14:1. A 20 m section [SX 8398 9122] near Kingsford shows yellowish-brown-weathered, grey shales with sporadic sandstones up to 0.2 m thick, dipping at 80° to 177°. Thicker sandstones have laminated tops and well-developed sole markings, with grooves oriented east–west. The shale:sandstone ratio is estimated to be between 4:1 and 5:1. Joints dip at 68° to 271°.

A quarry [SX 8513 9224] near Kingswell shows about 10 m of shales with sandstones, dipping at 86° to 173° (right way up). The shales are yellowish-brown-weathered, grey; a few grey, spheroidally weathered, plant-rich, sandy silt beds are present. Three beds contain soft, orange-brown, silty, ferruginous nodules. The sandstones are hard, grey, ranging mainly between 1 and 7 cm; one bed is 23 cm thick and shows a basal surface with flute casts indicating flow from the east. The shale:sandstone ratio is 3.4:1.

Exposures [SX 8613 9134] behind Longdown Village Hall show 4.5 m of grey shales with interbedded grey sandstones between 1 and 6 cm thick; the strata are stained orange-brown and reddish brown at the western end. The inverted sequence dips at 68° to 355°. The shale:sandstone ratio is 4.6:1.

A 30 m-long cutting [SX 8654 9208] in Bondhouse Copse on the southern side of the Springdale landfill site north of Longdown shows a 5 to 6 m-high section in folded shales and sandstones. Several fairly open folds have axial planes dipping north at about 60°; axial-planar cleavage is locally developed. The shales are dark to pale grey, locally carbonaceous, interbedded with hard, grey, fine-grained sandstones between 2 and 20 cm thick. The shale:sandstone ratio is 2.5:1.

Riley (1991) recorded *Reticuloceras* sp. from a locality [SX 8507 9000] in a track in Cotley Wood, near Culver.

Two cuttings along the old railway are present on the south side of the Culver Tunnel. In the north cutting [SX 8520 8993 to 8516 8986], shales are interbedded with hard, grey, turbidite sandstones up to 0.2 m thick; the shale:sandstone ratio is estimated to be between 3:1 and 4:1. The shales locally show the development of intense 'pencil' cleavage. A synform adjacent to the tunnel mouth on the west side plunges at 27° to 085°. The axial surface dips at 64° to 183°. Joints at the same locality dip at 68° to 089°. About mid-way along the cutting, bedding dips at 82° to 175° (inverted). Cleavage dips at 50° to 342°. There is a well-developed intersection of bedding and cleavage, giving 'pencil' shales. The lineation formed by this intersection plunges at 6° to 270°. Poorly developed sole markings on the base of the sandstones trend at 088°. The general lithology in the south cutting [SX 8510 8980 to 8505 8977] is similar to that in the north cutting. At the south end, bedding dips at 84° to 185°; cleavage dips at 45° to 012°. The intersection of bedding and cleavage produces a strong 'pencil' cleavage; the associated lineation plunges at 4° to 093°. The general trend of poorly developed linear sole markings is 126°.

Exposures [SX 8506 9001] in a track west of the Culver Tunnel yielded fossils indicative of the Reticuloceras coreticulatum (R$_{1c4}$) Marine Band (Table 3, locality 9).

A 15 m section [SX 8527 9070] near Hill Farm, Culver, shows grey to dark grey shales with interbedded grey sandstones between 1 and 6 cm thick. The shale:sandstone ratio is 4.6:1. The inverted sequence dips at 68° to 355°. Fossils indicate the R$_{1b}$ Zone (Table 3, locality 8).

In Eastern Cotley Wood, north of Perridge House, cuttings [SX 8639 9070] show a section in folded shales and sandstones up to 4 m high over a length of about 20 m. The axial surface of an anticline at the south-western end of the exposure dips at 63° to 353°. In the central part of the exposure, the axial surface of a syncline dips at 67° to 348°; the fold axis plunges at 10° to 265°. At the north-eastern end of the exposure, the axial plane of a partly faulted anticline dips at 60° to 348°; the fold axis plunges at 12° to 075°.

West of the western entrance to the Perridge Tunnel, exposures [SX 8580 9028] on the south side of the cutting show about 3 m of grey to dark grey, locally 'paper', shales interbedded with hard, grey, fine-grained sandstones up to 0.1 m thick; the dip is 57° to 168°. The shale:sandstone ratio is 2.2:1. Exposures [SX 8577 9029] on the north side of the cutting show an anticline with the axial surface dipping north at 50°, and the axis plunging at 35° to 262°.

Fossils from the Perridge Tunnel include *Bilinguites bilinguis*, indicating the highest zone (R$_{2b}$) so far recognised in the district (Butcher and Hodson, 1960).

Cuttings in the disused railway line east of Perridge show sporadic exposures for about 1.5 km [to SX 8834 9043]. Exposures [SX 8690 9044] beneath a railway bridge on the south side of the cutting show shales with sporadic sandstones dipping at 80° to 171°. On the north side of the cutting, an anticlinal axis plunges at 8° to 259°. The axial surface dips north at 40°. Exposures [SX 8741 9034] on the south side of the railway cutting near Fordland Farm, about 1 km east of Perridge House, show grey to olive shale, silty shale, and siltstone with interbedded hard sandstones up to 10 cm thick; the shale:sandstone ratio is 2.4:1. Nearby exposures [SX 8745 9034] show groove casts and flute casts on a sandstone base; the flutes indicate flow from the east. The south side of the cutting at Ide Brake [SX 8780 9035] shows dark to very dark grey, possibly carbonaceous, shales with sporadic sandstones to 0.15 m thick dipping at 46° to 172°. Flute casts indicate current flow from 090°.

Roadside exposures [SX 8660 8986] of grey shales with interbedded sandstones at the entrance to Darnaford dip at 30° to 328°; the axis of a small syncline plunges at 15° to 270°. Riley (1991) recorded fragments of an indeterminate ?homoceratid ammonoid from track exposures [SX 8666 8988] near the entrance to the farm.

A cutting [SX 8730 8994] about 800 m east-north-east of Darnaford shows about 10 m of weathered grey shales and sandstones, faulted at the southern end. Soft, dark brown, rotted nodules contain ammonoids preserved as three-dimensional moulds; they indicate the R$_{1c}$ Zone (Table 3, locality 11; Plate 2 (9)). Bedding dips at 40° to 008°; cleavage dips at 86° to 165°. The intersection of bedding and cleavage gives rise to 'pencil' shales.

An exposure [SX 8643 8887] in a cutting in Ramridge Plantation, north of North Wood, shows about 3 m of shale with interbedded sandstones between 0.1 and 0.25 m thick. The thickest sandstone shows clear grading, with very coarse- sand-grade material at the base; the top shows parallel lamination. The sequence dips at 28° to 315°. Exposures [SX 8635 8870] in a track in North Wood show shales and interbedded sandstones on the north side, dipping at 47° to 209°. The axial plane of a minor anticline dips at 78° to 190°; the fold axis strikes at 101° and is subhorizontal. Axial-planar cleavage is poorly developed. An adjacent syncline has an axial plane dipping at 70° to 190°.

Wheatley

The southern part of a 20 m-long section [SX 8892 9170] near West Wheatley shows an inverted sequence of about 12 m of yellowish-brown-weathered pale grey shales with interbedded sandstones and siltstones up to 12 cm thick (mostly less than 5 cm) dipping at 84° to 359°. This sequence is separated by a fault with associated gouge zone from the northern part of the section, which consists of a right-way-up sequence of about 8 m of grey to dark grey, locally paper shales, with interbedded hard, grey, fine-grained sandstones and siltstones, locally laminated and cross-laminated, dipping at 60° to 355°. The shale:sandstone ratio is 3.4:1. *Posidonia* sp. was noted (Riley, 1992 a).

Exposures [SX 8899 9208] in the bed of the Alphin Brook near Wheatley show 20 to 30 m of inverted shales with sandstones dipping at 80° to 007°. Grey and dark grey, silty shales are interbedded with hard, grey, fine-grained sandstones between 1 and 15 cm thick, together with some thin beds of sandy siltstone. The shale:sandstone ratio is 1.3:1. Groove casts on the base of a sandstone bed strike at 272°.

A cutting [SX 8961 9158 to 8960 9154] near Wheatley shows grey to dark grey, rusty-brown-weathering shales with interbedded sandstones mainly less than 0.1 m thick, exceptionally 0.4 m thick. At the northern end, the beds dip at 82° to 358° (inverted); at the southern end of the exposure, they dip at 49° to 335° (right way up). These exposures suggest a syncline with a north-dipping axial plane, the northern limb being steep and inverted, the southern limb less steep and right way up. Axial-planar cleavage is locally developed in the hinge zone. Fossils indicate the R_{1a} Zone (Table 3, locality 13).

Bude Formation

The Bude Formation (Edmonds et al., 1968) has an outcrop area of about 30 km² north of the Crediton Trough (Figure 3). The formation consists mainly of a folded and faulted sequence of massive, thickly bedded sandstones interbedded with siltstones and shales. Units of thinly bedded siltstone and claystone interbedded with thinner sandstones are present in the western part of the outcrop. The thickness of the formation is uncertain owing to the structural complexity of the sequence, but about 200 m are thought to be present within the district.

Areas in which thickly bedded sandstones predominate are shown on the published map. They form east–west ridges and have been quarried in places. The sandstones are massive, fine to medium grained, and show little internal structure and no recognisable sole markings. The associated siltstones are massive, blocky-weathering, and locally micaceous. Massive sandstones are particularly abundant in the formation west of a north–south fault that extends through Newbuildings [SS 797 035] in the north-west of the district. It is uncertain whether the fault had any influence on sedimentation.

The thinly bedded siltstones are commonly grey to olive-grey, with thin interbeds of dark grey, fissile claystone; both lithologies are commonly micaceous and, locally, finely laminated. Carbonaceous material is present in some siltstone beds. Interbeds of hard, fine-grained sandstone up to about 5 m thick are scattered throughout, and locally form the dominant lithology. They show poorly developed sole markings and graded bedding.

Much of the formation in the district is red or purple stained. The association of the staining with adjacent Permian outcrops and the preferential staining of sandstones suggest that it is post-depositional, formed by interaction with iron-rich groundwaters during the Permian.

The base of the formation is not exposed in the district. Durrance (1985) suggested that the Crackington Formation had been thrust northwards over the Bude Formation during the Variscan Orogeny, and that this thrust had been reactivated to form the Crediton Trough, thus explaining the coincidence of the boundary between the two formations with the trough along much of its length.

No macrofauna was found in the Bude Formation of the district. Elsewhere in south-west England, the formation has a diachronous base within the Langsettian Stage (Westphalian A), and ranges up into the Bolsovian Stage (Westphalian C) (Ramsbottom et al., 1978). Samples from the district yielded a few palynomorphs indicative of a Carboniferous, post-Tournaisian age. *Lycospora pusilla*, *Calamospora* spp., *Densosporites* spp., *Granulatisporites* spp. and abundant fragments of inertinite and vitrinite were recorded (Turner, 1991).

The depositional environment of the Bude Formation has been the subject of considerable debate (Higgs, 1984) and interpretations have ranged between deltaic and deep-water fan. Lack of exposure and the small outcrop area of the formation preclude detailed facies analysis in the present district. Sedimentary features of the thinner sandstones in the district suggest that they may be turbidites. The origin of the thick sandstones is uncertain; similar sandstones in the adjacent Okehampton district were thought by Edmonds et al. (1968) to have been deposited in water shallower than that of the Crackington Formation, possibly in transition to deltaic conditions. It has been suggested (e.g. by Higgs, 1986, 1991) that deposition of the Bude Formation took place in a lake, with intermittent marine incursions. Palaeomagnetic evidence suggests that the area lay within 5° of the equator (Scotese et al., 1979).

Newbuildings–Sandford–Chilton

A quarry [SS 789 046] near Lower Linscombe, about 1.3 km north-north-west of Newbuildings, formerly exposed the following sandstone sequence dipping at 16° to 033°: (top) massive and flaggy 5.5 m; shaly, micaceous 1.8 m; grey, lenticular 1.8 to 3.0 m; shaly, micaceous 0.8 m; and green, lenticular 4.0 m seen (base).

At East Henstill Quarry, about 1.4 km north-north-east of West Sandford, a face [SS 8158 0413] shows massive, olive-grey sandstones in beds up to 1.05 m thick, some with load-casted bases, interbedded with blocky, and locally nodular, greyish brown siltstone, with carbonaceous flecks in places. The sequence dips at 20° east and is intersected by near-vertical joints (Plate 1d).

Disused and mainly filled sandstone quarries occur near Bawdenhayes, about 1 km north of Sandford; a section [SS 8257 8353] shows 2.3 m of sandstone with siltstone interbeds dipping east at 12°. Nearby quarries [SS 8303 0336; 8312 0340; 8322 0332; and 8337 0324] between Hill Copse and Land Quarry Plantation show small sandstone exposures.

About 600 m north-east of North Creedy, exposures [SS 8349 0442 to 8355 0450] in the lane up Priorton Hill show olive-grey siltstones dipping at 15–30° north-east or north-north-east; near the top of the hill, flaggy and carbonaceous siltstone dips at 45° south [SS 8353 0473].

A section [SS 8440 0397] 150 m south of Bremridge Farm, Upton Hellions, shows, in downward sequence: sandstone, massive, purple-brown, micaceous, medium-grained 2.8 m; siltstone, reddish brown, weathered 0.7 m; siltstone, massive, purple-brown 0.6 m; siltstone, reddish brown, very weathered 2.3 m; and sandstone, hard, fine-grained 0.6 m seen.

Chilton Quarry [SS 8662 0449] shows 0.6 m of purple-brown siltstone in beds up to 50 mm thick, overlying massive (but broken on irregular joints), purple, micaceous, fine-grained sandstone, dipping at 40° to 010°.

Ramspit Quarry [SS 8625 0317], about 1 km south of Chilton, was worked in a sequence of northward-dipping turbiditic sandstones with subordinate beds of shale and siltstone. The following section is present:

	Thickness m
Sandstone, hard, locally micaceous, fine-grained	0.75
Shale, purple, with nodules of ferruginous siltstone	0.26
Sandstone, as above	0.30
Shale, as above	0.16
Sandstone, fine-grained, with vermiform load casts on the base	0.24
Shale, purple	up to 0.10
Sandstone, hard, purple, fine-to medium-grained	(seen) 0.57

Cadbury

Sporadic stream-bed exposures [SS 8923 0461 to 8915 0448] in Lynch Plantation, north of East Coombe, show up to 3 m of olive-grey to brownish grey shale and siltstone with a few beds of purple, fine-grained sandstone up to 0.3 m thick, dipping north at 50–60°.

Exposures [SS 9007 0408] in a cutting west of Bowley show about 6 m of olive-brown siltstone and silty shale weathering spheroidally into flattened ovoid forms up to 0.2 m along the long axis. Scattered mica and dark grey, very fine-grained comminuted plant debris are present. The dip is 40° to 144°.

The stream in Bowley Wood shows sporadic exposures [SS 9014 0393 to 9051 0386] in a strike section of grey to olive-grey finely micaceous silty shale and siltstone.

A stream section [SS 9089 0437 to 9090 0423] about 300 m north-north-east of Bowley shows the following strata:

	Thickness m
Sandstone, hard, brownish grey, fine-grained, dip 33° to 007°. Joints dip 68° to 096°	1.5
Clay, shale and siltstone in ?fault zone	1.5
Shale, purple, silty, dip 53° to 201°	5.0
Gap in section	c. 20.0
Sandstone, hard, greyish red, fine-grained, with scattered mica; locally weathered to soft sandstone. Dip 70° to 319°	6.5
Gap in section	c. 5.0
Sandstone, hard, greyish red, micaceous, fine-grained. Near-vertical, strike 306°	1.5
Siltstone and silty shale, greyish red, micaceous	10.0
Sandstone, greyish red, soft, micaceous, very silty, fine-grained	0.6
Gap in section	c. 17.0
Sandstone, hard, greyish red, fine-grained	2.0
Shale, greyish red, silty, dip 45° to 004°	4.0

About 150 m south of East Bowley, stream exposures [SS 9124 0365] show up to 2.0 m of greyish red, structureless siltstone and sandy siltstone, in beds up to 0.3 m thick, with pronounced spheroidal weathering at some levels; the beds dip at 18° to 154°; in nearby exposures [SS 9128 0364], the dip is 75° to 208°.

Sections [SS 9379 0452] on the River Exe near Chitterley show about 50 m of purple, massive, fine-grained sandstone in beds up to 2 m thick, with some interbeds of flaggy sandstone, overlying 3 m of purple mudstone. The dip is 36° to 184°. Farther south, exposures [SS 9377 0441] of purple shale and mudstone dip at 32° to 276°. Nearby [SS 9379 0438] are exposures of massive, spheroidally weathered, purple, micaceous sandy siltstone and silty fine-grained sandstone.

Silverton

A quarry [SS 9585 0470] 300 m south-west of Land Farm shows up to 1.8 m of massive, purple, hard, fine-grained sandstone. Prominent joints, spaced 0.8 to 0.9 m apart, dip at 86° to 034°; a second set is near-vertical, striking at 030°.

A quarry [SS 9595 0438 to 9604 0430] about 350 m north-north-east of Ash Farm, Silverton, shows 0.9 m of purple, massive, fractured fine-grained sandstone, overlying 0.5 m of purple shale with thin interbeds of sandstone, on 0.4 m of purple sandstone. The dip is 21° to 218°.

A borehole [SS 9629 0422] at Higher Roach Farm, Silverton, penetrated the following sequence (driller's log):

	Thickness m	Depth m
Topsoil etc.	2.13	2.13
Sandstone, coarse-grained	4.57	6.70
Shale, grey	2.14	8.84
Mudstone, dark brown	0.91	9.75
Sandstone, pale brown	12.81	22.56
Shale, grey	1.52	24.08
Sandstone, pale brown	14.32	38.40
Siltstone, pale brown to red	4.27	42.67

A quarry [SS 9707 0418] east of Roach shows up to 6 m of purple, fine-grained sandstone, dipping at 8° to 060°.

Shale and mudstone extends from near Tedbridge Copse [SS 970 050] eastwards to Rode Moors [SS 987 050]; grey shale is poorly exposed [SS 9737 0492] in the banks of the Burn River, Tedbridge.

A quarry at Forward Green, near Tedbridge, shows [at SS 9815 0473] up to 1.0 m of purple, massive to rubbly, hard, fine-grained sandstone; another quarry [SS 9782 0467 to 9801 0463] shows up to 1.0 m of purple, fine-grained sandstone with a few mudstone beds.

Westcott

In the Nag's Head Cutting [ST 022 051] on the M5 Motorway, Sherrell (1971) recorded a syncline containing grey, massive, thickly bedded, cross-bedded, coarse-grained sandstone in beds 0.6 to 1.0 m thick, together with more thinly bedded (average 0.15 m thick) fine-grained sandstone, and reddish brown thinly bedded mudstone and siltstone.

THREE

Lower Carboniferous igneous rocks and associated metamorphism

Intrusive dolerites and extrusive basalts and tuffs occur in association with the Lower Carboniferous sequence in the south-west of the district (Figure 5).

DOLERITES

Porphyritic and aphyric albite dolerites occur as folded and tectonically disrupted intrusions mainly within the Lower Carboniferous Combe Shale in the Teign Valley. The dolerites were intruded prior to the main period of folding and were deformed together with the enclosing strata; they locally form ridges which can be traced around the limbs of the major eastward-plunging folds. The thickest and laterally most extensive of these intrusions, the Ryecroft Sill, outcrops mainly in the adjacent Newton Abbot district (sheet 339) but has a small outcrop [SX 856 860] south of Doddiscombsleigh in the present district (Figure 5). Thinner, less continuous sills occur at or close to the junction between the Combe Shale and the Teign Chert. One of these, the Harehill Sill, extends from Harehill Plantation [SX 857 861], Doddiscombsleigh, to about 350 m north of Higher Barton [SX 859 852] (Newton Abbot district). Dolerite is also present near

Christendown Clump [SX 851 864], Doddiscombsleigh. In the Bridford area, dolerite sills are present in the limbs and the axial region of the Bridford Anticline and the Bridford Syncline (Figure 33); the most laterally persistent, the Bridford Sill, extends eastward from the centre of Bridford.

Most of the intrusions are less than 50 m thick, although locally they may be over 100 m thick. They are laterally continuous bodies with chilled intrusive and sometimes vesicular margins parallel with bedding in the host formation. They have long been regarded as sills emplaced during the interval that preceded deposition of most or all of the Teign Chert (Ussher, 1902). Whole-rock K/Ar measurements for the Ryecroft Sill at a locality [SX 839 846] near Lower Ashton, in the Newton Abbot district, gave an age of 344 ± 27 Ma (Miller and Mohr, 1964; recalculated using the constants of Steiger and Jäger, 1977).

The larger sills exhibit differentiation from mafic albite-dolerite at their bases, through albite-dolerite, to leucocratic quartz-bearing and highly feldspathic dolerite varieties near their tops. Enclaves of quartz-leucodolerite, 0.3 to 1.5 m in size, occur within the upper part of the Ryecroft Sill (Morton, 1958) and the Bridford Sill. Vertical lithological changes, including minor cryptic mineral

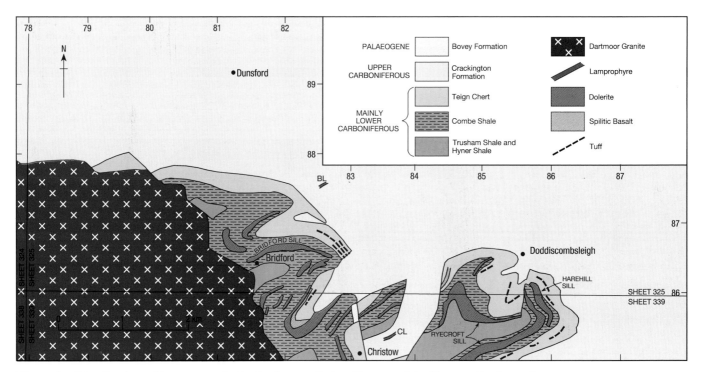

Figure 5   Distribution of igneous rocks in the Lower Carboniferous of the Exeter district and part of the Newton Abbot district, showing also the outcrop of the Dartmoor Granite and the location of the Bridford Lamprophyre (BL) and the Christow Lamprophyre (CL).

variations, in the upper part of the Ryecroft Sill of the adjacent Newton Abbot district have been described by Morton (1958) and Morton and Smith (1971), and summarised in Selwood et al. (1984). Differentiation may account for lithological variations in the Bridford Sill, but the mineralogical variation is partly masked by later thermal metamorphism that occurred during emplacement of the Dartmoor Granite.

Petrography

The typical mineral assemblage of the dolerite sills is albite-augite-biotite-chlorite-apatite-ilmenite. Textures are inequigranular and hypidiomorphic; albite forms euhedral tabular laths up to 10 mm long in porphyritic varieties, and contains abundant inclusions of sericite and chlorite. Trachytic textures are reported from the upper part of the Ryecroft Sill by Selwood et al. (1984). Augite, or titanaugite, forms large euhedral crystals and anhedral plates, 2 to 5 mm in size, that enclose or partly enclose albite to form ophitic or subophitic textures. Biotite occurs as relict fragments and sparse euhedral plates, and is a dark reddish brown variety. Chlorite forms interstitial fibrous aggregates that replace biotite or pyroxene, and occupies the core of many albite laths. Ilmenite and some magnetite occur as large anhedral and skeletal crystals extensively altered to secondary sphene ('leucoxene'). Apatite occurs as numerous acicular crystals up to 2 mm in length and cored by chlorite. Pyroxene, biotite, apatite and opaque minerals decrease, and albite increases, in modal abundance with increase in vertical height within the Ryecroft Sill (Morton, 1958).

Vesicular margins contain amygdales that range up to 25 mm in size. Outside the granite aureole, they consist largely of quartz (Selwood et al., 1984), but they consist of colourless amphibole and clinozoisite within the aureole. Quartz-leucodolerite enclaves contain anhedral interstitial patches of quartz that formed late in the crystallisation history.

Geochemistry

Tables 4 and 5 contain whole-rock analyses of the Ryecroft Sill taken from Morton and Smith (1971) and Chesher (*in* Selwood et al., 1984); partial analyses of mineral phases within the sill are given by Morton and Smith (1971). These data show that the albite dolerites contain high abundances of total alkalies (about 6.8 per cent at 50 per cent SiO_2) and fall within the alkaline field of Irvine and Barager (1971). They also contain high abundances of Al_2O_3 (14.5 per cent), TiO_2 (2.7 per cent) and P_2O_5 (2.2

Table 4 Analyses of the Scanniclift Copse Spilite and associated tuffs, and the Ryecroft Dolerite (data from Morton and Smith, 1971).

Major elements (weight per cent oxide)							
	1	2	3	4	5	6	7
SiO_2	55.40	58.79	70.48	69.73	74.13	44.15	46.58
TiO_2	1.99	2.20	0.87	0.74	0.19	4.36	3.36
Al_2O_3	17.53	13.47	10.27	12.77	13.55	13.74	14.13
Fe_2O_3	2.10	3.08	4.50	3.76	0.90	1.72	1.51
FeO	7.51	9.90	4.81	2.87	0.96	11.54	10.61
MnO	0.35	0.27	0.40	0.27	0.10	0.24	0.25
MgO	3.27	2.65	2.08	2.15	0.25	5.10	3.77
CaO	1.75	0.81	0.68	0.83	0.19	8.07	6.98
Na_2O	3.50	2.66	0.67	0.60	2.63	3.54	4.18
K_2O	3.45	3.67	4.61	3.87	4.82	1.72	1.57
P_2O_5	0.73	0.73	0.37	0.48	0.10	1.91	1.51
Loss	1.72	1.01	1.20	1.80	2.01	3.47	5.11
Total	99.28	99.24	100.94	99.87	99.83	99.56	99.87

1 and 2. Spilitised basalt, Scanniclift Cope [SX 843 861], Doddiscombsleigh; 3 and 4. Volcanic wackes, Scanniclift Copse [SX 843 861]; 5. Crystal-vitric tuff, 430 m south-east from Woodah Farm [SX 8498 8632]; 6. Dolerite, Ryecroft Sill, c. 3.7 m above quarry floor [SX 8432 8475]; 7. Dolerite, Ryecroft Sill (average of seven analyses) [SX 8432 8475].

Table 5 Analyses of the Ryecroft Dolerite, Scanniclift Copse Spilite and Christendown Clump Tuff (data from Chesher *in* Selwood et al., 1984), and the Dartmoor Granite (data from Darbyshire and Shepherd, 1985).

Major elements (weight per cent oxide)					
	1	2	3	4	5
SiO_2	49.92	49.60	50.80	73.10	71.16
Al_2O_3	15.00	14.08	14.03	11.86	14.31
Fe_2O_3	0.93	1.44	1.14	1.70	3.05
FeO	11.65	11.86	11.46	1.50	
MgO	3.19	3.04	3.41	3.23	0.49
CaO	5.48	5.69	7.82	0.01	1.33
Na_2O	4.60	4.80	6.00	4.42	3.24
K_2O	1.98	2.10	1.30	4.11	5.03
TiO_2	3.07	2.40	1.52	0.23	0.41
P_2O_5	2.24	2.05	0.73	0.07	0.19
MnO	0.14	0.10	0.05	0.01	0.07
Total	98.20	97.16	98.26	99.96	99.28
Trace elements (parts per million)					
Ba	796	551	1416	229	
Co	9	11	5	46	
Cu	68	64	137	6	
Cr	36	75	283		
Ga	26	25	25		
Li	60	80	75		
Ni	20	33	209	5	
Nb				19	
Rb	33	12	4	367	
Sr	331	271	45	96	
Y	55	46	52	33	
Zr				138	

1. Ryecroft Dolerite, Ryecroft Quarry [SX 8432 8475], average of seven analyses at 6.1 to 9.2m-thick vertical intervals;
2. Ryecroft Dolerite, Ryecroft Quarry [SX 8432 8475], chilled top;
3. Spilitic pillow lavas, Scanniclift Copse [SX 8448 8628 to 8439 8606]; 4. Quartz-keratophyre vitric-flow tuff, Christendown Clump [SX 8508 8635], Doddiscombsleigh; 5. Dartmoor Granite, Blackingstone Quarry [SX 784 857].

per cent) and, together with high Y contents (50 ppm), are comparable to modern mafic volcanic rocks described from within-plate tectonic settings (Pearce and Cann, 1973).

Morton and Smith (1971) and Selwood et al. (1984) used the analyses to demonstrate an upward increase in abundances of SiO_2, Al_2O_3 and alkalies, and a corresponding upward decrease in FeO, CaO, MgO, TiO_2, P_2O_5 and Fe_2O_3 within the Ryecroft Sill. Systematic geochemical variations of this type were attributed by these authors to processes of crystal fractionation.

Contact metamorphism and metasomatism

The intrusion of dolerite sills into the Lower Carboniferous sedimentary sequence resulted in contact metamorphism and metasomatism of the sediments immediately adjacent to the sills. Within the granite aureole, the effects of later thermal overprinting by the granite in large part mask the visible evidence of earlier thermal metamorphism at sill margins; spotted albite-hornfelses ('spilosites') are, however, preserved adjacent to the thick Bridford Sill.

Outside the aureole, spilosite commonly grades into hard splintery albite-hornfels ('adinole') adjacent to the sill contact. The sequence of mineralogical changes is described in Selwood et al. (1984, pp.78–79), who state that: 'It appears that dolerites with many vesicles and hence probably abundant volatiles (e.g. Ryecroft and Hill Copse sills), have given rise to pronounced sodium metasomatism of the shales, whereas those with less vesicular margins and relatively free from volatiles produced negligible metasomatic effects (e.g. Crockham sill)' [SX 849 808] (Newton Abbot district). The pronounced sodium metasomatism adjacent to the more vesicular intrusion margins is probably due to the higher permeability of such lithologies, which facilitated the incursion of sodium-bearing sea-water into the cooling igneous rock.

Typically, the metamorphic alteration extends for only a few metres into the adjacent shales, but up to 15 m of adinole is developed [SX 8573 8596] south of Harehill Plantation, Doddiscombsleigh. This exceptional thickness is probably due to the fact that the affected rocks are sandwiched between two dolerite intrusions, the Ryecroft Sill and the Harehill Sill. The adinole at Harehill Plantation consists of a microcrystalline mass of xenoblastic sutured grains of albite and quartz, with abundant weakly pleochroic, pale yellow mica and very pale green fibrous chlorite; a few granular aggregates of rutile, ilmenite and iron pyrites are present. Pseudomorphs after possible andalusite, which occur in the Newton Abbot district (as at Hill Copse, near Lower Ashton) are absent; instead, round spots up to 0.2 mm in diameter, a little coarser-grained than the matrix and without mica, are present (Morton, 1958).

Bridford area (within the granite aureole)

Aphyric metadolerite and metaquartz-dolerite form a folded sill, 30 to 40 m thick and 850 m long, in the northern limb and axial region of the Bridford Syncline (Figure 33), approximately 50 m below the base of the Teign Chert. An exposure

0.5 km from the granite margin [SX 8145 8689], west of Windhill Gate, shows greenish black, fine-grained metadolerite with pegmatitic segregations of metaquartz-dolerite. Veins and fracture coatings of amphibole and chlorite occur throughout.

A thick mafic sill occupies the core of the Bridford Syncline, 150 to 300 m from the margin of the granite. Brash of hard, dark green, fine-grained amphibolite with a granoblastic texture occurs on both sides of Seven Acre Lane, Bridford [e.g. SX 8112 8682]. The intrusion may represent a tectonically disrupted fragment of the Bridford metadolerite sill located north and north-east of the village and described below.

Quarries east of Bridford expose both basal and roof contacts of a metamorphosed and differentiated porphyritic albite-dolerite sill 1300 m long, the Bridford Sill (Figure 5). The sill is approximately 30 to 80 m thick, and typically composed of spheroidally weathered feldsparphyric metadolerite. Fine-grained aphyric metadolerite occurs above the floor and southern margin of the sill [SX 8230 8660], subvertical and slightly overturned toward the south. Adjacent hornfelsic, possibly tuffaceous, mudstones locally [SX 8226 8659] contain millimetre-sized rusty weathering spots of white mica and chlorite. The main quarry [mid-point SX 8220 8662] exposes 25 m-high faces of brown-weathering, greenish grey metadolerite that contains 2 mm long, white feldspar laths. The upper part of the sill is a greyish green, coarse-grained metaquartz-dolerite rich in albite; euhedral tabular laths of albite average 3 to 4 mm in length and are up to 10 mm in part. The lithology is hard, massive and contains abundant quartz veins. The roof (northern) margin of the sill is weathered, but appears to dip at about 80° towards the north. Spotted hornfelsic mudstones of a similar type to those in the basal aureole occur in loose material on the north side of the quarry [SX 8224 8667].

Quarries south of Pound Lane, Bridford expose metadolerite in the southern limb of the Bridford Anticline. Near South Wood [SX 8260 8644], extensive brash of brown weathering, pale grey or greenish grey metadolerite with white feldspar laths and a pyroxene-bearing matrix (2 mm grain size) occurs in a quarry. An identical lithology was worked 80 m farther south [SX 8261 8638], and although the two workings are separated by silicified hornfelses, they appear to have been excavated in the same intrusion which here divides into two sills. This intrusion, together with a 20 m-thick sill between a limekiln [SX 8247 8643] and Pook's Cottages [SX 8221 8621], was emplaced into the middle of the Combe Shale succession, at approximately the same stratigraphical level as the large porphyritic metadolerite within the north limb of the Bridford Anticline; the sills have been tectonically disrupted and there is no continuity around the nose of the fold.

Small, laterally discontinuous, poorly exposed sills occur between Poole Bere Cottage [SX 8192 8605] and Poole Grove [SX 8213 8593]; black or very dark green, fine-grained, extensively amphibolitised dolerite crops out [SX 8174 8594] near the road to Middle Hole, within the largest and most westerly of these intrusions.

Doddiscombsleigh area (outside the granite aureole)

Dark green, coarse-grained ophitic or porphyritic dolerite forms much of the Ryecroft Sill north of Great Leigh [e.g SX 8472 8599] and east of Down Lane, where the sill expands to more than 100 m thick. Extensive hydrothermal alteration of the margins occurs locally, and is well seen in the Doddiscombsleigh Anticline [SX 8461 8589] east of Scanniclift Copse, where ferruginous and in part bleached, very pale brown or buff dolerite with a relict ophitic texture is present.

Brown-weathered, ferruginous and extensively altered, highly feldspathic albite dolerite forms a small sill north of Down Lane [mid-point SX 8494 8637]. The sill is about 20 m thick, and is emplaced close to the axial surface of the Doddiscombsleigh Anticline. In a quarry [SX 8512 8646] near Christendown Clump at the eastern end of the sill, 5 to 8 m of hard, massive, dark greenish grey dolerite are exposed. Morton (1958, pp.142–143) noted columnar jointing, the columns being roughly tabular or hexagonal in cross-section; their long axes are oriented parallel to each other, and dip at 70° to 350°. Morton inferred that the sill dips at about 30° to 176°. Two samples from top and bottom of the sill showed marked differences, as follows:

	Top	Base
Plagioclase:	$Ab_{99}An_1$ cloudy and porphyritic	$Ab_{42}An_{58}$ pellucid and non-porphyritic
(Titan) Pyroxene:	Anhedral and mauve	Anhedral and very deep mauve
Biotite:	No trace	Remnants of formerly abundant biotite
Colour Index:	Around 45	Over 45

Morton (1958) considered that the upward decrease in mafic minerals was not unusual, but the change from labradorite to albite was abnormal, and perhaps due to albitisation.

Dolerite exposures [SX 8577 8609] in the Harehill Sill west of Harehill Copse are present in about 2 m of small crags beneath tree roots. The rock is fractured and largely weathered to yellowish brown and orange-brown; the less weathered rock is medium to dark grey. Morton (1958, pp.137–138) noted that dolerite from this sill is similar to albite dolerite from the Upper Beardon area (south of the present district), except for the following minor variations: the albite ($Ab_{98}An_2$) forms smaller phenocrysts; two varieties of chlorite are present, an abundant radiating fibrous pale green and weakly pleochroic chlorite with a low birefringence, and rare tiny subhedral crystals of colourless clinochlore (variety sheridanite); epidote is commonly seen as euhedral crystals of pistacite in vugs with chlorite and quartz; calcite occurs with the epidote; rarely, a colourless zeolite (thomsonite) is present in tiny veins and cavities; a very small proportion of quartz is seen with the epidote; a little iron pyrites is present; and the titanaugite is altered in part to chlorite and epidote.

SPILITIC BASALT

Albitised and extensively altered basaltic lava crops out in two places in the district. The first is a pillowed to massive flow 300 m long within the lower part of the Teign Chert in Scanniclift Copse [SX 8435 8606 to 8447 8633]. The second occurrence of spilitic basalt lies within the aureole of the Dartmoor Granite at Stone Copse Quarry, Bridford [SX 8241 8602], where the spilitic basalt is part of a more extensive, thick, massive flow outcropping mainly in the adjacent Newton Abbot district (Figure 5). In the quarry, a metamorphosed albite-dolerite sill emplaced into the Combe Shale forms the base of the flow, and volcanic mudstones, tuffs and cherts ascribed to the Teign Chert overlie the spilitic horizon.

The spilitic basalt is a greyish-brown-weathering, fine-grained, pale to dark green or greenish grey vesicular rock with sparse feldspar phenocrysts and, in part, extensive ferruginous and carbonate veining. Elongate pillow-like structures [SX 8444 8625] in pale green spilite at Scanni-

clift Copse are defined by fine-grained, dark green and irregular pillow margins that are gradational with non-pillowed material. Intervening sedimentary rock is absent, although there is a small amount of interstitial silica. Morton (1958) noted an adit of the former Scanniclift Copse Mine manganese workings which exposed [near SX 8447 8631] 7.6 m of bluish green pillow lavas in which pillows are up to 2.7 m long and 1.2 m wide (average 1.2 m long and 0.9 m wide) and have chilled, vesicle-free margins, concentric zones of white amygdales in the interior, and a large central cavity. The pillowed sequence is overlain by thin volcanic mudstones. The spilitic basalt at Scanniclift Copse thickens southwards, and about 12.2 m of massive, medium or dark green, locally highly vesicular lava crops out adjacent to the Scanniclift Thrust [SX 8447 8633] (Figure 33). Highly carbonated and ferruginous spilite varieties are present in old excavations in the vicinity.

The spilitic basalt at Stone Copse Quarry is a pale grey, fine-grained, vesicular lava with a faint greenish lamination and cross-cutting pink veins. Altered varieties that consist of pale greyish green basalt with a ferruginous stockwork are also present adjacent to colour-banded and carbonate-bearing cherts. The very fine grain-size of the cherts overlying the spilitic unit suggests a zone of contact metamorphism.

Petrography

Pillowed spilitic basalt from Scanniclift Copse [SX 8442 8621] consists of abundant small albite laths, 0.1 to 0.3 mm (rarely 0.6 mm) in length in a fine-grained, cloudy and altered matrix of chlorite and calcite; an opaque phase, sphene and quartz occur as accessory minerals, and there are abundant cross-cutting fractures infilled with iron oxide (E 66096). Amygdales are typically 1 to 2 mm in size, rimmed by iron oxide and filled with calcite and vermiform chlorite. Massive spilitic basalt from the southern part of the same flow [SX 8439 8613] is more extensively carbonated and, in addition to small albite laths, contains sparse euhedral albite phenocrysts up to 3 mm in length and abundant large and irregularshaped amygdales that have chlorite rims and coarse calcite interiors (E 66097). Some of the chlorite displays growth-controlled zones of vivid green colour, possibly related to the presence of trace metals such as Cr, Ni or V. Morton (1958) described two other basalt varieties from this locality, one with essential quartz and no calcite, and the other a non-vesicular, coarse-grained variety with euhedral pyroxene phenocrysts pseudomorphed by a mosaic of calcite, quartz, chlorite and amphibole.

Vesicular spilite from Stone Copse Quarry [SX 8238 8603] has a fine-grained, altered and turbid matrix in which chlorite, epidote, sphene and feldspar (probably albite) can be resolved (E 65527 and 65528). Amygdales, up to 12 mm in size, contain either intergrown prehnite and epidote, partially replaced by chlorite and carbonate, or acicular tremolite prisms and orthoclase. Narrow veins of very fine-grained tremolite are present throughout, and hairline veinlets of prehnite and fissures that contain sprays of epidote prisms with interstitial chlorite are also present. Highly altered varieties (E 9182) are dominated by aggregates of very fine-grained pale biotite which surround minute cores of axinite. Albite shows complex partial alteration to sericite, turbid clinozoisite, biotite (grown along fractures) and sparse pockets of a distinctive brown phyllosilicate mineral (?zinnwaldite).

Geochemistry

Analyses of the Scanniclift Copse Spilite taken from Morton and Smith (1971) and Chesher (*in* Selwood et al., 1984) are given in Tables 4 and 5. The data are closely comparable with analyses of albite dolerites in adjacent ground; like the dolerites, the spilite contains high concentrations of alkalies (6.3 to 7.3 per cent), Al_2O_3 (13.5 to 17.5 per cent), TiO_2 (1.5 to 2.1 per cent) and Y (52 ppm), but has rather less CaO and larger abundances of Na_2O, Ba and base metals. The similarities in mineralogy and geochemical composition led Morton (1958), Morton and Smith (1971) and Selwood et al. (1984) to suggest that the spilitic basalts are extrusive equivalents of the albite-dolerite sills.

TUFFS

Beds of tuff occur throughout the Teign Chert, but are more common in the lower to middle part of the formation, where an impersistent horizon can be traced throughout the middle Teign valley (Selwood et al., 1984). They also occur locally at the base of the spilitic basalt horizon. The main outcrops in the Bridford area are at the Bridford Barytes Mine and in Stone Copse Quarry; in the Doddiscombsleigh area, tuffs outcrop at Scanniclift Copse, in the Down Lane–Christendown Clump area, at Mistleigh Copse and at Harehill Plantation (Figure 5).

The tuffs range from a few millimetres thick to mappable units up to about 40 m thick. Three types can be recognised, depending on the predominant type of particle present: crystal tuffs, lithic tuffs, and vitric tuffs. The last-named predominate in the district. Detailed petrographical accounts of the tuffs are contained in Morton (1958, pp.56–66) and in Selwood et al. (1984, pp.62–63).

Most of the tuffs contain small pyroclasts of 'chequerboard' albite and the high-temperature, bipyramidal variety of quartz (ß-quartz), extensively replaced by sericite and secondary quartz. Morton (1958) and Morton and Smith (1971) have described graded bedding and radiolaria from these intervals, and suggested deposition in a marine sedimentary environment.

Geochemistry

Major-element analyses (Morton and Smith, 1971; Selwood et al., 1984) of the vitric and crystal-vitric tuffs southeast of Woodah Farm, at Down Lane [SX 8498 8632] and Christendown Clump [SX 8508 8635], and of 'volcanic wackes' or coarse altered tuffs from the Scanniclift Copse

horizon [SX 843 861] are given in Tables 4 and 5. According to these authors, the samples have rhyolite or soda rhyolite compositions, and represent the deposits of a volatile-rich, quartz-keratophyre magma that erupted explosively on the sea floor.

Bridford area (within the granite aureole)

West of the River Teign, vitric tuffs form three mappable intervals near the old Bridford Barytes Mine Quarry [mid-point SX 8296 8656]. The south-western edge of the quarry exposes about 10 m of pale grey, crumbly tuffs with black lithic fragments up to 150 mm in size. The basal 2.4 m of the unit is relatively fine grained and is overlain by a thin agglomeratic interval of welded lithic tuff fragments, which in turn is overlain by an upward-fining sequence of vitric and crystal tuffs with sparse fragments of mudstone, chert and vesicular lava. The welded lithic tuff fragments are composed of granules and very coarse sand-grade clasts of lava, mudstone, chert and feldspar grains in a flow-banded, fine-grained, sericitic matrix (E 65524). Fine-grained vitric and crystal tuffs of a similar type form mappable units south-west of the quarry [SX 8272 8665 to 8290 8650] and at the north-east quarry edge [SX 8293 8664 to 8299 8658].

The south-eastern face of Stone Copse Quarry [SX 8243 8600] is partly covered by fallen blocks of banded chert interbedded with very dark grey mudstone and pale grey vitric tuff containing sparse broken crystals of albite and quartz, together with small angular fragments of chert, hornfels and lava or pumice. The banded lithology is extensively fractured and veined; the vein assemblage is dominated by mats of fine-grained mica that contain patches of amphibole, epidote and axinite.

Medium grey, fine-grained and very finely laminated tuffs occur interlayered with basal cherts on Copplestone Down [SX 8159 8745], 1 km north of Bridford.

Doddiscombsleigh area (outside the granite aureole)

East of the River Teign, vitric and crystal tuffs form narrow layers within a mudstone interval, 0.6 to 1.8 m thick, in the inverted roof of the Scanniclift Copse spilitic basalt [SX 8446 8632 to 8439 8615], and vitric tuffs or vitric-crystal tuffs, approximately 1 m thick, occur in the floor of Down Lane [SX 8498 8632]; both localities are described by Morton (1958). Tuffs in the former locality are grey and finely banded, and consist of a very fine-grained matrix of plagioclase, quartz and chlorite with angular fragments of albite and quartz; some layers contain abundant subangular fragments of devitrified glass, while others contain sparse, 0.4 mm-size, fragments of lava. The Down Lane tuffs are bluish grey, and consist of a spongy aggregate of shards altered to microcrystalline quartz and sericite, and some pumice fragments replaced by subopaque material in a subordinate, fine-grained chloritic matrix. Crystals including albite, ß-quartz and zircon, and angular fragments of chert, shale and lava, were also recorded (Morton, 1958).

Small crags [8504 8637] at Christendown Clump consist of about 1 m of grey tuff dipping at 35° to 132°.

A quarry [SX 8540 8587] south of Mistleigh Copse contains exposures of grey, silicified, vitric tuff with chert lenticles, dipping at 50° to 163°. Selwood et al. (1984, p.63) recorded radiolaria in the groundmass.

Exposures of medium grey crystal tuff are present in a track [SX 8583 8619] just east of Harehill Plantation. About 200 m south-east, J A Chesher (MS fieldslip) recorded coarse crystal tuff.

FOUR

Permian and Triassic rocks (New Red Sandstone): introduction

The term 'New Red Sandstone', introduced by Conybeare and Phillips (1822), has long been used in south-west England for the thick succession of red, aeolian, fluviatile and lacustrine sediments deposited during the Permian and Triassic periods. This rests with marked unconformity on folded Carboniferous rocks, and has been divided into four groups, of which the lowest three, the Exeter Group, Aylesbeare Mudstone Group and Sherwood Sandstone Group, are represented by up to 1400 m of strata in the Exeter district. The fourth group, the Mercia Mudstone Group, outcrops in the adjacent Sidmouth district to the east. Selected stages in the evolution of the nomenclature of the New Red Sandstone of the district are shown in Tables 6 and 7. The nomenclature used in this memoir is summarised in Table 8.

The New Red Sandstone sequence crops out over about two-thirds (about 380 km²) of the Exeter district (Figure 2). In the northern half, a partly fault-bounded trough between Crediton in the west and Silverton in the east (the Crediton Trough) contains up to 800 m of Exeter Group strata. Rocks of this group also crop out around and south-west of Exeter city. The Exeter Group is overlain in the east by gently (mostly less than 3°) eastward-dipping Aylesbeare Mudstone Group strata which are in turn succeeded by rocks of the Sherwood Sandstone Group.

The Exeter Group consists predominantly of breccias and sandstones. It is divisible into two parts. The lower includes the Exeter Volcanic Rocks with lamprophyres dated at between 291 and 282 Ma, and is thus probably mainly Early Permian in age, though the lowest beds may be as old as latest Carboniferous. This sequence is largely confined to the Crediton Trough. The upper part of the Exeter Group, both in the Crediton Trough and around Exeter, unconformably overlies the lower part and has been dated as Late Permian (about 250–260 Ma) on the evidence of fossil pollen (Warrington and Scrivener, 1988; 1990). An implied hiatus of at least 20 Ma between the two sequences (Table 10) accommodates the emplacement of the Dartmoor Granite at 280 Ma and, elsewhere in south-west England, subsequent quartz-porphyry intrusions and tin mineralisation (Darbyshire and Shepherd, 1985, 1987; Halliday, 1980). This interval also saw isostatic uplift and regional denudation which followed the intrusion of the Dartmoor Granite (Warrington and Scrivener, 1990; Edwards et al, 1997).

The Exeter Group is succeeded unconformably by the Aylesbeare Mudstone Group; the latter consists pre-

Table 6 Selected stages in the evolution of the nomenclature of the New Red Sandstone of the Crediton Trough.

Hutchins (1963)	Edmonds et al. (1968)	This memoir		
		Western Crediton Trough	Eastern Crediton Trough	
St Cyres Beds	Crediton Conglomerates	Newton St Cyres Breccia	Shute Sandstone	
Crediton Beds		Crediton Breccia	Yellowford Formation	
Bow Beds	Knowle Sandstones	Creedy Park Sandstone		
		Knowle Sandstone	Thorverton Sandstone	
	Bow Conglomerates	Bow Breccia		
Cadbury Beds	Cadbury Breccia		Higher Comberoy Formation	
		Cadbury Breccia		
			Upton Formation	

Ussher, 1902	Ussher, 1913	Laming, 1966	Laming, 1968	Henson, 1970, 1972	Smith et al. (1974) Warrington et al. (1980)	Selwood et al. (1984)	Laming (1982)	This memoir
Upper Sandstone	Upper Sandstones	Upper Sandstone	Otter Sandstones	Otter Sandstone Formation	Otter Sandstone Formation	Otter Sandstone	Otter Sandstone	Otter Sandstone
Pebble Beds	Pebble Beds	Budleigh Salterton Pebble Beds	Budleigh Salterton Pebble Beds	Budleigh Salterton Pebble Beds	Budleigh Salterton Pebble Beds	Budleigh Salterton Pebble Beds	Budleigh Salterton Pebble Beds	Budleigh Salterton Pebble Beds
					SHERWOOD SANDSTONE GROUP	SHERWOOD SANDSTONE GROUP		SHERWOOD SANDSTONE GROUP
Lower Marls	Sandstones and Marls	Exmouth Beds	Littleham Beds	Littleham Formation (1970) / Littleham Mudstones (1972)	Littleham Mudstones	Littleham Mudstones	Littleham Mudstone	Littleham Mudstone / Aylesbeare Mudstone Group (where undivided)
	Straight Point sandstones		Exmouth Sandstones	Exmouth Formation (1970) / Exmouth Sandstones and Mudstones (1972)	Exmouth Formation	Exmouth Sandstone and Mudstone	Exmouth Formation	Exmouth Mudstone and Sandstone / Clyst St Lawrence Formation
	Lower Marls				AYLESBEARE GROUP		AYLESBEARE GROUP	AYLESBEARE MUDSTONE GROUP
Exminster breccias, Heavitree Breccia (Exeter area) (in part)	Langstone Point and Exmouth Shrubbery breccias		EXE GROUP		Clyst Sands / Langstone Breccias / Exminster Breccia / Dawlish Sands	Exe Breccia	Langstone Breccias / Clyst Sands (north) / Dawlish Sands (south)	Dawlish Sandstone / Poltimore Mudstone / Bussell's Member / Monkerton Formation
Teignmouth and Dawlish breccias, Heavitree Breccia (in part)	Dawlish sandstones and breccias		Langstone Breccia / Kennford Breccias / Dawlish Sands		Heavitree and Alphington Breccias		Heavitree and Kennford Breccias (north) / Coryton Breccias (south)	Heavitree Breccia
Lower Sandstone	Dawlish-type Breccia	Dawlish Sands	Heavitree and Alphington Breccias	Dawlish Sandstones	Dawlish Sandstone	Dawlish Sandstone		Alphington Breccia
	Teignmouth-type Breccia	Teignmouth Breccias	Teignmouth Breccias	Teignmouth Breccias	Teignmouth Breccias	Teignmouth Breccias	Teignmouth Breccia	Whipton Formation
Breccia and Conglomerate		TEIGNHEAD GROUP	TEIGNHEAD GROUP					Knowle Sandstone
								EXETER GROUP

Table 7   Selected stages in the evolution of the nomenclature of the New Red Sandstone of the district outside the Crediton Trough.

Table 8 New Red Sandstone stratigraphy of the district.

	Group	Exeter area to Bicton	Crediton Trough	
			Crediton to Silverton	Clyst Hydon area
TRIASSIC	SHERWOOD SANDSTONE GROUP	Otter Sandstone Budleigh Salterton Pebble Beds *unconformity*		
TRIASSIC	AYLESBEARE MUDSTONE GROUP	Littleham Mudstone Exmouth Mudstone and Sandstone *u n c o n f o r m i t y*		Aylesbeare Mudstone Group (where undivided) Clyst St Lawrence Fm.
PERMIAN	EXETER GROUP	Dawlish Sandstone Monkerton Formation Heavitree Breccia Alphington Breccia Whipton Formation *u n c o n f o r m i t y*	Dawlish Sandstone Poltimore Mudstone Busssell's Member Shute Sandstone Newton St Cyres Breccia Crediton Breccia Yellowford Fm. Creedy Park Sandstone	
PERMIAN	EXETER GROUP	Knowle Sandstone* *u n c o n f o r m i t y*	Knowle Sandstone* Thorverton Sandstone * Bow Breccia* Cadbury Breccia*	Higher Comberoy Formation Cadbury Breccia Upton Formation

* Formations containing isolated volcanic members (Exeter Volcanic Rocks).

dominantly of reddish brown mudstones with lenticular sandstones at some levels, and is overlain by pebble beds and sandstones of the Sherwood Sandstone Group.

BIOSTRATIGRAPHY

Prior to the present survey, the oldest stratigraphically useful fossils known from the New Red Sandstone of Devon were the vertebrate remains from the Otter Sandstone. Indigenous fossils from formations lower in the succession comprised only the trace fossils noted below, and others from near the base of the Exeter Group in the Torbay area (Pengelly, 1864; Ussher, 1913; Laming, 1966, 1970; Ridgway, 1974; Pollard, 1976; Selwood et al., 1984; Mader, 1985 b).

The majority of the New Red Sandstone formations of the district have been examined for palynomorphs (Table 9; Figure 6), with 121 samples analysed from 60 sites; 17 samples (all from the Exeter Group) proved productive. Miospores, principally pollen, have provided the first satisfactory biostratigraphical dating of the New Red Sandstone formations older than the Otter Sandstone. The most significant assemblage, from the Whipton Formation at Shilhay [SX 9195 9205], Exeter, is indicative of a Late Permian (Kazanian–Tatarian) age

(Warrington and Scrivener, 1988, 1990) (Table 10). It is dominated by bisaccate pollen, including the taeniate *Lueckisporites virkkiae* (Plate 3a), and is comparable with assemblages from the Late Permian Zechstein sequence (e.g. Clarke, 1965). Poorer bisaccate pollen assemblages, including taeniates, have been recovered from the Whipton Formation elsewhere at Exeter, and from the Creedy Park Sandstone, Crediton Breccia and Newton St Cyres Breccia in the Crediton Trough (Figure 6). Possible specimens of *Lueckisporites* from the Creedy Park Sandstone and the Crediton Breccia indicate that these units are of Late Permian age, like the Whipton Formation.

Specimens of *Crassispora kosankei*, *Lycospora* sp. and *Florinites pumicosus* from the Creedy Park Sandstone, Newton St Cyres Breccia and Whipton Formation are probably reworked from Carboniferous deposits. *Potonieisporites novicus* from the Creedy Park Sandstone may be in situ or reworked, as this taxon ranges from the Late Carboniferous (Stephanian) to the Late Permian (e.g. Clarke, 1965; Clayton et al., 1977).

Spores recovered from the Bow Breccia and Knowle Sandstone include *Convolutispora* sp.?, *Crassispora kosankei* and *Lycospora* spp., and are characteristic of Carboniferous sequences of Namurian to Stephanian age, though *Lycospora* ranges into the Early Permian (Clayton et al., 1977).

Table 9
Palynological study of the New Red Sandstone, with numbers of (I) sample sites, (II) samples, and (III) productive samples by lithostratigraphical unit in (A) the Crediton Trough succession and (B) the succession outside the Crediton Trough. The locations of palynology samples are shown in Figure 6.

III	II	I	A	B	I	II	III
				Otter Sandstone	2	2	
				Budleigh Salterton Pebble Beds	1	1	
				Aylesbeare Mudstone Group	2	18	
				Clyst St Lawrence Formation	2	2	
	4	4	Dawlish Sandstone	Dawlish Sandstone	1	1	
				Monkerton Formation	3	4	
			Shute Sandstone				
2	3	3	Newton St Cyres Breccia	Heavitree Breccia			
			Yellowford Formation				
1	9	5	Crediton Breccia	Alphington Breccia	5	6	
5	19	10	Creedy Park Sandstone	Whipton Formation	11	37	5
	3	2	Thorverton Sandstone				
	5	4	Knowle Sandstone	Knowle Sandstone	1	1	1
3	5	3	Bow Breccia				
			Higher Comberoy Formation				
	1	1	Cadbury Breccia				
			Upton Formation				
11	49	32	Totals		28	72	6

The specimens are translucent and yellow to mid brown in colour. They differ markedly in character from the dark brown or black, commonly opaque, spores recovered from known Carboniferous deposits in Devon, and thus appear not to have been affected by metamorphism associated with the Variscan Orogeny, which involved deposits at least as young as Bolsovian (Westphalian C) (Ramsbottom et al., 1978; Freshney et al., 1979).

The Whipton Formation and equivalent units in the Crediton Trough succession are regarded as Late Permian (Kazanian to Tatarian) in age on palynological evidence; vertebrate tracks from the Dawlish Sandstone also indicate a later Permian age (Table 10). The vertebrate fauna from Otter Sandstone sites in the Sidmouth district is indicative of an Anisian (early Mid Triassic) age (Milner et al., 1990; Benton et al., 1994).

Trace fossils from the Dawlish Sandstone near Broadclyst reflect elements of a contemporary land fauna. Shapter (1842) recorded 'tracings of annelides, the claw-like feet marks of two species of small crustaceans and obscure impressions of other objects' from a quarry at Broadclyst [?SX 989 982] (Figure 6, locality B). Clayden (1908 a, b) reported footprints from the same unit at Poltimore [SX 971 971] (Figure 6, locality P); Haubold (1973) identified these as *Laoporus ambiguus*, and attributed them to a pelycosaurian reptile. McKeever and Haubold (1996) consider *Laoporus* Lull 1918 to be a junior synonym of *Cheilichnus* (Jardine, 1850), which they regard as indicative of a later Permian age. Specimens from Poltimore (Plate 3i) are referable to *C. bucklandi*, the smallest of the *Cheilichnus* species now recognised (written communication, Dr P J McKeever, 1995).

In the Newton Abbot district, bioturbation and burrow structures occur in the Aylesbeare Mudstone in coast sections some 6 km south of the Exeter district (Henson, 1970; Selwood et al., 1984; Mader, 1985 a, b). To the east, in the adjoining Sidmouth district, coast sections of the Otter Sandstone within 5 km of the Exeter district show abundant traces of plant root systems (Mader, 1990; Purvis and Wright, 1991), and have yielded a varied macrofossil assemblage (Whitaker, 1869; Huxley, 1869; Johnston-Lavis, 1876; Metcalfe, 1884; Walker, 1969; Paton, 1974; Spencer and Isaac, 1983; Milner et al., 1990; Benton, 1990; Benton et al., 1994). This comprises plant debris, including fragments of an equisetalean (*Schizoneura*), and remains of arthropods, including an insect and branchiopod crustaceans, but is dominated by vertebrates, including fish (*Dipteronotus cyphus*), amphibians (*Eocyclotosaurus* sp., *Mastodonsaurus lavisi*, indeterminate capitosaurids and other remains) and reptiles (*Rhynchosaurus spenceri*, *Tanystropheus* sp., and remains of indeterminate procolophonids, rauisuchians and a possible ctenosauriscid) (Benton et al., 1994).

CORRELATION

The position of the Permian–Triassic boundary remains poorly resolved, but must lie in the sequence of some 560 m of beds between the Exeter Group and the Otter Sandstone (Table 10). For descriptive convenience, it is taken at the base of the Aylesbeare Mudstone Group. The base of the New Red Sandstone at Exeter is Late Permian in age, rather than around the Carboniferous– Permian

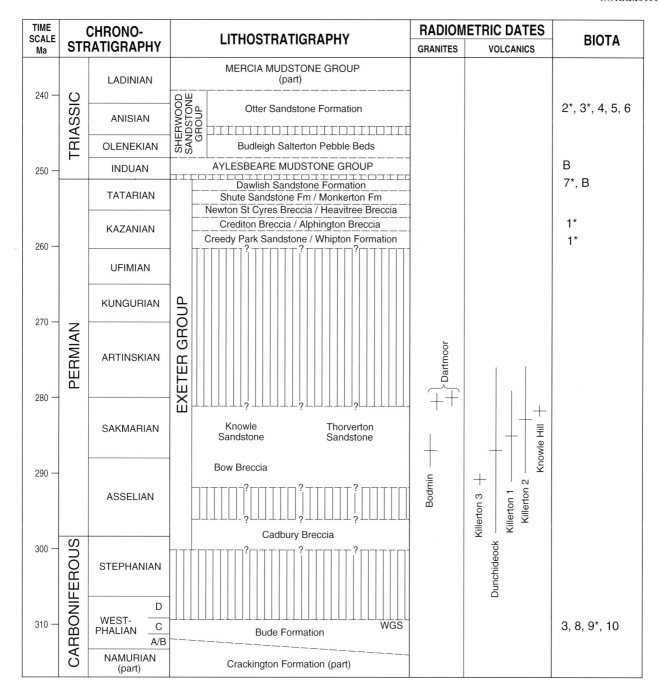

Table 10 Variscan and post-Variscan events in south-west England in relation to the Exeter Group of the district and contiguous areas.

WGS — Warren Gutter Shale (Bolsovian: Westphalian C, Ramsbottom et al., 1978). Vertical ruling indicates a gap in sequence. Biota: 1 — pollen, 2 — land vertebrates, 3 — fish, 4 — crustaceans, 5 — insects, 6 — plants, 7 — footprints, 8 — bivalves, 9 — ammonoids, 10 — orthocone nautiloid, B — burrows/bioturbation, * — significant for dating.

boundary as previously held (e.g. Laming, 1965). However, in the Crediton Trough the New Red Sandstone includes older Permian deposits, dated by their associated volcanic rocks, and the basal Cadbury Breccia may be latest Carboniferous (Stephanian) or earliest Permian in age (Table 10). The relationship of the New Red Sandstone deposits to the timing of other events in the Late Carboniferous to

Mid Triassic history of the Exeter region (Table 10) has been reviewed by Warrington and Scrivener (1988, 1990) and Edwards et al. (1997).

Correlations within the New Red Sandstone are rendered difficult owing to the marked lateral and vertical lithological variation present in the sequence, and the scarcity of fossils. However, late Permian palynomorphs suggest correlation of the Whipton Formation of the Exeter area with the Creedy Park Sandstone and Crediton Breccia of the Crediton Trough (see above). General correlations between the breccia formations are made on the basis of clast assemblages; of particular significance is the presence of distinctive cleavage fragments of potassium feldspar (termed 'murchisonite' by Levy, 1827), which occurs both in the Newton St Cyres Breccia of the Crediton Trough and in the Heavitree Breccia around Exeter, and thus indicates a correlation between the two breccias. There is a general upward increase in the variety of clast types in the breccias, reflecting the progressive erosion of folded Carboniferous rocks, the metamorphic aureole of the Dartmoor Granite, minor intrusions, and the granite itself (Table 11). Correlation within the Exeter Group has also been aided by a study of the geochemistry of the fine fraction (less than 0.106 mm) of samples of breccias and sandstones (pp.35, 38). Suggested correlations between formations in the Crediton Trough and those of the area around and south-west of Exeter are given in Figure 7.

The division between the Early Permian and Late Permian parts of the sequence is reinforced by data from rare earth analyses, together with Rb/Sr and Sm/Nd isotope studies. These studies demonstrate that rhyolite and felsite clasts from the Bow Breccia have geochemical affinities with the rhyolite and felsite occurrences of the Cawsand and Withnoe areas of east Cornwall, and with the Bodmin Moor Granite. In contrast, felsite and rhyolite pebbles from the Crediton and Newton St Cyres breccias have Sm/Nd isotopic properties close to those of the Dartmoor Granite, and the rare earth analyses correspond closely with those of typical Dartmoor Granite and its late-stage aplite intrusions.

It is notable that, while the acid igneous clasts in the Bow Breccia are not tourmalinised, those in the Crediton and Newton St Cyres breccias are; the tourmalinised material increases in abundance upwards from the middle part of the Crediton Breccia, where it first appears. The nature of this material suggests that boron-rich fluids were active in the lower part of the parent volcanic sequence, close to the granite pluton. There is, therefore, an inverted stratigraphy revealed by study of the breccia clasts, from untourmalinised volcanic rocks in the lowest part of the Late Permian breccias through tourmalinised rhyolite and felsite to granite.

SUMMARY OF DEPOSITIONAL ENVIRONMENTS

The compressive mountain-building phase of the Variscan Orogeny, culminating in the Late Carboniferous, was succeeded by a period of lithospheric extension, during which thick sequences of sediments were deposited in partly fault-controlled basins. Volcanic activity associated with extension is represented by the Exeter Volcanic Rocks, dated radiometrically at about 291 to 282 Ma. The Variscan mountains began to be rapidly eroded in the Early Permian, and possibly from the Late Carboniferous (Stephanian) onwards. Huge volumes of debris from the disintegration of the highlands were deposited on alluvial fans adjacent to the mountain front. Deposition was mainly by debris flow and sheet-flooding during occasional heavy rainstorms in a semi-arid climate. These coarse-grained deposits are now preserved as the breccias of the Exeter Group. The breccias are overlain by, and interfinger with, sandstones of the Dawlish Sandstone Formation, which were formed as wind-blown desert dunes, with some interdune deposits.

The Early Permian extensional phase, with associated rift-valley topography, was followed by gradual peneplanation, and by the end of the Permian (here taken at the top of the Exeter Group) a period of regional thermal relaxation had led to an increased area of deposition (Holloway, 1985). As a result, the basal Triassic Aylesbeare Mudstone unconformably oversteps from Permian strata on to Variscan 'basement', as at the eastern end of the Crediton Trough. The Aylesbeare Mudstone was probably deposited in a continental sabkha–playa lake complex.

The Budleigh Salterton Pebble Beds, which sharply overlie the Aylesbeare Mudstone, represent a sudden influx into the district of coarse gravels containing exotic clasts derived from an external source to the south, and were deposited by low-sinuosity, highly braided rivers. A thin bed containing wind-faceted pebbles caps the Pebble Beds, and indicates a hiatus of uncertain length in the Late Scythian, during which the region was a wind-blasted stony desert. The succeeding Otter Sandstone was largely deposited on a sandy braidplain, with some aeolian dunes (Mader and Laming, 1985).

The microfloras from the Whipton Formation, and from the Creedy Park Sandstone to Newton St Cyres Breccia sequence in the Crediton Trough, comprise gymnosperm pollen that reflects a parent flora of conifers. As in contemporary volcanic areas, soils derived from the weathering of the Exeter Volcanic Rocks may have favoured establishment of this vegetation locally, although no plant macrofossils are known.

The existence of a land fauna in Late Permian times is shown by trace fossils, including vertebrate (reptilian) tracks, which are consistent with the continental, largely aeolian facies of the Dawlish Sandstone. Root system traces in the Otter Sandstone reflect an indigenous vegetation dominated by conifers (Mader, 1990). Purvis and Wright (1991) considered some vertical traces to represent tap roots of phreatophytic plants. These traces are the main evidence of the existence of a significant flora in the area during the Mid Triassic; only indeterminate plant debris and fragments referable to the equisetalean *Schizoneura* have been noted (Benton et al., 1994). The Mid Triassic macrofauna of the region, represented solely by remains from the Otter Sandstone, included insects, aquatic crustaceans and other arthropods, and fish, in addition to a terrestrial tetrapod fauna that included carnivorous rauisuchians and herbivorous rhynchosaurs and procolophonids. The presence of fish and amphibians that required aquatic habitats, and of the specialised semi-aquatic reptile *Tanystropheus,* is noteworthy.

More detailed interpretations of the depositional environments of the Exeter Group, Aylesbeare Mudstone Group, and Sherwood Sandstone Group, are given in chapters 5 and 6.

FIVE

Permian rocks: Exeter Group

INTRODUCTION

The term Exeter Group is introduced in this memoir to include all the formations in Devon between the unconformity at the base of the New Red Sandstone and the overlying Aylesbeare Mudstone Group (Table 8). The group consists predominantly of breccias, with some sandstone and mudstone units; these lithologies commonly weather to gravel, sand and clay. The component formations are listed in Table 8 and their outcrops shown on Figure 6. Because of the structural separation between the Exeter area and the Crediton Trough, local formational names are necessary; probable correlations, based on similarities in clast content, geochemistry and limited palaeontological evidence, are given in Figure 7. The stratigraphical relationships of the Exeter Group formations in the Crediton Trough are shown in Figure 8. Lavas locally interbedded with the sedimentary rocks (Figure 7) are not individually named, but are referred to collectively as the Exeter Volcanic Rocks; they are described in Chapter 7.

Hutchins (1963) and Edmonds et al. (1968) have shown that in the Crediton Trough, each breccia formation contains a characteristic assemblage of fragment types reflecting the erosional history of the area. There is a general tendency for the clast assemblages to become more diverse as erosion cut down into a wider variety of rocks in the source areas. The earliest formation, the Cadbury Breccia, consists almost exclusively of clasts derived from the Carboniferous Bude Formation, with scarce pebbles from the Devonian of north Devon. The succeeding Bow Breccia contains, in addition to Culm sandstone, clasts of shale, slate and hornfels, together with a minor fraction of pebbles of acid igneous origin; these igneous types predate the intrusion of the Dartmoor Granite. The younger Crediton Breccia and the equivalent Alphington Breccia at Exeter have a mixed clast assemblage, including Culm slate and sandstone, rhyolite fragments, large clasts of quartz-porphyry, and tourmalinised slate and hornfels. This assemblage indicates erosion of the roof zone and aureole of the Dartmoor Granite. The succeeding Newton St Cyres Breccia (and the Heavitree Breccia at Exeter) contain abundant fragments of the potassium feldspar locally known as 'murchisonite', in addition to clasts of granite which can be matched with the Dartmoor Granite (Dangerfield and Hawkes, 1969), indicating that erosion had by this time reached the main mass of the granite itself. Table 11 shows the succession of clast types recognised in the Exeter Group breccias of the district.

Analysis of the sedimentary features of the Exeter Group during the resurvey (Jones, 1992 a, b) confirms earlier observations (e.g. Laming, 1966) that the breccias were deposited on alluvial fans, and the Dawlish Sandstone Formation was largely aeolian in origin. Detailed facies analysis is given by Jones (1992 a, b).

Chemical analyses of the fine-grained (less than 0.106 mm) fractions of Exeter Group breccias and sandstones from the district and the adjacent Okehampton

Table 11 Succession of clast types in the Exeter Group of the district.

Classification	Clast content	Common	Rare
	Culm sedimentary rocks · Chert · Devonian sedimentary rocks · Quartz-porphyry (untourmalinised) · Tourmalinite · Hornfels · Rhyolite and acid tuff (untourmalinised) · Lamprophyre · Basalt · Vent breccia · Tourmalinised acid igneous rocks	Murchisonite	Granite
DAWLISH SANDSTONE			
NEWTON ST CYRES BRECCIA / HEAVITREE BRECCIA / SHUTE SANDSTONE / MONKERTON FORMATION			
CREDITON BRECCIA / ALPHINGTON BRECCIA / YELLOWFORD FORMATION			
CREEDY PARK SANDSTONE / WHIPTON FORMATION			
KNOWLE SANDSTONE / THORVERTON SANDSTONE			
BOW BRECCIA			
CADBURY BRECCIA			

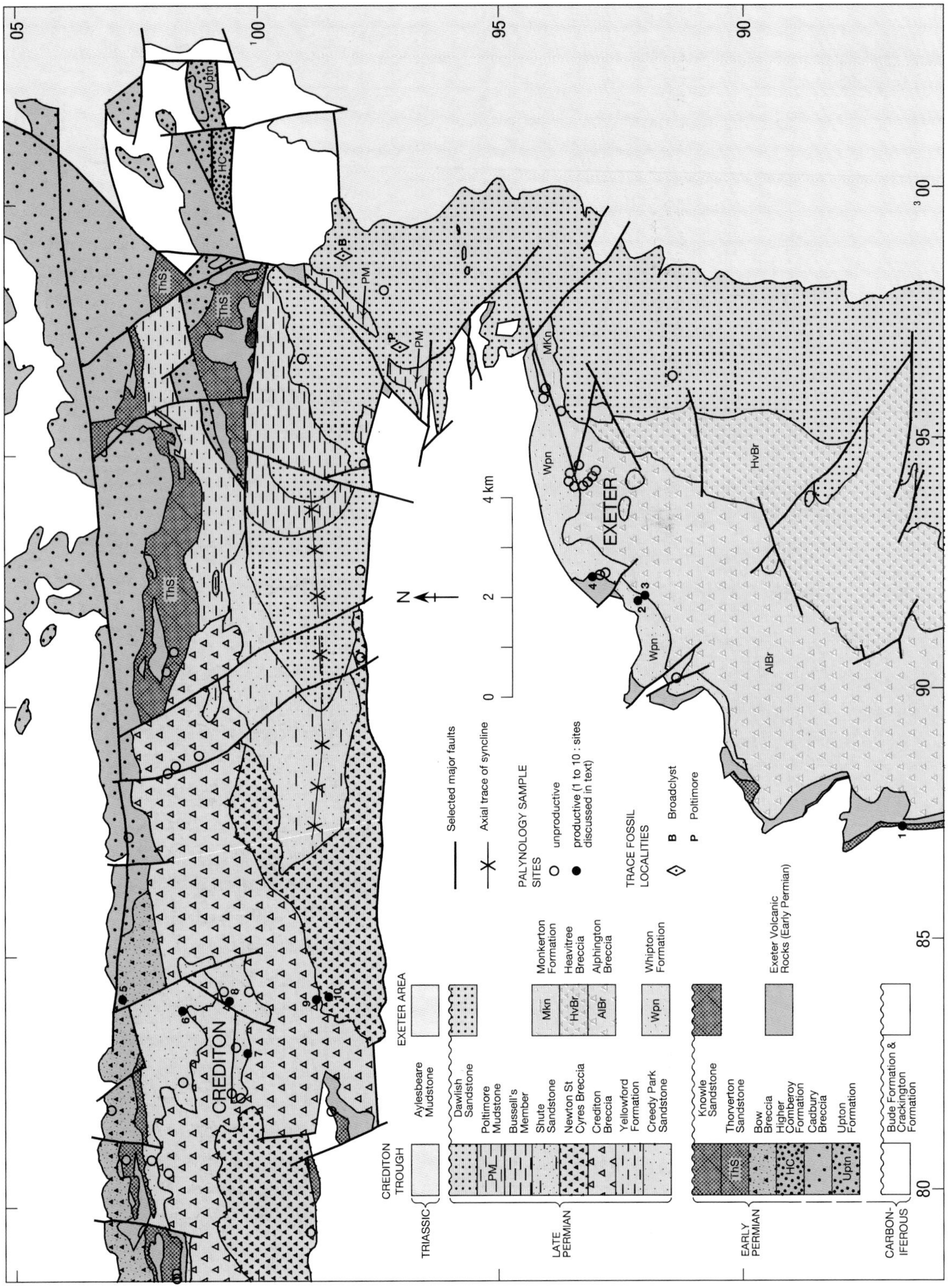

Figure 6
Distribution of the Exeter Group formations of the district, showing also palynology sample sites and trace fossil localities. Productive palynology sites (1–10) are discussed in the text.

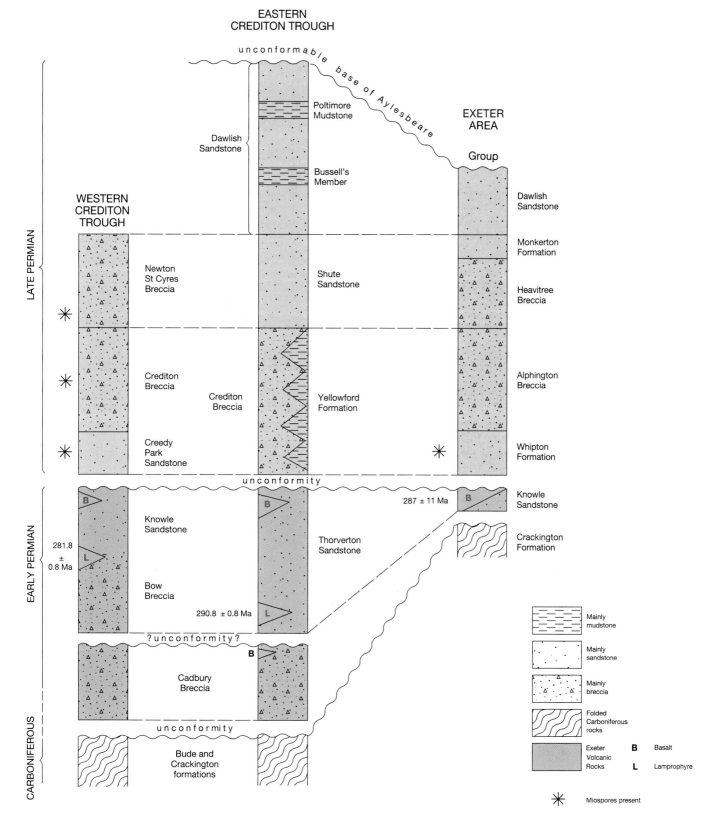

Figure 7 Correlation of Exeter Group sequences in the district.

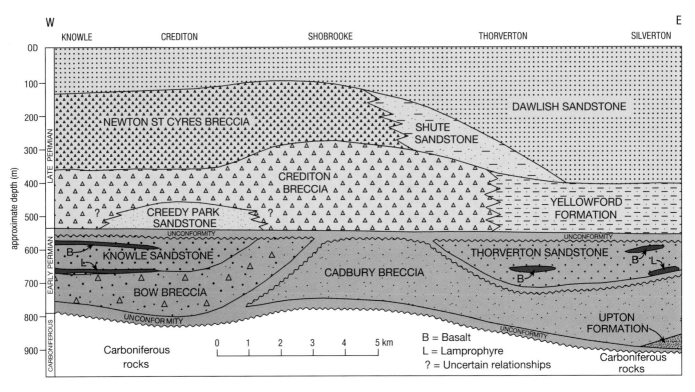

Figure 8   Diagram of stratigraphical relationships of the Exeter Group in the Crediton Trough.

district were carried out using the techniques of Peachey et al. (1985). The results are given in Ault et al. (1990, 1993). Comparisons between samples were aided by plotting ratios of selected elements. There are marked geochemical contrasts between different parts of the Exeter Group sequence, and these contrasts are present both in the Crediton Trough and in the sequence around and south-west of Exeter, confirming correlations made on other criteria.

The Early Permian formations—the Cadbury Breccia, Bow Breccia, Knowle Sandstone and Thorverton Sandstone—are characterised by broadly similar ratio plots (Figure 9) and show geochemical characteristics typical of mature sediments with generally low values of the mobile elements Mg, Ca, Sr and Ba. Correlation of the sandstones and breccias beneath the basalt lavas of the Dunchideock area south-west of Exeter with the Knowle Sandstone of the Crediton Trough is supported by the close similarities in the geochemistry.

The lower part of the Late Permian sequence—Crediton Breccia and Yellowford Formation in the Crediton Trough, and Alphington Breccia in the Exeter area—shows a marked geochemical contrast with the Early Permian sequence that lies unconformably below it. These breccias postdate the emplacement of the Dartmoor Granite at 280 Ma, and contain a considerable variety of clasts (Table 11). This variety is reflected in the geochemistry. The relatively unweathered lithic material gives rise to generally higher values of the mobile elements (Mg, Sr and Ba, but not Ca), and, in particular,

many samples are characterised by high values of V and Sr, giving distinctive ratio plots (Figure 9).

The Whipton Formation of the Exeter area, at the base of the Late Permian sequence, and its correlative the Creedy Park Sandstone of the Crediton Trough, vary in their geochemistry. The Creedy Park Sandstone generally resembles the Crediton Breccia, into which it is believed to pass laterally (though the Sr values are lower, resulting in a different ratio plot), rather than the underlying Knowle Sandstone, from which it differs in having higher Li, B, V, Cr, Sr and Zr and lower Be.

In the eastern part of the area, near Exeter (the Heavitree Breccia, Monkerton Formation and Dawlish Sandstone), and around Silverton at the eastern end of the Crediton Trough (the Dawlish Sandstone), the appearance of clast material, including potassium feldspar (murchisonite) and granitic material, indicating that the Dartmoor Granite had been unroofed and was contributing to the New Red Sandstone sequence, is reflected in the geochemistry by a pronounced increase in the values of K and also by higher values of K_2O/Al_2O_3, which are indicators of an increase in the proportion of potassium feldspar in the analysed fine-grained fraction of the rock. These formations are also characterised by low values of Li and Sr, as illustrated in the ratio plots (Figure 9). In formations of similar age in the western part of the Crediton Trough, these geochemical features are present, in a subdued form, in some samples from the Newton St Cyres Breccia and Dawlish Sandstone, but not in the Shute Sandstone.

Figure 9 Plots of ratios of selected elements in the fine-grained (< 0.106 mm) fractions of samples from the Exeter Group of the district, showing variation at different stratigraphical levels in the Crediton Trough and in the Exeter area.

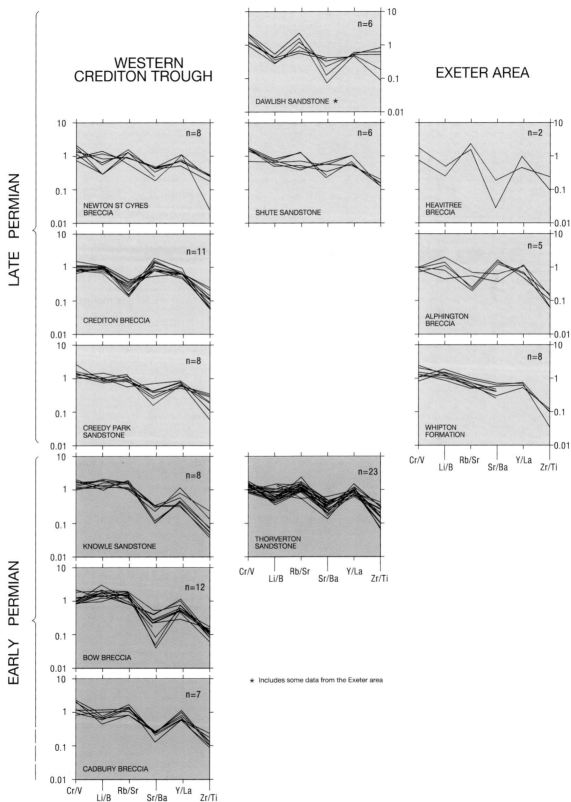

EASTERN CREDITON TROUGH

WESTERN CREDITON TROUGH

EXETER AREA

LATE PERMIAN

EARLY PERMIAN

DAWLISH SANDSTONE ★

NEWTON ST CYRES BRECCIA

SHUTE SANDSTONE

HEAVITREE BRECCIA

CREDITON BRECCIA

ALPHINGTON BRECCIA

CREEDY PARK SANDSTONE

WHIPTON FORMATION

KNOWLE SANDSTONE

THORVERTON SANDSTONE

BOW BRECCIA

CADBURY BRECCIA

★ Includes some data from the Exeter area

STRATIGRAPHY

Exeter area

KNOWLE SANDSTONE

Outside the Crediton Trough (p.29) the Knowle Sandstone consists of reddish brown, moderately well-cemented sandstone and fine-grained breccia, and is up to about 20 m thick. A discontinuous outcrop extends from Halscombe [SX 882 898], near Ide, southwards through Idestone and Dunchideock to the southern boundary of the district (Figure 6). The Knowle Sandstone lies unconformably on the Crackington Formation, and includes basalts of the Exeter Volcanic Rocks at the top (Figure 7). The basalts locally rest directly on Crackington Formation, suggesting that erosion of the Knowle Sandstone occurred before extrusion of the lavas. The correlation of these strata with the Knowle Sandstone of the Crediton Trough is supported by their similar lithologies and geochemistries.

No direct evidence of the age of the formation is available. South-west of Exeter, the Knowle Sandstone lies beneath, and is therefore older than, basaltic lavas which at Dunchideock have been dated using the potassium-argon method at 287 ± 11 Ma (recalculated from Miller and Mohr, 1964, using the constants of Steiger and Jäger, 1977). A sample from beds exposed beneath the lava in School Wood Quarry [SX 8750 8708], Dunchideock (Figure 6:1), yielded the spores *Lycospora* sp. and *L. pusilla*, probably reworked from Carboniferous rocks (p.31).

The main clast types are Culm sandstone and shale, lava, and porphyry. An exposure [SX 8823 8806] near Idestone shows a few subrounded pebbles of grey fine-grained limestone, possibly of Devonian age. At School Wood Quarry [SX 8750 8708], dolomite pebbles are present in breccias beneath basaltic lava.

Halscombe

Surface material [SX 8795 8986] near Halscombe includes pale red, moderately well-cemented, medium-grained sandstone, and pale red to greyish red, moderately well-cemented, fine-grained breccia with clasts of Culm sandstone and siltstone and fine-grained ?lava. A sample [SX 8811 8979] is moderate red, moderately well-cemented, fine- to medium-grained sandstone. A thin section (E 65347) shows it to be a well-sorted quartz-arenite composed of well-rounded 'millet seed' quartz grains, with chert and lithic clasts, in a cherty cement. Grains have early diagenetic iron oxide coatings. The cherty cement also forms early grain-coating veneers as well as later pore-fillings.

Idestone

An excavation [SX 8789 8837] at Idestone showed 4 m of red, fine-grained breccia, locally hard and well cemented, with clasts of Culm sandstone and shale, and some lava. Impersistent beds of reddish brown, coarse-grained sandstone are present. The beds dip at 18° to 035°. An old quarry [SX 8823 8806], about 500 m south-east of Idestone, shows vesicular basalt overlying breccia, although the contact is not exposed. Dips average 20° to 075°. The breccia is hard, red, moderately well cemented, mainly fine grained, with impersistent beds of pebbly sandstone and a few subrounded clasts of medium grey, very fine-grained ?Devonian limestone.

Dunchideock

South of Webberton Wood, a cutting [SX 8730 8693] shows up to 2 m of reddish brown, fine-grained breccia, with clasts mainly less than 3 cm, some to about 7 cm, of cream-weathering porphyry (angular to subrounded), Culm sandstone, shale, and chert, in a gravelly clayey sandy silt matrix, resting unconformably on Crackington Formation.

In the western corner [SX 8750 8708] of the southern quarry in School Wood (Plate 10c), sandstone and breccia are visible beneath lava, as follows:

	Thickness m
Basalt, purple, with white amygdales, complexly intermingled with reddish brown, micaceous, fine-grained sandstone, dip 20° to 065°; sandstone is dominant in the lowermost 0.6 m	2.8
Sandstone and fine sandy breccia, pale red, hard, massive	1.2
Breccia, pale red, hard, well-cemented, with subrounded dolomite clasts to 7 cm	0.6

In an exposure [SX 8720 8667] about 400 m west-north-west of Penhill Farm, Dunchideock, up to 2 m of pale red, moderately to weakly cemented, fine-grained breccia with clasts mainly of angular Culm sandstone, fine-grained lava, and some white ?feldspathic material dips at 23° to 073°.

WHIPTON FORMATION

From Pinhoe [SX 970 943] westwards through the centre of Exeter to Ide [SX 895 903], the basal unit of the Exeter Group is the Whipton Formation, which rests on the Crackington Formation except where local patches of basalt intervene at Rougemont [SX 921 930], Barley Lane [SX 900 922], Pocombe [SX 899 915], and West Town Farm (Ide) [SX 891 903] (Figure 6). South of West Town Farm, the Whipton Formation is absent and the overlying Alphington Breccia rests on basalt.

The formation, first named from the Whipton area of Exeter by Scrivener (1984), consists predominantly of reddish brown, weakly cemented, silty, fine-grained sandstone, with local developments of siltstone, claystone and breccia. These lithologies weather to sand, silt, clay and clayey gravel.

In the type area, the formation is estimated to be about 60 m thick. It is about 20 m thick around Ide, and at least 25 m thick in Exeter city centre. East of Whipton, the outcrop narrows eastwards and the formation either thins or is overstepped by the Monkerton Formation in that direction. At the most easterly recorded outcrop [SX 970 943], near Pinhoe, the thickness is about 10 m.

The type section [SX 9435 9419 to 9449 9394] is in a stream along the western boundary of the grounds of Northbrook School, Whipton. At the north end, reddish brown, gravelly clay and clayey sand with beds of breccia rest unconformably on a southerly dipping (about 15° dip) surface cut across silicified and manganese-stained Crackington Formation shales dipping north at about 50°. A basal breccia, 0.3 m thick, consists of subangular cobbles of sandstone and shale in a ferruginous cement, and is overlain by poorly sorted sandstone and mudstone, with thin beds of breccia and some coarser breccia units. The sequence continues to the south-east, poorly exposed,

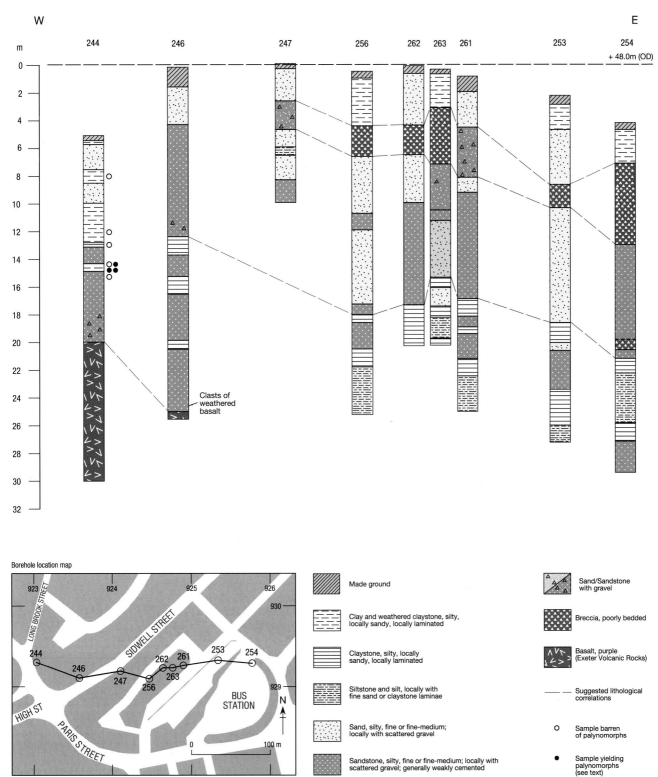

Figure 10 Cross-section through boreholes in the central area of Exeter city to illustrate lithological variation in the Whipton Formation. Borehole numbers are those of BGS records for sheet SX 99 SW.

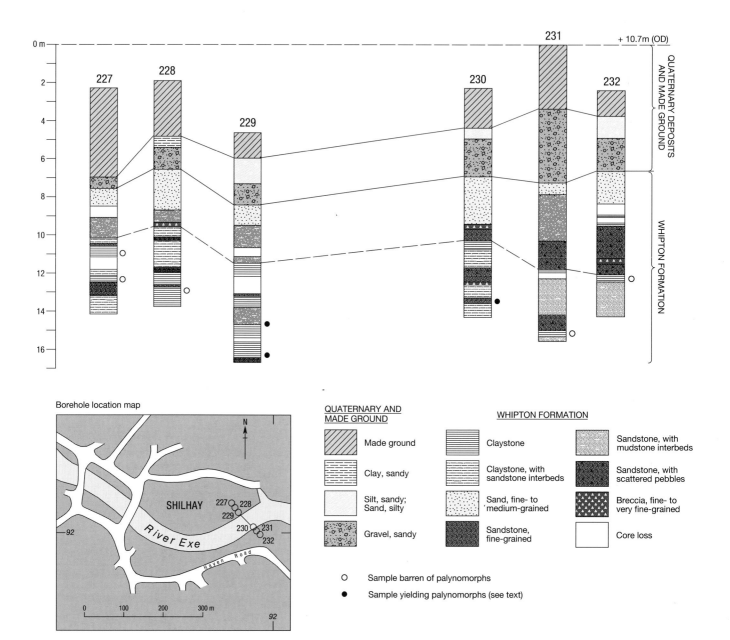

Figure 11 Cross-section through Whipton Formation and Quaternary deposits in the Shilhay area of Exeter. Borehole numbers are those of BGS records for sheet SX 99 SW.

in gently south-dipping mudstone. Where the Whipton Formation rests on basalt, as at Pocombe Quarry [SX 8994 9133], the basal 2 m or so consist of a rubbly deposit comprising irregular clasts of purple lava in a matrix of hard, reddish brown, fine-grained sandstone.

The overall lithology is constant along the outcrop of the Whipton Formation, but there are considerable small-scale local variations in the proportions of sandstone, silt-stone, claystone and breccia. Examples are shown in Figures 10 and 11.

The sandstones are mostly very weakly cemented at outcrop, red or reddish brown, and vary in grain size

from coarse to silty or, in places, clayey fine-grained sand. Some clay is commonly present, although relatively clean sands occur locally. The degree of sorting varies from poor to moderate. Bedding is typically parallel; small-scale cross-bedding is rarely developed. The breccias form the basal parts of sandstone units and commonly grade upwards into coarse-grained sandstones. The thick-ness of the breccia units seldom exceeds 1 m, and is com-monly much less. Basal contacts are sharp and may show channelling into the underlying finer-grained units. Clast size is variable, with coarse-grained breccias being subor-dinate to finer breccia units. Culm sandstone and slate

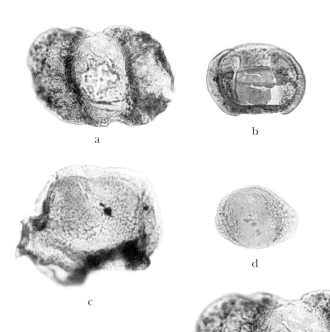

a

b

c

d

e

f

g

h

Plate 3 Late Permian fossils, Exeter Group.

a–h Pollen, Whipton Formation, Shilhay [SX 9195 9205], Exeter. All specimens × 500; specimens held in the palynology collection, British Geological Survey, Keyworth, registered in the MPK series.

a   *Lueckisporites virkkiae*, var. A of Clarke, 1965 (MPK 5926).
b   *L. virkkiae*, var. B of Clarke, 1965 (MPK 5928).
c   *Perisaccus granulosus* (MPK 5921).
d   *Klausipollenites schaubergeri* (MPK 5934).
e   *Taeniaesporites ortisei* (MPK 5922).
f   *Protohaploxypinus microcorpus* (MPK 5920).
g   *L. virkkiae* var. B of Clarke, 1965 (MPK 5929).
h   *K. schaubergeri* (MPK 5935).
i   Footprints (*Cheilichnus bucklandi*) from the Dawlish Sandstone, Poltimore [SX 971 971]; part of trackway found on 19 May, 1908 (Clayden, 1908 a). Photograph by courtesy of the Royal Albert Memorial Museum and Art Gallery, Exeter; registered specimen number EXEMS 287/1908 Slab A.

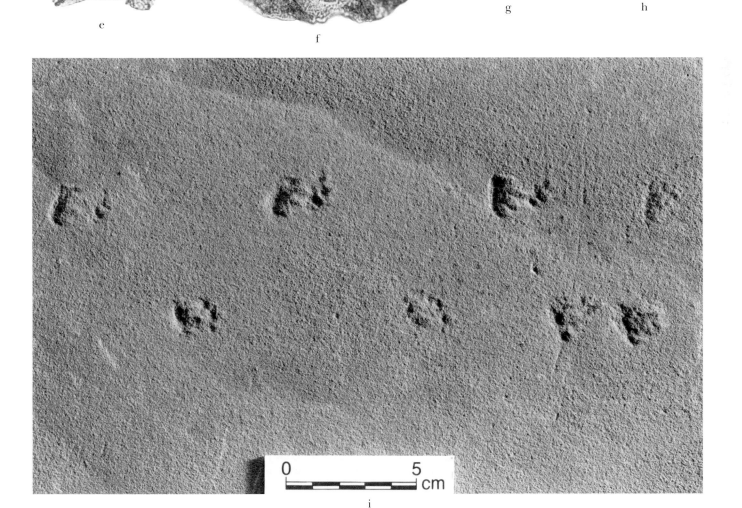

i

clasts predominate, with a very minor amount of quartz-porphyry and other igneous debris; in this, they differ from the overlying Alphington Breccia. The breccia matrix may be red, rather coarse-grained sand or brownish red, more argillaceous material. The breccias are matrix supported except for the coarser-grained units. The red or purplish brown mudstones may form units up to several metres in thickness, with prominent green and grey-green reduction zones. More usually, the argillaceous beds are relatively thin and locally occur as discontinuous lenses within the sandstones.

At Dunsford Hill [SX 903 914], fining-upwards units up to 22 cm thick have pebbly bases 1 to 2 cm thick and are overlain by finely laminated, ?plane-bedded, or massive fine-grained sandstone, and probably represent tractional sheetflood events. Sharp bed boundaries indicate that deposition was in separate events. Other nearby exposures [SX 9025 9139] of very fine- to fine-grained, massive sandstone in beds up to 10 cm thick may indicate aeolian deposition; however, the bedded nature of the sandstone could also be explained by individual sheetflood events. The massive nature of the sandstone possibly indicates rapid deposition from turbulent sheetfloods, which would suppress the generation of sedimentary structures.

Samples from a borehole [SX 9195 9205] at Shilhay on the northern bank of the River Exe (Figure 6, locality 2) yielded a moderately rich miospore assemblage comprising *Lycospora* sp., indeterminate trilete spores, *Perisaccus granulosus*, *Lueckisporites virkkiae* (vars A and B of Clarke, 1965), *Lunatisporites* sp., *"Taeniesporites" ortisei*, *Protohaploxypinus microcorpus*, *P. samoilovichii*, *Jugasporites* sp., *J. delasaucei*, *Falcisporites zapfei*, *Klausipollenites schaubergeri* and indeterminate bisaccate pollen (Plate 3a–h). Indeterminate bisaccate pollen were also recovered from a borehole [SX 9198 9203] at Haven Banks, Exeter (Figure 6, locality 3). A borehole [SX 9230 9293] in central Exeter (Figure 6, locality 4) yielded a solitary trilete spore, cf. *Punctatisporites*, and bisaccate pollen including *Lunatisporites* sp. and indeterminate taeniate and non-taeniate specimens. These assemblages indicate a Late Permian age for the formation (Warrington and Scrivener, 1988, 1990).

Pinhoe

A borehole [SX 9701 9412] drilled for the M5 Motorway proved 6.6 m of Whipton Formation beneath 2.1m of drift. The sequence consisted of 0.3 m of medium-grained sandstone, overlain by 3.1 m of sandy siltstone, overlain by 3.2 m of silty and sandy clay.

The following section, typical of the Whipton Formation of the area, was proved in a borehole [SX 5545 9373] adjacent to Pinhoe Road:

	Thickness m	Depth m
Clay, silty, sandy, red, firm, with occasional hard clay fragments	5.0	5.0
Clay, silty, fine-grained, sandy; with some thin beds of silty sandstone	2.9	7.9
Sandstone, silty, clayey, red	0.7	8.6
Clay, silty, sandy, red, stiff, with weak sandstone fragments	1.5	10.1

	Thickness m	Depth m
Clay, sandy, red, stiff to very stiff	1.5	11.6
Sand, silty, slightly clayey, red	0.6	12.2
Silt, clayey, hard, friable, with layers of clayey sand and thin weak sandstone	1.6	13.8

Whipton

The type section of the Whipton Formation is the stream section [SX 9435 9419 to 9449 9394] along the western boundary of Northbrook School (p.40). A stream section [SX 9417 9342 to 9419 9332] at Whipton exposes the highest beds of the Whipton Formation (clayey sand and mudstone), and the junction with the overlying Alphington Breccia (breccia with beds of purplish brown, gravelly mudstone).

Exeter city centre

A site investigation borehole [SX 9275 9328] adjacent to Sidwell Street proved 8.7 m of Whipton Formation beneath 1.8 m of drift and made ground. The lowest 3.5 m proved is laminated sandy siltstone with sand beds, overlain by 2.5 m of very sandy, clayey siltstone with sand beds, overlain by 2.7 m of very silty, fine- to medium-grained sand with sandy silt beds.

Site investigation boreholes [SX 9230 9293 to 9258 9293] between Longbrook Street and Paris Street proved up to 25 m of Whipton Formation, consisting mainly of silty, fine- to medium-grained sandstone and sand, with subordinate thicknesses of breccia, clay, claystone and siltstone (Figure 10). The westernmost boreholes show the formation resting on basalt (Exeter Volcanic Rocks). Three samples from another borehole [SX 9230 9293] yielded sporadic miospores (Figure 6, locality 4; Figure 10), including, from 9.32 m depth, a poorly preserved trilete spore cf. *Punctatisporites*, from about 9.8 m depth a vestigial specimen of a taeniate bisaccate pollen, possibly referable to *Lunatisporites*, and from 9.81 m depth, specimens assignable to *Lunatisporites*. The association comprises forms which are also present in the richer assemblage at Shilhay.

South-west of the city centre, a trial borehole [SX 9166 9231] in Beedles Terrace, near the southern end of Fore Street, proved 2.1 m of made ground and head, on 6.7 m of Whipton Formation consisting of reddish brown and grey clayey fine-grained sand with thin breccia beds, resting on weathered and reddened Crackington Formation shale.

Shilhay area, Exeter

Boreholes near Shilhay penetrated up to 8.5 m of Whipton Formation beneath made ground and drift (Figure 11). The top 3 m of the formation in the northernmost four boreholes is sand and sandstone, overlying a mainly mudstone sequence. In the southernmost two boreholes, there is possibly a lateral passage from the mudstone sequence into sandstones, or a fault between the two sets of boreholes. A sample from 9.9 to 10.1 m depth in one borehole [SX 9195 9205] (Figure 6, locality 2) yielded a moderately rich Late Permian miospore assemblage.

Dunsford Hill, Exeter

Boreholes and excavations for a reservoir at Dunsford Hill [SX 903 914] showed the formation to consist predominantly of reddish orange-brown, silty, fine- to medium-grained sand, with some breccia, and a bed of reddish brown mudstone 1.15 to 1.5 m thick in the lower part, dipping at 5° to 087°. The sequence proved in a borehole [SX 9026 9131] is as follows:

	Thickness m	Depth m
Soil and ? head	3.00	3.00

WHIPTON FORMATION

Sand, slightly clayey, silty, fine to medium; locally with fine to medium gravel	1.00	4.00
Sand, silty, fine to medium, locally with fine to medium gravel and dark brown, fine sandy silt and coarse sand laminae	2.60	6.60
Sand, partially cemented, slightly clayey, silty, fine, medium and coarse, locally with angular gravel	0.90	7.50
Sand, slightly clayey, silty, fine to medium, with a few bands of silty clay	1.50	9.00
Sand, silty, fine to medium	1.20	10.20
Mudstone, very thinly bedded and laminated, silty; locally clay	1.15	11.35
Sand, silty, fine to medium	1.45	12.80

Pocombe area, Exeter

A cutting [SX 8994 9133] at the entrance to Pocombe Quarry shows 1 to 2 m of reddish brown, irregularly bedded, fine-grained sandstone, with a gradational base, on 1.5 to 2 m of an admixture of hard, reddish brown, fine-grained sandstone and purple lava, with an ill-defined base, on 2 m of moderately massive, irregularly fractured, red-purple and greyish purple basalt.

Ide to West Town

A section [SX 8997 9037] in Ide, near the top of formation, shows 1 m of reddish brown, clayey, fine- to very fine-grained breccia (clasts mostly less than 1 cm), locally passing into sandy breccia and silty fine-grained sand, overlying 0.6 m of reddish brown, sandy silt, on 0.6 m of breccia.

The probable contact with the overlying Alphington Breccia is exposed [SX 8902 9021] in the railway cutting south of West Town Farm; soft reddish brown, fine- to medium-grained sand and fine-grained breccia dip east at about 10°, and are overlain by reddish brown, weakly cemented breccia (?Alphington Breccia) with clasts of purple-weathered lava and pale grey ?dolomite.

A section [SX 8904 9029] at West Town Farm, Ide, shows up to 2.5 m of reddish brown, soft, silty, fine- to very fine-grained sand, with irregular centimetre-scale beds, possibly partly brecciated, of black-stained sand and very weakly cemented sandy siltstone. About 1 m above the base, a 5 cm bed of gravelly fine-grained sand contains weathered clasts of purple and grey, rotted vesicular lava, and overlies 0.3 m of soft, gravelly, fine-grained sand which probably contains much derived volcanic material.

ALPHINGTON BRECCIA

The name Alphington Breccia has been used in the present survey to describe breccias that crop out around and to the south of Exeter (equivalent to the lower part of the Heavitree and Alphington Breccias of Smith et al., 1974). The Heavitree Breccia is readily distinguished by the characteristic presence in it of abundant fragments of a distinctive potassium feldspar (murchisonite).

The Alphington Breccia has an outcrop 2 to 3 km wide extending from the Heavitree area of Exeter, south-westwards through the centre of Exeter to Alphington, and thence through Shillingford St George [SX 904 878] to the southern boundary of the district (Figure 6). At the northern end of the outcrop, it is faulted against the Monkerton Formation. The low degree of cementation and the high shale content (which causes it to weather to a clay), give rise to subdued topography. South of Alphington, dips are generally less than 10°, giving rise to long dip slopes in the Kennford and Shillingford Abbot areas. Because it is poorly cemented, the Alphington Breccia has not been quarried, and there are few exposures.

In the Exeter city area, as far south-west as West Town Farm, Ide [SX 883 899], the Alphington Breccia rests on the Whipton Formation; from West Town to the southern edge of the district near Lawrence Castle [SX 875 860], the breccia rests on basalts at the top of the Knowle Sandstone. The junction of the Alphington Breccia with the Whipton Formation is gradational over a vertical interval of about 10 m, the boundary being drawn where breccia becomes the dominant lithology.

The formation consists predominantly of reddish brown, poorly sorted, fine-grained breccias containing a wide variety of angular to subrounded clast types in a matrix of reddish brown, clayey sand or silty clay. The clasts form a characteristic suite of Culm-derived shale and sandstone, together with hornfels, chert, quartz-porphyry and lava. A few granitoid fragments and murchisonite cleavage fragments are present in the higher part of the formation. The size of the clasts is variable: most are less than 4 cm in diameter but some range up to 10 cm, and boulders of quartz-feldspar porphyry are locally present. In general, the beds in the lower part of the sequence are coarse-grained, with interbeds of finer grained breccia, siltstone or claystone, or less commonly, red sandstone; the upper part of the sequence is finer grained and better cemented. The breccias are rather poorly bedded, but bedding generally is indicated by the alternation of coarse and fine-grained breccia beds, and locally by impersistent lenses of reddish brown or greyish green mudstone. Individual beds of breccia are graded, and may channel into the underlying bed.

There appears to be little major lithological variation along the outcrop of the Alphington Breccia. At Dunchideock Barton [SX 8772 8747 to 8776 8746], where the formation rests on basalt, the basal few metres consist of interbedded reddish brown, moderately well-cemented, fine-grained breccia and pebbly mudstone. A similar clayey facies probably extends to Webberton Cross [SX 880 874]. Elsewhere along the outcrop, pebbly mudstones appear to be absent from the basal beds. At The Quay, Exeter [SX 921 920], for example, beds close to the base are very coarse-grained breccia with clasts up to small boulder size, including abundant quartz-porphyry (Plate 5a).

The formation reaches its maximum thickness, estimated to exceed 240 m, in the area south of Alphington; in the Exeter city area, the thickness is about 160 m. North-east of Heavitree, the Alphington Breccia either thins out northwards or is overstepped by the Monkerton Formation.

No palynomorphs have been recorded from the Alphington Breccia. The Crediton Breccia, with which the Alphington Breccia is correlated, has yielded pollen assemblages indicating a Late Permian age (p.68).

The sedimentary features of the basal part of the formation at Dunchideock Barton [SX 8776 8746] have been described by Jones (1992 b). Up to 2 m of poorly cemented, unbedded and ungraded, matrix-supported breccia contains subrounded to subangular and poorly to very poorly sorted clasts, with maximum particle size approximately 0.25 m, distributed evenly throughout the bed. The breccia lacks grading or bedding, and clasts are generally flat-lying, although locally, vertical clasts are present. These features are characteristic of subaerial debris flows (Gloppen and Steel, 1981; Lowe, 1982).

Whipton to central Exeter

A stream bank [SX 9422 9326] at Whipton shows 2.7 m of head resting on 3.0 m of fine- to medium-grained breccia in beds up to 0.20 m thick, fining upwards into reddish brown sandy clays up to 0.25 m thick. Greenish grey reduction streaks, pods and irregular patches are common throughout, but especially in the breccia units. In the same stream, a section [SX 9448 9301] near the end of Madison Avenue shows 0.44 m of brownish red mudstone, with abundant fine gravel, overlying 0.7 m of reddish brown clayey breccia passing down into coarse-grained breccia with irregular subrounded clasts including some quartz-porphyry up to 0.10 m, on 0.74 m of brownish red, fine-grained silty sandstone, locally gravelly and with common reduction patches. The sequence dips south at 6°.

A degraded face [SX 9318 9289] in the Clifton Hill brickpit shows about 4 m of fine-grained breccia with angular fragments of sandstone, quartzite, shale and slate with quartz-porphyry and minor chert in a purple-red clayey matrix. Green and greenish grey reduction patches are common.

Beds of very coarse breccia, with clasts up to small boulder size, occur in the river cliff [SX 9214 9203] at The Quay (Plate 5a). The general dip of about 5° south-east is locally obscured by channelling of the coarse base of one bed into the underlying finer top of another. Clasts consist of Culm sandstone, shale, slate, hornfels, abundant quartz-porphyry, chert and lava. Some quartz-porphyry clasts show rhyolitic texture and flow-banding.

Shillingford Abbot

An exposure [SX 9120 8912] north of Shillingford Abbot shows 2.5 m of fine- to coarse-grained breccia with thin (1 cm) beds of fine- to medium-grained sandstone towards the top. The clasts include abundant shale, some slate, a fragment of rotted lava, and some granitic fragments up to 12 cm across with kaolinised feldspars. Another exposure [SX 9098 8890] shows 4 m of fine- and coarse-grained breccia with clasts generally up to 0.15 m across, and a 0.1 m-thick bed of fine-grained sand. A third exposure [SX 9088 8893] shows 3 m of mostly fine-grained breccia, but with one bed of clasts up to 0.3 m across. Blocks of fine-grained quartz-feldspar-biotite porphyry up to 0.3 m across are common west [SX 905 889 and 902 891] and south-west [SX 901 883] of Shillingford Abbot.

Ide

An exposure [SX 8907 9019] in the railway cutting south of West Town Farm shows reddish brown, sandy, fine-grained breccia and sand, containing clasts of soft, pale greyish purple-weathered vesicular lava and purple-weathered, non-vesicular basalt. Another section [SX 8904 9021] in the cutting shows 4 m of reddish purple-brown, weakly cemented breccia with clasts mainly less than 6 cm predominantly of Culm sandstone in gravelly, silty

sand, together with clasts up to 0.1m of reddish brown lava weathered to clay with white vesicles, and pale purple-grey, weathered basalt. The dip is very approximately 5° east.

A cutting [SX 8861 8953] behind a barn at Whiddon Farm, about 1.5 km south-west of Ide, shows up to 2.5 m of reddish brown, poorly bedded breccia with angular clasts of Culm sandstone, shale, moderately abundant, pale cream-weathering porphyry, and scattered chert (some up to 0.1 m across), in a matrix of gravelly, clayey sand. The section includes some large (0.4 m) boulders of porphyry. Bedding is picked out by the alternation of coarse and fine units, and by sporadic beds of fine (less than 1 cm) gravelly, silty clay, and some impersistent, greyish green bands.

Idestone–Dunchideock

At Marshall Farm, about 0.7 km east of Idestone, a section [SX 8857 8850] shows 4 to 5 m of reddish brown breccia, mainly unbedded to very poorly bedded, with clasts mainly less than 0.1 m across, but with some boulders (up to 0.4 m across) of quartz-feldspar porphyry. The clasts are predominantly subangular sandstone, shale, porphyry and chert, in a gravelly, clayey, silty sand matrix. Large (up to 0.2 m) clasts of lava are present about 1.5 m above the base.

At Manstree Cross [SX 8871 8800], south of Marshall Farm, exposures show up to 2 m of reddish brown clayey breccia with generally small (less than 4 cm) subangular clasts of Culm shale and sandstone, fairly common porphyry, chert and hornfels. A few large (0.2 m) boulders of weathered cream-coloured quartz-feldspar porphyry are present. Locally, the proportion of fine matrix increases to give pebbly clay.

Cuttings for a barn adjacent to Biddypark Lane, Dunchideock Barton, reveal fine-grained breccias and mudstones, at the base of the Alphington Breccia, resting on lava. Sections up to 2 m high beside the barn [from SX 8772 8747 south-eastwards] reveal interbedded reddish brown, moderately well-cemented, fine-grained breccia and mudstone. At the southernmost corner of the barn [SX 8776 8746] up to 2.5 m of reddish brown, fine- to very fine-grained, unbedded and ungraded, matrix-supported breccia, and mudstone with only sporadic large clasts of porphyry and rotted clasts of purple vesicular lava are seen. The mudstone commonly contains scattered small angular clasts of sandstone. The beds dip at 12° to 015°.

Shillingford St George–Clapham

About 2 m of reddish brown, clayey, fine-grained breccia dipping about 8° east are exposed [SX 8976 8767] on the north side of Pond Wood, Shillingford St George. Cream-weathered angular clasts of porphyry, mainly less than 4 cm, are locally concentrated in layers which pick out bedding. Other clasts present include Culm sandstone, shale, chert and hornfels.

North-east of Clapham, laneside sections [SX 8971 8705 to 9000 8719] show sporadic exposures of breccia. A typical section [SX 8980 8702] shows about 2 m of very weakly cemented, reddish brown, fine-grained breccia, with clasts mostly less than 3 cm but with a few scattered larger clasts. A few impersistent thin (about 1 cm) greyish green horizons are present. Clasts are predominantly sandstone and shale, with chert, very scarce porphyry and acid igneous rocks. The breccias dip at 13° to 168°.

Exposures [SX 8915 8745] near Yeo's Farm, about 150 m north-west of Clapham, show about 2 m of reddish brown, unbedded to very poorly bedded, clayey, silty, fine, matrix-rich breccia. The clasts (generally less than 4 cm) are of subangular sandstone, shale, and chert; some larger clasts include scattered quartz-feldspar porphyry.

Heavitree Breccia

The breccias termed Heavitree Breccia in this memoir were first noted by De la Beche (1839), who referred to the 'conglomerates of Heavitree', a name which was modified to Heavitree Conglomerate by Murchison (1867), and to Heavitree Breccia by Ussher (1902). The nomenclature of the formation has been reviewed by Bristow (1984 a).

The base of the formation is defined by the marked influx of murchisonite feldspar fragments. The breccia is locally well cemented and the basal beds form a prominent scarp above the Alphington Breccia over the whole length of the outcrop within the district (Plate 4a).

The Heavitree Breccia outcrop reaches its maximum width in the area between Kennford and Exminster. From there, the outcrop narrows northwards across the River Exe into the Countess Wear area towards the type locality at Heavitree [SX 9498 9214]. North of Heavitree, the breccia thins rapidly and passes laterally into the dominantly sandy Monkerton Formation. The northernmost outcrops of the Heavitree Breccia are present in the Hill Barton area [SX 954 930] of Exeter. The formation gives rise to a characteristic topography, of steep scarp faces and longer dip slopes, which is more rugged than that of the Alphington Breccia. The outcrop of the formation is broken by faulting, particularly south of the River Exe.

The Heavitree Breccia consists of well-cemented, granule- to pebble-sized clasts, commonly matrix-supported, set in a matrix of poorly sorted, clay-rich, fine- to coarse-grained sandstone. It weathers to gravelly clayey sand or gravelly sandy clay. The clasts include sandstone, slate, vein quartz, hornfels, porphyritic felsite, porphyritic granite, fine-grained tourmaline granite, microgranite, rhyolite, acid lava, and fragments of murchisonite. A bulk sample from Matford Park Quarry [SX 936 890], north of Exminster, had a calcium carbonate content of 5.3 per cent by weight; a sample from a nearby quarry [SX 9384 8875] had a carbonate content of only 1 per cent.

Thin interbeds of sand and sandstone occur at many localities, but only in the Exminster Hill area [around SX 945 869] are they thick enough to map. There, three interbeds of medium-grained sandstone occur within the breccia.

The thickness of the Heavitree Breccia is estimated to reach a maximum of 300 m in the south of the district. Between Countess Wear and Heavitree, the thickness is estimated to be about 135 m. From Heavitree northwards, the formation thins rapidly to disappear as a mappable unit just north of Hill Barton.

There is no direct evidence for the age of the Heavitree Breccia; it overlies the Alphington Breccia, of probable Late Permian age, and is overlain by the Dawlish Sandstone, of possible Late Permian age.

Typical sedimentary features of the Heavitree Breccia are seen in deep motorway cuttings [SX 934 882 to 931 879] west of Exminster, and are described in Jones (1992b). Thickly bedded breccia forms the bulk of the sequence and overlies thinly bedded breccia (Figure 12; Plate 4b). Individual beds in the thinly bedded facies are typically erosively based, from 0.2 to 0.8 m thick, and

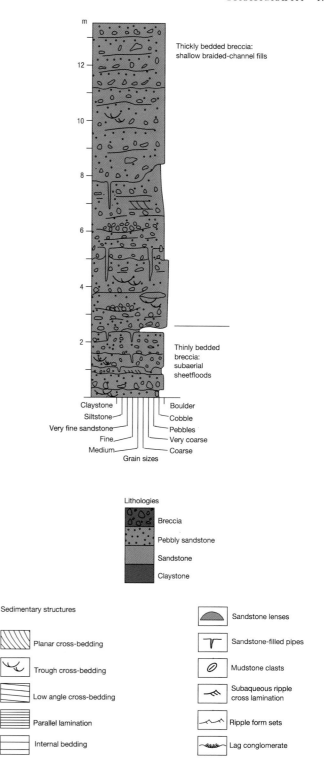

Figure 12 Representative graphical sedimentary log and interpretation of the Heavitree Breccia in the M5 Motorway cuttings [SX 931 880], near Exminster.

extend laterally for at least 100 m. Internally, the beds are either massive or crudely bedded, with thicker units showing evidence for the amalgamation of thinner beds of breccia; bedding is picked out by variations in clast size

Plate 4a View southwards from a point [SX 9125 8914] near Shillingford Abbot, towards Shillingford Plantation, showing the well-defined wooded scarp feature of the Heavitree Breccia on the Alphington Breccia (A 15328).

Plate 4b Thickly bedded Heavitree Breccia dipping 7° north-east in cuttings [SX 9306 8792] at the western end of the M5 Motorway. Individual beds are from 2.3 to 3.5 m thick and are generally laterally persistent for more than 500 m. They contain crudely defined units up to 0.15 m thick, with impersistent internal erosion surfaces; bedding is picked out by variations in clast size and concentration (A 15299).

and concentration. Trough cross-bedding occurs in isolated sets up to 0.1 m thick, directed towards the north-north-east. Rare sandstone lenses are up to 0.02 m thick and 0.2 m long. Thickly bedded breccias are distinguished from the thinly bedded facies by more distinctive channelled bases, greater bed thicknesses, and more complex internal geometry. Beds are commonly from 2.3 m to 3.5 m thick, and are generally laterally persistent for more than 500 m. Channel margins are occasionally preserved as steep-sided erosion surfaces that pass laterally into thin wings up to 1 m long. The upper bed boundaries are marked by erosional contacts with the overlying beds. Internally, the breccia beds contain crudely defined units up to 0.15 m thick, with impersistent internal erosion surfaces. Bedding is picked out by variations in clast size and concentration. Trough cross-bedding is common, forming isolated sets and rare cosets up to 0.16 m thick, directed towards the east-south-east

The thinly bedded breccias probably represent the deposits of a series of unconfined sheetfloods, and the thickly bedded breccias were deposited in fluvial channels (Jones, 1992 b).

Within both types of breccia are irregularly spaced, vertical to near-vertical, sandstone-filled cracks up to 0.12 m wide and 0.4 m long which taper downwards and are filled by a mixture of poorly sorted, fine-grained sand-

Plate 4c Detail of bedding in Heavitree Breccia, A30 Link Road cutting [SX 9292 8797]. The breccia is well cemented and matrix-supported. Clasts are mainly poorly sorted, granular to small pebble sized, with some medium to large pebble sized. The white to flesh-coloured angular clasts are murchisonite feldspar. The scale is 30 cm long (A 15296).

Plate 4d Detail of sand-filled pipes, probably desiccation cracks, in Heavitree Breccia at Heavitree Quarry [SX 9498 9214], Exeter. The hammer is 0.3 m long (A 15295).

stone and some pebbles (Plate 4d); they are probably desiccation cracks (Jones, 1992 b).

Heavitree area, Exeter

In Woodwater Lane, an exposure [SX 9504 9164] shows the following Heavitree Breccia succession:

	Thickness m
Breccia, reddish brown, coarse-grained, very sandy, with beds of poorly cross-bedded red sand up to 0.2 m thick	5.00
Sand, reddish brown, fine-grained and cross-bedded	0.40
Breccia, fine- to coarse-grained, feldspathic, with reddish brown fine-grained sand matrix	2.10
Sand, reddish brown, fine-grained, soft and friable	1.10
Breccia, seen	0.10

Heavitree Quarry [SX 9498 9214], the type locality for the formation, shows reddish brown, well-cemented, fine- to medium-grained breccia with subrounded to angular clasts seldom exceeding 4 cm, of sandstone, chert, feldspar (murchisonite), vein quartz, hornfels, quartz-porphyry, microgranite, tourmalinite and basic lava. The matrix is poorly sorted, fine-, medium- and coarse-grained, slightly clayey sand. Beds are laterally persistent and 0.2 to 1.0 m thick. Lenses and thin persistent beds of red sandstone occur sporadically. Small-scale cross-bedding is locally common. Vertical, sand-filled pipes, up to 1 m high and 10 cm across at the top, are common at several levels; they are probably desiccation cracks (Plate 4d).

Countess Wear area, Exeter

A cutting [SX 9406 9057] adjacent to the Country House Inn, Topsham Road, shows (beneath 1.1 m of drift) about 3 m of red, sandy, feldspar-rich breccia, in beds 0.4 to 0.6 m thick, dipping gently south-east. The breccias are interbedded with red silty sands, gravelly in places. A road cutting [SX 9499 9048] north of Countess Wear shows 2.4 m of well-cemented red breccia with abundant clasts of feldspar and some coarse granite fragments. An exposure [SX 9463 8940] at Lower Wear shows 5 m of fine to medium breccia in beds between 0.1 and 0.8 m thick; clasts are up to 10 cm across, and include abundant feldspar and granitic material.

Exminster–Kennford

A quarry [SX 936 890] near Matford Park Farm exposes 20 m of breccia. Beds 0.1 to 1 m thick are defined by relatively pebble-free layers and also coarser layers which contrast with the beds of poorly sorted, but dominantly fine, breccia. Most clasts are less than 1 cm across, but pebbles up to 5 cm across are common; rare clasts up to 15 cm across also occur. The clasts, which include granite, lava, vein quartz, fine-grained sandstone and murchisonite, are matrix supported in clayey sand. A sample had a calcium carbonate content of 5.3 per cent by weight. In a quarry [SX 9384 8875] opposite the former Limekiln Lane Cottages breccias dip at 10° to 110°; a sample here contained 1 per cent carbonate.

A section [SX 9305 8862] in the sunken part of Matford Lane shows up to 2 m of cross-bedded, dominantly fine-grained, breccia in beds about 0.8 m thick. Pebbles, generally less than 6 cm across, of granite, tourmalinised quartz-porphyry and lava are the commonest constituents; broken feldspar crystals are common. The dip is 5° to 145°.

South of Exminster, sands of the overlying Dawlish Sandstone interdigitate with the Heavitree Breccia; the junction of such a sand bed with overlying breccia is exposed at Crablake Farm [SX 9456 8626]. About 300 m south of the farm, a quarry [SX 9462 8595] exposes 10 m of breccia in beds 0.2 to 1.0 m thick. Common clasts of granite, tourmalinised quartz-porphyry and feldspar phenocrysts occur in a matrix of fine- to medium-grained sand. Interbeds of cross-bedded, medium-grained sand up to 0.1 m thick are also present. The dip is 15° to 035°.

Shillingford St George

A quarry [SX 9065 8746] in Shillingford Wood shows 12 m of well-cemented, mainly fine-grained breccia, but with scattered cobbles up to 0.2 m across, in beds 0.8 to 2 m thick. Feldspar fragments up to 2 cm long are common, together with angular and subangular clasts of granite, quartz-porphyry, lava, chert and slate. The dip is 14° east. Near-vertical joints trend north-east.

M5 Motorway and A38 road cuttings (Exminster to Kennford)

The M5 Motorway cutting [SX 934 882 to 931 879] about 1.5 km west-north-west of Exminster exposes about 30 m of breccias dipping 7° north-east (Figure 12; Plates 4b, c). The breccias are dominantly fine grained, but local coarse beds may include fragments up to 0.2 m across. The clasts include quartz-feldspar porphyry, rhyolite, granite, basalt, dolerite, tuff, agglomerate, lava, lamprophyre and siltstone.

Sections [SX 916 870] in the A38 road cuttings near The Wobbly Wheel, north of Kennford, expose well-bedded breccia with beds varying in thickness from 0.6 to 2 m. The breccias are dominantly fine, but with some clasts up to 15 cm across. The clasts, which include granite, quartz-feldspar porphyry, basalt, shale, sandstone, vein quartz and murchisonite, are mostly matrix supported in clayey, fine-, medium- and coarse-grained sand. Locally they are clast supported. Sand-filled pipes, up to 0.5 m high and up to 8 cm across, cut the upper part of one bed [SX 9160 8714]. They have an irregular spacing of, on average, one every 2 m. A K/Ar radiometric age determination on biotite from a granite clast from this cutting gave a date of 281 ± 7 Ma (Rundle, 1981).

MONKERTON FORMATION

Between Heavitree and Pinhoe, the Heavitree Breccia passes north-eastwards laterally and upwards into a sequence predominantly of clayey fine-grained sandstones, which Bristow and Williams (1984) called the Monkerton Member. It is here termed the Monkerton Formation, a component formation of the Exeter Group.

The formation has a small outcrop area north-east of Exeter, from Hill Barton in the west [SX 956 928], through Monkerton [SX 965 938] (the type area), to just beyond the M5 Motorway in the east [SX 974 939] (Figure 6). The few exposures indicate that it consists of a variable sequence of clayey sandstone, dominantly fine grained but locally medium or coarse grained, with sandy and silty claystone, siltstone, and thin beds of breccia. These lithologies commonly weather to sand, clay, silt and gravel.

In the Hill Barton area, the Monkerton Formation overlies the Heavitree Breccia. Hereabouts there appears to be a fairly rapid upward passage from breccia into a dominantly argillaceous sequence; the boundary is not marked by a feature. North of the Hill Barton Fault, the Heavitree Breccia passes laterally into the Monkerton Formation, which there rests on the Whipton Formation. The formation is absent east of the Pinhoe Fault. The maximum thickness of the Monkerton Formation is about 40 m.

There is no direct evidence of the age of the Monkerton Formation; however, its probable lateral equivalence to the Heavitree Breccia suggests that it too is probably of Late Permian age.

Lack of exposure precludes detailed sedimentological studies of the formation; however, it seems probable that it represents the deposits of a more distal part of the Heavitree Breccia alluvial fan system.

Monkerton

Boreholes [e.g. SX 9691 9371] for the M5 Motorway proved up to 18 m of weakly cemented, reddish brown, clayey, fine-grained sandstone. Sandy mudstone was exposed [SX 9649 9380] near Monkerton Farm. Exposures [SX 9630 9400 to 9610 9404] in the Pin Brook, near the base of the formation, are of red and greyish green, silty and fine-grained sandy clay, and thin clayey, fine-grained sandstones.

Clayey, fine-, medium- and coarse-grained sands, with minor amounts of clean sand and sandstone, were augered in the low-lying ground [SX 960 937] west of Monkerton. Probably one of these clay-free sandstones was formerly worked in the quarry [SX 9584 9363], now largely filled, at Quarry Gardens.

Whipton–Hill Barton

A temporary excavation [SX 9567 9363] next to Hart's Lane, Whipton, showed nearly 10 m of Monkerton Formation, dipping at 14° to 165°, as follows:

	Thickness m
Sand, reddish brown, silty, fine-grained, locally cemented; in basal 0.4 m, 3–4 cm beds of reddish brown, silty clay with grey-green reduction spots	7.50
Breccia, fine-grained (clasts to 1 cm)	0.15

Plate 5a Beds of coarse Alphington Breccia in river cliffs at The Quay [SX 9214 9203], Exeter. Clasts include Culm sandstone, shale, slate, hornfels, quartz-porphyry and rhyolite (A 15304).

Plate 5b Dawlish Sandstone near Lake Bridge [SX 9275 9780], Brampford Speke. The hammer (0.3 m long) rests on a bed of pebbly sandstone and breccia deposited by fluvial sheetfloods in an interdune environment. Aeolian dune-bedded sandstones overlie and underlie the interdune deposits (A 15288).

	Thickness m
Sand and weakly cemented sandstone, reddish brown, fine-grained; very silty near top	0.30
Sandstone, ferruginous and pebbly, hard, locally well-cemented	0.50
Sand, reddish brown, fine- to medium-grained, locally with small pebbles	0.35
Sand, brown, fine-grained, with subrounded pebbles to 1 cm	0.12
Sand, dark brown, clayey, silty, fine-grained	0.13
Silt, clayey, and silty clay, reddish brown, small grey-green lenses in top 0.2	0.60

In Hospital Lane, Whipton, an exposure [SX 9538 9355] shows a 1 m bed of cross-bedded, fine- and medium-grained sandstone within clayey, fine-grained sandstone. Farther south, stratigraphically higher exposures [SX 9540 9353] show red clay, clayey, fine-grained sand and clayey breccia. In the Hill Barton area [around SX 955 928] red, sandy, and locally pebbly, clays were augered.

DAWLISH SANDSTONE

The Dawlish Sandstone is the equivalent of the greater part of Ussher's (1902) 'Lower Sandstone'. The name 'Dawlish Sandstones' was introduced by Ussher (1913). In this account, it includes a sequence of sandstones with some mudstones that has been referred to wholly or in part by a variety of local names, including Dawlish Sands, Exminster Sandstone, Clyst Sands, and Exeter Formation (Table 7; Bristow et al., 1985).

The outcrop of the Dawlish Sandstone occupies an approximately north–south-trending belt about 3 km wide, extending from Broadclyst in the north to the Exe estuary in the south (Figure 6). Northwards from Broadclyst, the formation expands in thickness westwards into the Crediton Trough where two mudstone members, the Bussell's Member and the Poltimore Mudstone, have been mapped.

Plate 5c   Aeolian cross-bedding in the Dawlish Sandstone in a quarry [SX 976 910] at Clyst St Mary. The hammer is 0.3 m long (A 15289).

The greater part of the formation consists of reddish brown, weakly cemented sandstones, mainly cross-bedded, with intercalated thin lenses and beds of reddish brown claystone or clayey siltstone and fine-grained breccia. These lithologies locally weather to sand, clay, silt and gravel. Sedimentological analysis (Jones, 1992 a, b) indicates that the sequence is made up of two main facies; wind-blown sand dunes, and interdune deposits (pebbly sandstones, sandstones, and mudstones) formed in the intervening low-lying areas.

In the Crediton Trough, the Dawlish Sandstone rests on the Shute Sandstone. South of the southern boundary fault of the trough, the Dawlish Sandstone thins over what was probably, at the time of deposition of the formation, an upstanding fault-bounded block of Carboniferous rocks. In this area, between Broadclyst and Pinhoe, the Dawlish Sandstone rests on the Crackington Formation, and overlaps outcrops of lava of probable Early Permian age. In the Monkerton area, the Dawlish Sandstone overlies the Monkerton Formation, and the boundary is marked by a prominent north-facing scarp. From south of Monkerton to Exminster, the formation overlies the Heavitree Breccia with little or no interdigitation. From Exminster southwards, the Dawlish Sandstone overlies and interdigitates with the Heavitree Breccia.

The thickness of the Dawlish Sandstone in the Crediton Trough is estimated at about 350 m. From Broadclyst southwards to the southern edge of the district, the formation is probably between 100 and 120 m thick.

In the easternmost part of the Crediton Trough (Figure 6), the lowermost 100 m or so of the Dawlish Sandstone consists mainly of reddish brown sandstone (74 to 93 per cent) with sporadic beds of mudstone, and is overlain by the Bussell's Member. The sequence in the type section of the member, the Bussell's Farm Exploration Borehole [SX 9529 9873], Rewe, is shown in Figure 16. The base is taken at an upward change from coarse-grained pebbly sandstone to clayey sandstone and mudstone. The Bussell's Member forms generally low-lying ground, much of which is mantled by drift deposits. The outcrop extends between Stoke Canon to Columbjohn [SX 958 996] and eastwards to near Budlake [SX 980 997]. There is no exposure, but lithologies encountered in the type borehole can be proved in auger holes. A mappable, relatively clay-free unit of fine-grained sandstone locally gives rise [e.g. SX 963 995, near Columbjohn] to dip and scarp features within the low-relief outcrop of the mudstone. The member thickens northwards from 24.4 m in a borehole [SX 9458 9788] at Huxham, to 50 m at Bussell's Farm 1 km north-east of Huxham, and to more than 69 m in the Columbjohn Borehole [SX 9584 9964] (Figure 16), a further 1 km north-east. The maximum thickness is uncertain, but could be up to 130 m.

The Bussell's Member is succeeded by about 80 m of reddish brown sandstones, mainly fine grained but locally medium and coarse grained to pebbly. The basal part of the sequence is apparently more pebbly than the higher part, and contains thin beds of breccia. The sandstones are overlain by the Poltimore Mudstone which consists of red sandy mudstone and clayey fine-grained sandstone outcropping around Poltimore village [SX 966 970] (Figure 6). Its base is cut out by faulting at Poltimore, but farther north-east, around Mooredge Cottages [SX 983 985], the mudstones rest sharply on sandstones. The thickness of the Poltimore Mudstone is estimated to be about 15 m.

Sandstones between the top of the Poltimore Mudstone and the base of the Aylesbeare Mudstone in the Broadclyst area are estimated to be about 80 m thick. The junction of the sandstones with the Poltimore Mudstone is marked by a prominent feature.

Between Broadclyst and the southern boundary of the district, the formation consists mainly of reddish brown, cross-bedded sandstone. Around Broadclyst, and near Sowton [SX 976 925], several beds of red clay and sandy clay are thick enough to be mapped. Fine-grained breccia beds were also mapped at two places [around SX 962 927 and 972 926] west of Sowton; soil brash suggests that thin lenses of breccia occur elsewhere but are not mappable. The sandstones in the Broadclyst area are mostly fine grained and poorly sorted. Sandstones around Sowton are mostly fine grained and poorly to moderately sorted.

Around Clyst St Mary, the sandstones are reddish brown, fine to medium grained, with rare isolated coarse grains; sorting varies from poor to well sorted, and the sand grains are moderately to well rounded. Bishop's Court Quarry [SX 963 914] contains the most extensive exposures of the formation in the district, and the sedimentary features are summarised below.

In the Topsham area, the formation consists of reddish brown, fine- to medium-grained sandstones and sand, commonly cross-bedded and with red mudstone and breccia lenses. A breccia lens [SX 981 883] at the top of the formation, beneath the Exmouth Mudstone and Sandstone, may correlate with the Exe Breccia of the Newton Abbot district (Selwood et al., 1984).

Around Exminster [SX 941 881], the formation consists of yellow, fine- to medium-grained, moderately well-sorted sandstone, with a few beds of breccia near the base. South of Exminster [around SX 940 860], a much-faulted outcrop of reddish brown sandstone contains numerous interbeds of breccia, indicating an eastwards and downward passage into the Heavitree Breccia. The sandstones are commonly cross-bedded, fine grained, and moderately sorted. Persistent beds of thick breccia [e.g SX 947 859 and 950 864] may equate with the Exe Breccia of the Newton Abbot district (Selwood et al., 1984).

Trace fossils indicating the presence of a contemporary land fauna have been recorded from the Dawlish Sandstone of the Broadclyst area. Shapter (1842) recorded tracks of small crustaceans and obscure impressions of other objects near Burrow Farm, Broadclyst [?SX 989 982] (Figure 6:B), and Clayden (1908 a, b) reported footprints from Poltimore [SX 971 971] (Figure 6:P). These are referable to *Cheilichnus*, an ichnogenus indicating a later Permian age (McKeever and Haubold, in press), and include *C. bucklandi* (Plate 3i) (written communication, Dr P J McKeever, 1995).

The sedimentary features of the Dawlish Sandstone are typified by exposures at Bishop's Court Quarry [SX 963 914], described by Jones (1992 b). Up to 30 m of sandstones with rare interbedded mudstones and pebbly sandstones are present (Figure 13) and belong to two main facies associations: aeolian dune, and interdune. Their geometry is illustrated in Figure 14. The aeolian dunes have well-developed foreset laminae dipping downwind at between 20 and 32°. Lamination types identified in the aeolian facies include grainfall, grainflow, and translatent lamination (Jones, 1992 b). The lamination types, together with the geometry of the deposits, the simple hierarchy of bounding surfaces and the absence of any channelised form, are indicative of aeolian dunes. The trough-like form of the cross-bedding suggests that dunes were the transverse type (Collinson, 1986). The distribution of cross-bedding azimuths (inset on Figure 14) indicates a

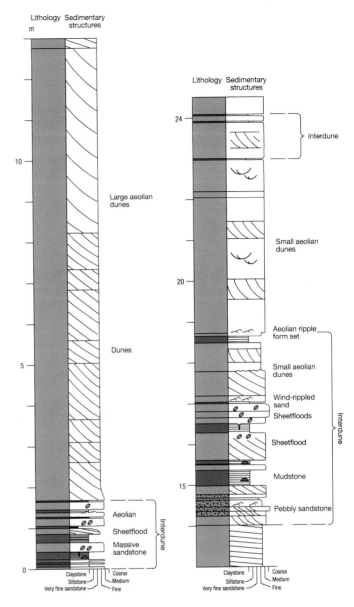

Figure 13 Representative graphical sedimentary log and interpretation of the Dawlish Sandstone at Bishop's Court Quarry [SX 963 914]. The log is a composite recorded from Faces A, E and F of Figure 14. The interdune sequence between 14 and 19 m corresponds to that seen in Face A of Figure 14.
See Figure 12 for key to graphical log symbols.

wind that blew from south-east to north-west. The close clustering of azimuths is considered to be a characteristic feature of transverse dunes (Glennie, 1970).

The interdune facies association formed in topographically low-lying areas between dunes. It is characterised by pebbly sandstone, small-scale cross-bedded sandstone, wind-rippled sand, thinly bedded mudstone, and massive sandstone (Jones, 1992 b). The interdune deposits are indicative of an alternation of wet and dry interdune periods. Wet interdune periods were marked by influxes

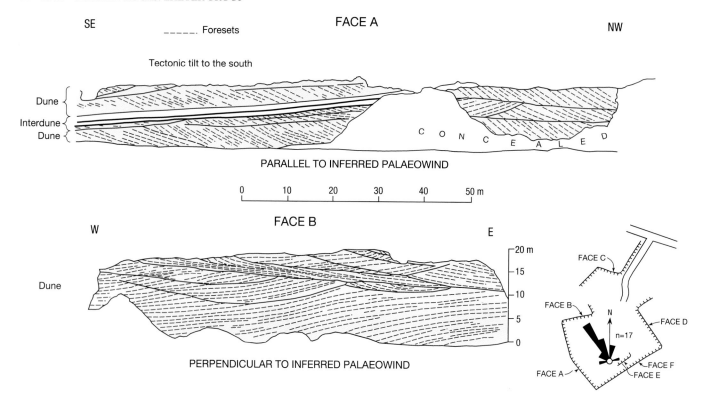

Figure 14 Field sketch of two faces within Bishop's Court Quarry [SX 963 914], to illustrate the geometry of aeolian dune and interdune facies in the Dawlish Sandstone. The inset shows a plan of the quarry with lettered faces identified, and a palaeowind rose diagram.

of fluvial sheetfloods which deposited cross-bedded, pebbly sand. Ponding of the flood waters led to the formation of small standing bodies of water in which fine-grained sediment was deposited from suspension, to form beds of mudstone. Subsequently, the interdune dried, and beds of mud were desiccated. Small aeolian dunes and wind ripples also migrated across the interdune during these periods. Subsequent sheetfloods ripped up the desiccated mud, incorporating mudstone clasts into the flow, depositing a layer of sand and infilling desiccation cracks in the underlying mudstone.

Brampford Speke–Rewe

Graphic logs of selected water boreholes in the Rewe and Brampford Speke areas are shown in Figure 15. Correlation of the boreholes is difficult owing to the existence in several cases of alternative logs, some of which differ substantially.

Starved Oak Cross No.1 Borehole [SX 9130 9883], about 1.5 km west-north-west of Brampford Speke, is recorded by Davey (1981) as having penetrated 91.4 m of clean medium- to coarse-grained sandstone (locally silty, especially near the base), with units of red and white clay (Figure 15). About 68 m of the total thickness of 91.4 m (74 per cent) was recorded as sandstone. The adjacent Starved Oak Cross No.2 Borehole [SX 9130 9879] was fully cored to a depth of 23.3 m; 3 m out of every 10 m was cored below that depth (Figure 15). The cores were described by Davey (1981), and by Scrivener (1983). The latter indicated that the cored sequence is predominantly sandstone to a depth of 52.4 m; below that to the final depth at 91 m, the

cored intervals are mainly of pebbly sandstones and breccias containing murchisonite fragments; this sequence can be classified mainly as Newton St Cyres Breccia, but might also include some Shute Sandstone.

The graphic log of the Sandy Lane Borehole [SX 9186 9824] (Figure 15) is based on the description by Davey (1981). Only scarce mudstones up to 1.6 m thick are recorded in a 91.4 m-thick sequence of which 93 per cent is sandstone.

The driller's log of the Fortescue Farm Borehole [SX 9287 9938] (Figure 15) indicates predominantly sandstone down to 44 m, underlain by alternations of sandstone with 'marl' and sandy 'marl' down to the final depth at 91.4 m.

The Burrow Farm Borehole [SX 9408 9958] (Figure 15), as interpreted from the driller's log, penetrated mainly mudstone, attributed to the Bussell's Member, to 39 m depth. Beneath, to 93 m depth, the sequence is mainly sandstones, which rest on pebbly sandstones to 114 m depth, and these on marly sandstones to the bottom of the borehole at 122.5 m.

An exposure [SS 9291 0022] in a river cliff of the Exe shows 6 m of reddish brown sandstone interbedded with breccia in cross-bedded units up to 1.0 m thick. Scattered clasts also occur sparsely distributed in a sandstone matrix; most of the clasts are Culm sandstone with vein quartz and some lava; the dip is 14 to 180°. Exposures [SS 9293 0015; 9294 008; 9293 0001] of reddish brown fine-grained sandstone dipping 15° to the south are present in the banks of the river. The river cliff [SX 9269 9910] south-south-west of Fortescue Farm exposes two beds of reddish brown sandstone separated by 0.06 m of reddish brown mudstone. The upper bed, 3.6 m thick, is moderately to poorly cemented, mostly medium to coarse grained, with some finer laminae and thin beds. Planar cross-bedding is well developed

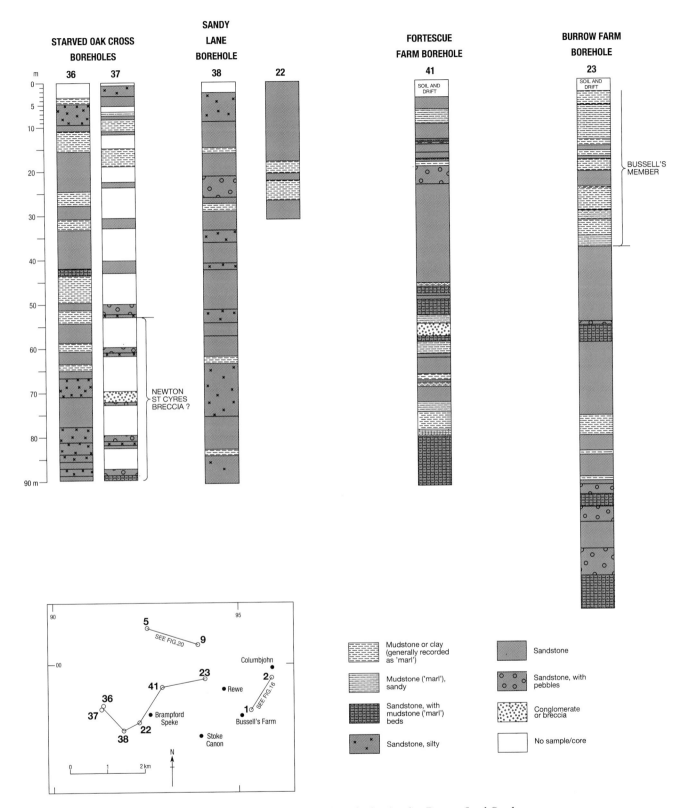

Figure 15 Graphic logs of the Dawlish Sandstone in boreholes in the Brampford Speke area. Borehole numbers are those of BGS records for sheet SX 99 NW. Boreholes 25, 22, 41 and 23 are based on the driller's record. Boreholes 36 and 38 are based on the lithological log of Davey (1981 a); borehole 37 is based on the lithological log of Scrivener (1983). See text for discussion of lithology and classification.

in places, in units up to 0.30 m thick. Foreset dips show current flow from the south-south-east. The lower, medium-grained sandstone, 2.2 m thick, contains laminae of coarser 'millet seed' sand up to 10 mm thick. Planar cross-bedding is parallel and dips around 14° to 005°.

Sections in a cutting [SX 9268 9850] in a river cliff at Brampford Speke expose 5.85 m of reddish brown, medium-grained, parallel- or massive-bedded sandstone dipping at 16° to 340°, and overlain by 0.15 m of weakly cemented breccia.

A river-cliff section [SX 9275 9780] at Lake Bridge, Brampford Speke shows 6 m of sandstone dipping 12 to 14° south (Figure 17; Plate 5b). Most of the aeolian dune sequence is reddish brown, weakly cemented, medium-grained sandstone, with sets of cross-bedding typically 0.3 m high up to a maximum of 1 m, with foresets dipping at up to 18°. The interdune deposits are well-bedded combinations of sandstone, pebbly sandstone, and breccia, containing clasts of sandstone, hornfels, rhyolite, and abundant murchisonite. A thin (0.03 m) bed of reddish brown mudstone overlies the interdune sequence in the middle of the section.

Nether Exe–Bussell's Farm (Rewe)–Columbjohn

The Bussell's Member, recorded as plastic red marl, was present beneath 3.05 m of drift, to 28.96 m depth in a borehole [about SS 941 001] near Barnfield Cottages, Nether Exe. Another borehole [about SS 947 001] to 46.33 m depth was recorded as being in plastic marl with occasional sandstone horizons.

Two boreholes were drilled at the same site [SX 9529 9873] near Bussell's Farm: the Bussell's Farm Exploration Borehole (1970), and the Bussell's Farm Production Borehole (1973) (Figure 16). The log of the Exploration Borehole by Professor A Stuart was based on descriptions of partial cores and bulk samples. However, the log differs in several respects from the driller's log; for example, the driller recorded 'red sandstone with pebbles and marl bands' between 51.36 and 61.72 m, not noted by Stuart. The base of the Bussell's Member is taken at an upward change from coarse-grained pebbly sandstone to clayey sandstone and mudstone at 48.46 m depth (Bristow, 1983).

The logs of the Production Borehole differ substantially from those of the adjacent Exploration Borehole. The uppermost 33 m of the Production Borehole is recorded as marl with interbedded sandstones; the lower 60 m is recorded as mainly sandstone and 'marly' sandstone with a few 'marl' beds. The base of the Bussell's Member cannot be identified with certainty, but may lie considerably higher than the base (48.46 m) in the adjacent Exploration Borehole.

Huxham

The Huxham Borehole [SX 9458 9788] showed 56.1 m of Dawlish Sandstone beneath 4.9 m of river terrace gravel. The sequence consists of 18.0 m of red sandstone, overlain by 24.4 m of brown mudstone (Bussell's Member), overlain by 13.7 m of red sandstone with a few thin breccia beds.

Belfield House–Danes Wood–Newhall Farm

Sandstones overlying the Bussell's Member form a feature from east of Bussell's Farm [SX 951 986], north-eastwards through Danes Wood [SX 964 990] (near Belfield House), to near Newhall Farm [SX 981 993]. Thin (up to 0.1 m) red mudstones occur locally, and a 3.8 m mudstone was proved beneath drift in a borehole [SX 9730 9814] 1 km east-south-east of Belfield House.

A section [SX 9692 9911] in Danes Hill Lane, east of Danes Wood, shows 4 m of weakly cemented, reddish brown, fine- to medium-grained sandstone, overlain by 4 m of reddish brown, poorly sorted, fine-, medium- and coarse-grained, locally pebbly sandstone, in weakly defined thin beds, and with some cross-bedding. The basal bed is 0.1 m of sandy breccia. Clay bands up to 10 cm thick occur in the stratigraphically higher parts of the cutting. Some of the sandstones are ferruginously cemented to form hard bands 1 cm thick, and similar material is common as brash at several localities [e.g. SX 9570 9865 and 9735 9910].

A roadside exposure [SX 9820 9935] north-east of Newhall Farm shows about 3 m of poorly sorted, thinly bedded, fine-, medium- and coarse-grained sandstone, pebbly in part and with thin interbeds of breccia. Most of the remaining exposures show less than 1 m of cross-bedded, fine-, medium- and coarse-grained friable sandstone [e.g. SX 963 981 and 9622 9803].

Broadclyst

A quarry at Broadclyst [SX 983 974] exposes 8 m of friable, poorly sorted, cross-bedded, fine-, medium- and coarse-grained sandstone dipping at 5° south-east, with cross-bedding foresets dipping at 24° to 030°. A quarry [SX 971 971] near Poltimore is reputed to have been worked in sandstone to a depth of 30 m. Ussher (1902) noted that the sandstone at this latter locality was evenly laminated in layers averaging 5 cm in thickness in the upper 1.2 m, 7 to 17 cm thick in the next 1.5 m, and in beds 30 cm thick below this level. Clayden (1908 a,b) recorded reptilian footprints in this pit (Figure 6:P). Ussher (1902, p.28) noted a bed of dark red-grey mottled marly clay in red laminated sands in a pit [SX 989 982] (now infilled) near Burrow Farm. This may be the pit in which Shapter (1842) noted ripple marks and annelid and crustacean tracks (Figure 6, locality B).

Mosshayne

In the Mosshayne area [SX 980 947], poorly sorted sandstone was formerly exposed in the pit [SX 979 945] north of the railway line. Red clay at a higher stratigraphical level is exposed in a pit [SX 9793 9540] and in the banks of a pond [SX 9798 9455]. The succeeding sandstone is exposed in a lane [SX 9811 9455]. There, 2 m of horizontally bedded, fine-, medium-, and coarse-grained sandstone are overlain by red sandy clay 2 to 5 m thick.

Two boreholes [SX 9858 9425 and 9884 9423] proved clay units beneath the alluvium of the River Clyst. The first proved 8 m of red, friable clay above weakly cemented sandstone; the second penetrated 2.3 m of reddish brown, silty clay.

Clyst Honiton

Reddish brown, fine- to medium-grained sands near the top of the formation are poorly exposed in road cuttings [SY 0000 9469 to 0008 9473] near South Whimple Farm. A nearby borehole [SY 0005 9451] penetrated the top 6.85 m of the formation which was very dense reddish brown fine- to medium-grained sand. A cutting [SX 9995 9465] along the Honiton Road revealed about 3 m of thinly cross-bedded, poorly to moderately sorted, fine- to medium-grained sand in units between 1 and 2 m thick. Cuttings [SX 9955 9325] in the road to Exeter Airport show up to 2.5 m of horizontal, thinly bedded, poorly to moderately sorted, fine-, medium- and coarse-grained sandstone.

A borehole [SY 0020 9305] 100 m south-south-west of Fair Oak Farm penetrated 88.69 m of Dawlish Sandstone beneath 47.25 m of Aylesbeare Mudstone.

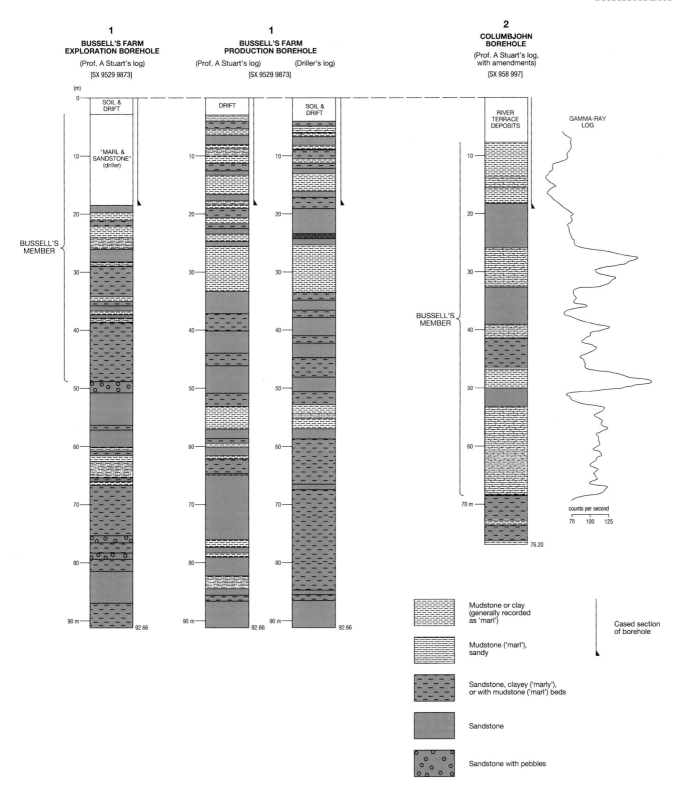

Figure 16 Graphic logs of water boreholes in the Dawlish Sandstone of the Bussell's Farm (Rewe) and Columbjohn areas. Borehole numbers are those of BGS records for sheet SX 99 NE. See Figure 15 for location map, and text for discussion of lithology and classification.

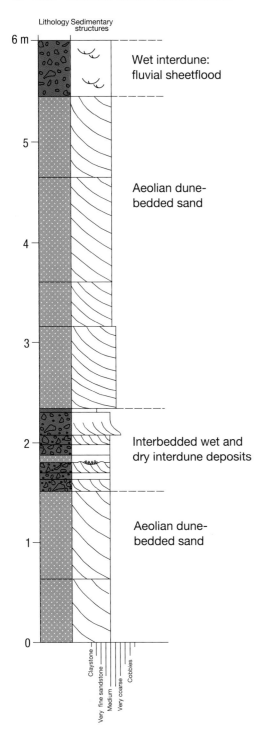

Lithology Sedimentary
structures

Wet interdune:
fluvial sheetflood

Aeolian dune-
bedded sand

Interbedded wet and
dry interdune deposits

Aeolian dune-
bedded sand

Figure 17   Representative graphical sedimentary log and interpretation of the Dawlish Sandstone at Lake Bridge, Brampford Speke [SX 9275 9780]. See Figure 12 for key to graphical log symbols.

Blackhorse

Exposures [SX 9810 9355; 9815 9345; 9825 9343] in the Blackhorse area show 2 to 3 m of thinly bedded, cross-bedded, fine-, medium- and coarse-grained sandstone. An exposure [SX 9822 9359] shows fine-, medium- and coarse-grained sandstone with thin beds of red clay and breccia; an 0.8-m bed at the top contained beds of breccia and clay up to 0.1 m thick. North of Blackhorse, 2.3 m of red, pebbly clay were proved in a borehole [SX 9797 9384].

Monkerton

In Pinn Lane [SX 9651 9357], 2 to 3 m of poorly sorted, fine-, medium- and coarse-grained, but dominantly fine-grained, cross-bedded sandstone are exposed. A few metres south, thinly bedded, fine-, medium-, and coarse-grained sandstone with a few scattered pebbles up to 1 cm across occurs; a red clay, 2 cm thick, is also present. A quarry [SX 9653 9327] at Onaway shows 8 m of yellow, poorly to moderately sorted, fine-, medium- and coarse-grained, cross-bedded friable sandstone in units 3 m thick. Foresets dip 6° north-north-west. Anastomosing, hard, ferruginous bands are present throughout much of the face. Boreholes for the M5 Motorway [e.g. SX 9688 9350] proved up to 23 m of fine- and medium-grained sandstone.

Sowton

The cutting in Sowton Lane [SX 9735 9292] shows about 3 m of thinly bedded, cross-bedded, fine-, medium- and coarse-grained sandstone; the foresets dip 10° north. Farther south-east, a small section [SX 9763 9267] shows fine-, medium- and coarse-grained sand with a 10-cm breccia bed.

Clyst St Mary

About 30 m of Dawlish Sandstone is exposed in Bishop's Court Quarry [SX 963 914], and the sequence is shown in Figures 13 and 14. The sedimentary features are summarised above (p.53), and described in detail in Jones (1992 b). Irregular, hard, iron-pan layers are scattered throughout the sequence and are particularly abundant in the basal 3 m. The beds in the working part of the quarry dip 2° to 300°.

A borehole [SX 9658 9116] for the M5 Motorway proved Dawlish Sandstone to 23.04 m depth, beneath 0.46 m of soil. Two boreholes [SX 9607 9104; 9607 9108] reached the Heavitree Breccia at 11.70 m and 11.00 m.

A sand pit [SX 976 910] in Clyst St Mary shows 3.35 m of pale reddish brown, cross-bedded, fine- to medium-grained sand and sandstone, with foresets dipping at up to 15° (Plate 5c).

Topsham

A quarry [SX 9627 8834] in Ferry Road, Topsham exposes 2 m of flaggy, laminated, pale reddish brown sandstone resting on 1.5 m of cross-bedded, fine-grained, reddish brown sandstone containing a wedge-shaped bed of breccia which is up to 1 m thick. Near Exton, railway cuttings contain 0.6 m of reddish brown, coarse- to medium-grained sandstone [SX 9771 8704] and 0.6 m of reddish brown, cross-bedded sand [SX 9776 8684].

Exminster

The rapid gradational passage from the Heavitree Breccia into the Dawlish Sandstone, dipping at 9° to 110°, was formerly exposed in the M5 Motorway cutting [SX 9375 8840 to 9410 8843], but the sections are now grassed over.

A cutting [SX 9422 8810] opposite Exe Vale Hospital shows 2.5 m of yellow, cross-bedded, fine-, medium-, and coarse-grained, weakly cemented (locally iron-cemented) sandstone. Cross-bedding foresets dip at 30° to 325°.

At Crablake Farm [SX 9456 8627], and in the nearby road cutting [SX 9444 8635], thinly bedded, moderately sorted, cross-bedded, fine-, medium- and coarse-grained sandstone is overlain by an impersistent bed of breccia, succeeded by red sandy clay. The dip in the road cutting is 15° north-north-east.

A thick bed of breccia overlies the Dawlish Sandstone south of Crablake Farm; a quarry [SX 9462 8595] exposes 10 m of breccia in poorly defined beds 0.2 to 1 m thick, with interbeds of cross-bedded medium-grained sand up to 0.1 m thick. Clasts of granite, tourmalinised quartz-porphyry, and feldspar phenocrysts occur in a matrix of fine- to medium-grained sand. Near Luccombs Farm [SX 9322 8641], 3 to 4 m of breccia overlie reddish brown, fine- and medium-grained sandstone dipping 10° north-east.

A quarry [SX 9285 8595] on the south side of the Old Dawlish Road south-west of Luccombs Farm exposes 3 m of weakly cemented cross-bedded, fine- to medium-grained sandstone. The foresets are inclined 20° west; bedding dips 15° north-east.

Crediton Trough

The inferred stratigraphical relationships of the Exeter Group in the Crediton Trough are shown in Figure 8. In the central and northern Crediton Trough, the Upton Formation, Cadbury Breccia and Bow Breccia are older than any New Red Sandstone elsewhere in the district. These formations are not found in the New Red Sandstone outcrops around Posbury on the southern edge of the Crediton Trough, nor south-west of Exeter, where the oldest New Red Sandstone formation is the Knowle Sandstone, which rests directly on the Crackington Formation.

UPTON FORMATION

The Upton Formation, named by Edwards (1988) after Upton Farm [ST 021 006], Clyst St Lawrence, occupies small, almost wholly fault-bounded outcrops between Paradise Copse [ST 010 010] and Clyst Hydon (Figure 6). A small outcrop [SS 971 051] is also present at the northern boundary of the district, about 0.5 km west of Tedbridge. The stratigraphical relationships of the formation are uncertain; around Upton Farm, it appears to rest unconformably on Crackington Formation. Exposures are few and small, and mainly confined to stream sections and sunken tracks.

The formation consists of orange-brown to purplish brown gravelly mudstone with scattered angular to subrounded Culm sandstone fragments, which at some localities are sufficiently abundant for the deposit to weather to a clayey gravel. The matrix consists of millimetre-sized angular fragments of shale which are highly weathered to a clay or 'shale-paste', and retain only a remnant platy shale-like appearance. In auger samples the deposit is generally a gravelly clay.

The thickness is uncertain, but is probably at least 10 m. The deposit possibly partly represents a redistribution of weathered regolith material developed on the Culm surface. Its age and relationship to the Cadbury Breccia are uncertain. The two formations may be contemporaneous, and the differences in their lithologies may reflect differences in provenance. The Cadbury Breccia was derived mainly from the north, and is rich in Bude Formation sandstone clasts; the Upton Formation contains abundant shale clasts, indicating possible derivation from the Crackington Formation.

There is no exposure in the fault-bounded outcrop between Paradise Copse and Higher Comberoy, where augering proved orange-brown to purplish brown, sandy, gravelly clay. A stream section [ST 0239 0091] east of Legars Upton, north of Upton Farm, shows orange-brown to purplish brown, gravelly clay, consisting of small scattered angular fragments of Culm sandstone in a shale-paste matrix; farther east, [ST 0271 0091], beneath 1.5 m of head, the Upton Formation consists of purplish brown, clayey gravel with angular to subround Culm sandstone, shale, and quartz, locally passing into sandy gravelly clay. Small exposures [e.g. ST 0244 0074] in a sunken track east-south-east of Legars Upton show about 1 m of orange-brown to purplish brown gravelly clay with scattered angular Culm sandstone, locally clayey gravel. A stream section [ST 0187 0068] shows up to 1 m of orange-brown gravelly clay, consisting of highly weathered shale fragments, with scattered angular Culm sandstone and siltstone clasts. Near Upton Farm, small exposures [ST 0207 0058] are of gravel consisting of angular to subround Culm sandstone clasts (less than 3 cm) in a matrix of granular silty clay.

CADBURY BRECCIA

The Cadbury Breccia (Edmonds et al., 1968) is named from the village of Cadbury [SS 910 050], about 3 km north-north-west of Thorverton. The formation is a brown to reddish brown, unbedded to very roughly bedded deposit consisting of red-stained subangular to subrounded pebbles and cobbles of sandstone in a matrix of poorly sorted silt and fine- to medium-grained sand (Plate 6a). The sandstone clasts are mostly derived from the Bude Formation, and are up to 0.3 m in diameter. The sandstone clasts may be quartz-veined, and clasts of vein-quartz occur sporadically in the breccia. Rare clasts of sandstone with Devonian fossils have been recorded (Downes, 1881; Martin, 1908). Also scarce are small (less than 30 mm) fragments of very angular cherty shale and chert, ranging in colour from white, through pale grey, to greyish pink and pink. They have not yet been matched definitely with any formation. Surface debris on the Cadbury Breccia outcrop north-east of Bradninch [e.g. ST 007 047] includes subrounded fragments of grey fossiliferous limestone, but no rock of this type has been seen in situ. The limestones are comparable to the Lower Carboniferous Westleigh Limestone (Thomas, 1963). In the Thorverton area, near Silverton and near Beare, a few localities near the top of the formation show small fragments, generally less than 40 mm, of quartz-porphyry.

Along the northern margin of the Crediton Trough, from the western boundary of the district to the north of Sandford [SS 8232 0314], Cadbury Breccia may be present beneath the Bow Breccia, but is too thin to be represented on the map. From a place [SS 845 030] near Lower Creedy, the formation extends along the northern side of the Crediton Trough, giving rise to the Raddon Hills and hill cappings (attaining about 250 m above OD) around Cadbury and Fursdon (Figure 6). The Raddon Hills Fault (p.128) extends along the southern flank of the Raddon Hills and thence eastwards to Bradninch.

Plate 6a   Cadbury Breccia at Raddon Hill Farm [SS 9041 0309]. The breccia is brown to reddish brown, unbedded to very crudely bedded, very poorly sorted, comprising angular to subrounded clasts almost exclusively of Culm sandstone from granule size to c. 0.15 m, in a matrix of very clayey sandy silt. Also present are rare clasts of very pale grey, angular cherty shale; one 10 cm clast of fossiliferous sandstone was recorded. The hammer is 0.3 m long (A 15309).

The breccia forms high ground from north of Silverton eastwards to Bradninch. Lower ground between Greenslinch [SS 969 035] and near Hele [SS 999 028] is underlain by silty and clayey fine-grained sand and sandy silty clay, apparently intercalated within the Cadbury Breccia, being overlain by breccias of Cadbury type between Stockwell [SS 976 029] and Hele. The outcrop is terminated in the east and west by faults.

The outcrop of the Cadbury Breccia is repeated by faulting in an area north and west of Killerton Park, forming the hill [SS 981 014] near Penstone, and the hill [SS 965 016] near Hayne House. Outcrops of Cadbury Breccia between Clyst Hydon and Beare [SS 990 009] are more disconnected and commonly fault-bounded.

Along the northern margin of the Crediton Trough, the thickness of the Cadbury Breccia is difficult to estimate owing to uncertainty about the dip and the amount of throw on the Raddon Hills Fault. South of the fault, near the western end of the outcrop, the Cadbury Breccia thins at the expense of the overlying Bow Breccia; the outcrop pattern suggests unconformity between the two formations. At least 70 m of Cadbury Breccia are present in the Raddon Hills area, and the thickness could be up

Plate 6b   Poorly bedded, reddish brown Crediton Breccia, Forches Cross [SS 8318 0088], Crediton. The breccia consists of small pebbles in a silty mudstone matrix; lenses of coarser-grained, clast-supported breccia are locally present. The clasts include Culm sandstone, shale, slate, quartz-porphyry, rhyolite, tourmalinite, and tourmalinised (hydrothermal) breccia. The hammer is 0.3 m long (A 15302).

to 150 m. Farther east, a thickness of up to 270 m is estimated in the Bradninch area.

The formation may be latest Carboniferous (Stephanian) or earliest Permian in age. Sandstone clasts with Devonian fossils were described from probable correlatives of the Cadbury Breccia in the Tiverton area by Downes (1881) and from a wider area, including that north of Thorverton and Silverton, and around Bradninch in the present district, by Martin (1908). All the fossils identified were stated by Martin to be known from the Pilton and Marwood Beds of north Devon. *Phacops latifrons*, recorded from the Tiverton district, is practically the only trilobite occurring in the Pilton Beds, where it is very common. A sandstone clast at a locality [SS 9239 0500] north of Fursdon yielded *Cyrtospirifer* sp., productoid? debris, a poorly preserved phacopid trilobite pygidium, and crinoid debris, indicating a Devonian age; the nearest sources for the lithology and fauna are likely to be the Pilton Formation in north Devon, or marine Devonian strata in the Quantock Hills, Somerset (Riley, 1992 b).

The Cadbury Breccia is broadly similar throughout its outcrop in the Exeter district. However, the maximum size of clasts in the formation decreases upwards from about 0.35 m at the base to about 0.07 m near the top, and the clast assemblage becomes somewhat more varied towards the top of the formation, notably in the presence at a few localities of clasts of porphyry.

Between Greenslinch [SS 969 035] and a place [SS 999 028] near Hele, silty and clayey, fine-grained sand and sandy silty clay, about 90 m thick, are apparently intercalated within the Cadbury Breccia, and overlain by breccias of Cadbury type between Stockwell [SS 976 029] and Hele. The outcrop is probably terminated in the west and east by faults. Exposures are few and the outcrop has been determined mainly by augering.

Igneous rocks occur within the Cadbury Breccia at a small number of localities north of Thorverton, near Silverton, and near Beare (see Chapter 7).

The matrix of the Cadbury Breccia is mostly a poorly sorted mixture of silt, fine- and medium-grained sand, with grains heavily coated with clay and iron oxide. Hutchins (1963) noted that the sand grade fraction also contains a high proportion of shale fragments. Hutchins (1963, p.113) also examined the heavy-mineral suite, which was poor in variety and derived mainly from pre-existing sediments. Rounded zircon was predominant, with rounded brown and green tourmaline and rounded rutile subordinate. Very rare staurolite was present. Opaque ore minerals were not normally abundant; detrital sphalerite appeared in most samples. Angular quartz and quartzite fragments were the most prominent transparent light detrital grains, feldspar being rare or absent.

Grain-size analyses of sands from within the Cadbury Breccia are from thin interbeds within the breccias, and from the thicker unit outcropping between Greenslinch and Hele (see above). Four samples from the first group are fine grained, moderately well sorted to poorly sorted, and positively to strongly positively skewed. Three samples representing the second group are finer grained than the first group (fine- to very fine-grained sand), moderately well sorted to poorly sorted, and strongly positively skewed.

Few sedimentary structures are apparent. The deposit is mostly unbedded to very crudely bedded, especially in the lower part of the sequence. The matrix-supported character, poor sorting and lack of grading within the breccias suggests that they were deposited by debris flows (Jones, 1992 a).

Coombe Barton

North of Coombe Barton, a section [SS 8746 0273] shows up to 0.9 m of blanket head and regolith on 3 m of reddish brown, unbedded, ungraded, matrix-supported breccia. The matrix consists principally of a poorly sorted mixture of silt and fine- and medium-grained sand with grains heavily coated with clay and iron oxide. The clasts are variably sized, subrounded to subangular, medium to large pebbles, mainly of Culm sandstone and sparse vein quartz, with maximum size of 0.16 m. The larger cobbles are relatively rare, and the fine gravel includes fragments of shale, angular cherty shale and siltstone.

Raddon Hills

Near Raddon Hill Farm, 3 to 4 m of Cadbury Breccia are exposed in a cutting [SS 9041 0309] (Plate 6a). The section consists of brown to reddish brown, unbedded to very roughly bedded, very poorly sorted breccia, comprising angular to subrounded clasts, almost exclusively of purple-stained Culm sandstone from granule size to about 0.15 m (the larger cobbles are somewhat subrounded), in a matrix of very clayey sandy silt. Also present are rare clasts of very pale grey angular cherty shale, and one 10 cm clast of reddish brown weakly cemented micaceous fine-grained sandstone with moulds of shelly fossils, mostly crinoid and chonetoid debris. In places, bedding is shown very indistinctly by alternations of slightly less pebbly and cobbly gravel.

Thorverton

Small exposures [SS 9281 0305] at Bidwell Cross, about 1 km north-north-east of Thorverton, show up to 0.8 m of unbedded, reddish brown breccia with angular to slightly rounded Culm sandstone to 0.1 m, some vein quartz, and rare angular fragments of white, very pale grey and pale red very siliceous shale.

Exposures at Bidwell [SS 9335 0321] showed about 2.5 m of reddish brown unbedded breccia, composed mainly of angular to subround Culm sandstone to 0.2 m (mainly less than 5 cm), with scarce vein quartz and chert, in a matrix of gravelly, very clayey silt, and with a few impersistent lenses of very clayey, silty, fine-grained sand. About 40 m farther north, exposures [SS 9332 0325] show 0.7 m of sand and clay, apparently underlying the breccias.

Cadbury

A cutting [SS 9113 0488] for a barn just east of Church Farm, Cadbury, showed 2.4 m of pale reddish brown to brown unbedded breccia, predominantly of angular to subrounded Culm sandstone pebbles and cobbles up to 0.2 m (most less than 0.1 m) in a matrix of ill-sorted, granular, clayey sandy silt, and with rare vein quartz pebbles and angular fragments of cream-coloured, cherty shale.

Fursdon

Between Fursdon Lodge and Fursdon House, the base of the formation is defined by a feature and by the emergence of springs. Cuttings [SS 9218 0465; 9254 0471] near Fursdon House and Fursdon Barton reveal up to 2.0 m of unbedded reddish brown, earthy gravel with angular to subrounded Culm sandstone up to 0.2 m across, some vein quartz and rare cream-coloured, very siliceous shale, in a clayey silt matrix.

Silverton

Laneside exposures [SS 9543 0235] south-west of Silverton, and cuttings [SS 9535 0222 to 9534 0214] in the lane at Kenson Hill are typical of the finer-grained, higher part of the Cadbury Breccia. They reveal 3 to 4 m of indistinctly bedded, reddish brown breccia, with clasts up to 0.1 m, predominantly of angular to subround Culm sandstone, vein quartz and scarce chert, in a matrix of gravelly clayey sandy silt. Cementation is variable; some units are well cemented. Dips are generally 12–14° towards the south.

Exposures [SS 9630 0363] on the north side of Livingshayes show about 3 m of reddish brown unbedded breccia with lenses of reddish brown, moderately well-sorted, fine-grained sand.

Stockwell–Hele

Exposures [SS 9700 0267] north of Park Farm, Stockwell, show about 1.5 m of reddish brown, roughly bedded breccia, composed mainly of angular to subrounded Culm sandstone clasts (generally less than 7 cm, commonly less than 4 cm, maximum size seen 10 cm) with scarce chert or cherty shale, and vein quartz, in a gravelly clayey silty sand matrix. Similar breccia is visible in cuttings up to 4 m deep [SS 9716 0273 to 9735 0266] north of Combesatchfield.

At Stockwell, roadside exposures [SS 9761 0291] show 1.1 m of interbedded reddish brown, fine-grained breccia and sand in beds 0.1 to 0.2 m thick.

A 0.4 m exposure [SS 9827 0301] at Moorland shows reddish brown, moderately sorted, very fine-grained, silty sand. Farther east in the same stratigraphical unit, an exposure [SS 9912 0288] in a silage-pit at Hele Payne shows, beneath 1.5 to 2.4 m of head, 0.6 m of reddish brown, finely micaceous, moderately well-sorted, fine-grained sand, on 0.3 m of reddish brown, silty to very silty clay (fine-grained breccia at the base), on 0.1 m of reddish brown, clayey, sandy silt.

A gamma-ray log (Davey, 1981 a) of a 48.77 m-deep borehole [about SS 992 026] at Hele Payne indicates a clay-rich sand sequence with mudstone beds up to 1 m thick at 17.5, 22.5 and 27 m depth.

Bradninch

A section [SS 9853 0440] close to the base of the formation at Rhode Farm shows up to 3 m of reddish brown, unbedded to very indistinctly bedded, very poorly sorted gravel, consisting mainly of angular to subround Culm sandstone clasts in a gravelly clayey sandy silt matrix. The maximum size of cobbles seen is 0.35 m; most cobbles are less than 0.1 m, but many are between 0.1 and 0.2 m. Other clasts include scarce angular chert, and vein quartz.

At Paceycombe Farm, north of Bradninch, exposures [SS 9990 0454] show up to 2.5 m of unbedded to roughly bedded, reddish brown gravel with angular to subround Culm sandstone and quartz and one clast of fossiliferous sandstone in a gravelly, clayey, sandy silt matrix.

A cutting [ST 0033 0339] near Kensham House shows 1.5 m of reddish brown, fine-grained breccia, the majority of clasts being less than 5 cm, of angular and subround Culm sandstone, in a gravelly silty sand matrix; vein quartz and chert are also present.

Roadside exposures [ST 0023 0433] at Cullompton Hill, Bradninch, show about 3 m of poorly-bedded breccia with interbeds up to 0.3 m thick of reddish brown, fine- to medium-grained sand.

Pottshayes–Washbeerhayes

A cutting [ST 0051 0203] at Pottshayes Farm shows 3 to 4 m of reddish brown, very poorly sorted, unbedded to very roughly bedded gravel, with angular to subround Culm sandstone pebbles and cobbles, mainly less than 0.1 m but some up to 0.3 m, especially in a bed about 0.8 m above the base of the section. The larger clasts are in a matrix of gravelly clayey sandy silt. Sandstone clasts form over 95 per cent of the total; other types include vein quartz and scarce chert.

HIGHER COMBEROY FORMATION

The Higher Comberoy Formation, named by Edwards (1988), consists of siltstone, sandstone and breccia, that weather to silt, sand and gravel. The formation is present only in the area about 2 km west-north-west of Clyst St Lawrence, where it has a small (about 0.6 km²) fault-bounded outcrop between Brook Hill [ST 002 006] and Higher Comberoy Farm [ST 016 005]. Owing to its fault-bounded nature, its stratigraphical position is uncertain. The sequence dips southwards (to 165°) at 5°. The exposed sequence is at least 55 m thick and consists of at least 30 m of siltstone and clayey very fine-grained sandstone, overlain by breccia 7–16 m thick, which is in turn overlain by at least 9 m of clayey very fine-grained sandstone. The central breccia unit forms a moderately well-defined scarp feature and has been mapped separately.

Very small exposures [ST 0024 0059] at Brook Hill, in the lower siltstone and sandstone unit, show reddish brown, micaceous, clayey, very fine-grained sand. Small exposures [ST 0143 0078] in the track north of Higher Comberoy Farm show reddish brown, poorly sorted silt; at one place [ST 0135 0083], 1.0 m of reddish brown, poorly sorted silt contains a 0 to 0.2 m-thick gravel bed. Small exposures of reddish brown, very fine-grained breccia are present north of Higher Comberoy Farm [e.g. ST 0155 0052; 0153 0059]. The upper sandstone unit is not exposed; around and south-west of Higher Comberoy Farm, auger holes penetrate clayey, very fine-grained sand.

BOW BRECCIA

The Bow Breccia (Bow Conglomerate of Edmonds et al., 1968) typically comprises clasts of subangular to well-rounded Culm sandstone, shale, slate, vein quartz and hornfels, and a minor fraction of subrounded pebbles of argillised and iron-stained, acid igneous rocks, and bleached and rotted lamprophyre, in a matrix of reddish brown silty sandstone. Beds and lenses of sandstone are common, and increase in number towards the top of the formation. The igneous pebbles are mostly of quartz-porphyry, acid tuff and rhyolite, the last of ignimbritic aspect. Pebbles of Middle Devonian (Givetian) marine limestone occur in the Okehampton district (Hutchins, 1958), but have not been observed in the Exeter district.

Lamprophyric lavas are interbedded with the formation near Woolsgrove [SS 792 028], and basalts occur at the junction with the overlying Knowle Sandstone. Near Neopardy, at the southern edge of the Crediton Trough, small exposures [SX 7938 9888] of rhyolite may rest on

Bow Breccia. These volcanic rocks are described in Chapter 7.

The formation crops out along the northern margin of the Crediton Trough (Figure 6). Between Elston [SS 7827 0294] and the valley of the River Creedy, it rests unconformably on the Bude Formation, although Cadbury Breccia, too thin to be represented on the map, may locally intervene, [e.g. SS 783 029 to 790 029]. From Lower Creedy [SS 8425 0272] to its most easterly outcrop near Great Gutton [SS 8612 0249], the Bow Breccia overlies the Cadbury Breccia, possibly disconformably. Where it occupies the whole of the Crediton Trough, west of Bow in the adjacent Okehampton district, the Bow Breccia is up to about 400 m thick; at the eastern edge of the Okehampton district, where the outcrop is restricted to the northern margin of the trough, a thickness of 76 m was estimated (Edmonds et al., 1968). In the present district, the formation is up to 140 m thick, decreasing to around 40 m at the eastern end of the outcrop.

Samples from the basal beds of the Bow Breccia at Lower Creedy House [SS 8424 0282] (Figure 6:5) yielded the spores ?*Convolutispora* sp., *Crassispora kosankei*, *Lycospora* sp. and *L. pellucida*. The Bow Breccia is considered to be Early Permian in age and the spores are regarded as reworked from deposits of Carboniferous age (p.31).

The sedimentary features of the Bow Breccia at some localities [e.g. SS 8160 0255, west of Sandford] — including matrix-support, imbrication of clast long axes parallel to flow, and inverse to normal grading — indicate deposition by debris flows and density-modified grain flows (Jones, 1992 a) (Figure 18). At other localities [e.g. SS 8107 0290, in West Sandford] cross-bedding within the breccia suggests that the debris flows were partly reworked by sheet-floods.

West Sandford–Sandford

A cutting [SS 8160 0255] between Sandford and West Sandford exposes 1.7 m of reddish brown breccia (Figure 18). A bed of breccia up to 0.82 m thick is overlain by thinly bedded, pebbly sandstones. Clasts include slate, sandstone and porphyry, present as poorly sorted clasts within a matrix-supported bed. They are up to cobble size, with the coarsest clasts 0.10 m above the base of the bed. These are generally rounded and show imbrication of clast long axes parallel to flow. The matrix is poorly sorted, fine- to coarse-grained sandstone. The bed shows moderately developed inverse to normal coarse-tail grading of clasts; the larger clasts increase then decrease in size upwards. Above this bed are thinly bedded, pebbly sandstones, averaging 0.08 m thick, which form sharply based sheets of poorly sorted, fine- to coarse-grained sandstone containing many small, isolated flat-lying pebbles up to 0.10 m. The beds are generally internally structureless, matrix-supported and ungraded.

Road cuttings [SS 8107 0290] north of West Sandford expose up to 2 m of thinly bedded, sharp or slightly erosively based sheets of reddish brown, pebbly breccia overlain by beds of fine-grained sandstone. The breccia is poorly-sorted and matrix- to clast-supported. The clasts are dominantly subrounded and flat-lying; common normal coarse-tail grading is developed. The pebbly breccia fines abruptly upwards into sandstone which contains occasional small pebbles and is dominantly massive, although rare ripple cross-lamination is present. Small-scale cross-bedding indicates flow towards 050°.

Lower Creedy

At Bradley Farm, an excavation [SS 8544 0236] showed, beneath 0.4 m of blanket head and regolith, 0.6 m of brownish red, silty and sandy breccia, with angular to rounded pebbles, poorly sorted, and up to 10 cm in diameter, of Culm sandstone with minor vein quartz, hornfels, acid tuff, quartz-porphyry and bleached lava, resting on 0.9 m of brownish red silty sand with a sparse clast population (as above) to 3 cm in diameter.

An exposure [SS 8605 0252] near Great Gutton, close to the boundary with the Cadbury Breccia, shows 0.6 m of coarse-grained clayey breccia with subrounded clasts of Culm sandstone and vein quartz, crudely bedded and dipping 5° due south. Pockets of creamy-white kaolinitic material apparently represent rotted lava or tuff fragments.

KNOWLE SANDSTONE

The Knowle Sandstone of this memoir is not exactly equivalent to the Knowle Sandstones of the Okehampton district (Edmonds et al., 1968), because the latter included sandstones which have now been separately mapped in the Exeter district as the Creedy Park Sandstone. Together with the overlying Creedy Park Sandstone, the formation occupies low-lying ground extending from the western boundary of the district at Knowle [SS 783 016], through Spencecombe, to West Sandford and Sandford (Figure 6). Small outcrops are present on the southern margin of the Crediton Trough between Uton [SX 825 986] and Posbury Clump, and in the Gunstone House area [SX 807 989]; hereabouts, the Knowle Sandstone rests on Crackington Formation and is overlain by basalt, as at Dunchideock south-west of Exeter (p.40).

The Knowle Sandstone comprises reddish brown, moderately- to well-cemented, well-bedded sandstone. Individual beds vary in particle size, content of fines, and degree of sorting. Interbeds of breccia are present throughout, but decrease in number upwards; breccia clasts are similar to those in the Bow Breccia. Lavas of the Exeter Volcanic Rocks are interbedded at two levels (Figure 7): lamprophyric microsyenites are present at or close to the junction with the underlying Bow Breccia, and olivine basalts occur in the middle to upper part of the formation. They are described in Chapter 7. $^{40}Ar/^{39}Ar$ dating of the lamprophyric lavas at Knowle·Hill Quarry [SS 789 022] gave an age of 281.8 ± 0.8 Ma (Chesley, personal communication, 1992). The sandstone may be ashy or scoriaceous in the vicinity of the lavas; in places, it has been baked by the lavas but elsewhere there is evidence of disturbance by the interaction of the lava with wet sediment during extrusion. Farther east, in the absence of the Exeter Volcanic Rocks, the base is gradational from the underlying Bow Breccia, while the top is marked by a strong disconformity overlain by the Creedy Park Sandstone (Figure 8). The Knowle Sandstone of the area south-west of Exeter is described above (p.40).

The thickness of the Knowle Sandstone at the western margin of the district was estimated at 120 to 150 m by Edmonds et al. (1968). In the Knowle to Sandford area, the thickness is up to 100 m. Around Posbury and Uton, up to about 30 m are estimated to be present.

The sedimentary features of the Knowle Sandstone indicate deposition mainly by unconfined tractional sheet-

a.   BOW BRECCIA

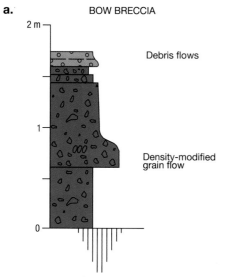

Debris flows

Density-modified
grain flow

b.   KNOWLE SANDSTONE

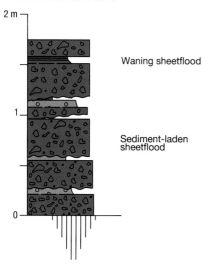

Waning sheetflood

Sediment-laden
sheetflood

c.   KNOWLE SANDSTONE

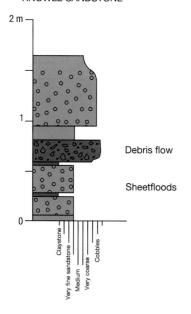

Debris flow

Sheetfloods

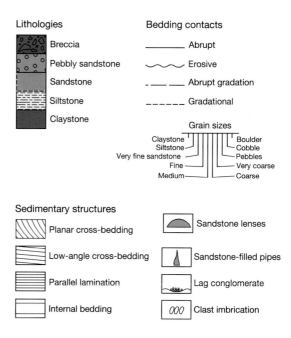

Lithologies

- Breccia
- Pebbly sandstone
- Sandstone
- Siltstone
- Claystone

Bedding contacts

- ———— Abrupt
- ∿∿∿ Erosive
- – — – — Abrupt gradation
- – – – – – Gradational

Grain sizes

Claystone — Boulder
Siltstone — Cobble
Very fine sandstone — Pebbles
Fine — Very coarse
Medium — Coarse

Sedimentary structures

- Planar cross-bedding
- Low-angle cross-bedding
- Parallel lamination
- Internal bedding
- Sandstone lenses
- Sandstone-filled pipes
- Lag conglomerate
- *000* Clast imbrication

Figure 18   Representative graphical sedimentary logs and interpretation of the Bow Breccia and the Knowle Sandstone: (a) Bow Breccia between West Sandford and Sandford [SS 8160 0255]; (b) Knowle Sandstone at West Sandford [SS 8107 0285]; (c) Knowle Sandstone at Brandirons Corner [SS 7876 0200].

floods (Jones, 1992 a). Matrix-supported, structureless breccias [e.g. SS 7876 0200, south of Brandirons Corner] were probably deposited by debris flow (Figure 18).

Details of the Knowle Sandstone in the area south-west of Exeter are given on p.40.

Knowle

A cutting [SS 7876 0200] in the lane leading south from Brandirons Corner shows 1.7 m of reddish brown pebbly sandstone, with a 0.2 m-thick breccia 0.9 m below the top of the section (Figure 18). The sandstones are fine grained and occur as thin beds with sharp bases and tops; they are generally poorly sorted, with angular granules and small pebbles up to a maximum of 0.15 m in size common towards the base. Normal coarse-tail grading is usually well developed. Grains are commonly coated with dark reddish brown clay. Breccia is present as a thin, poorly sorted, matrix-supported bed that lacks internal structure, but shows inverse coarse-tail grading at the base. The matrix is poorly sorted fine-grained sandstone, and clasts are generally pebble-sized, with a maximum size of 0.07 m. Clast types include slate, black mudstone, volcanic rocks, sandstone and quartz. Mudstones commonly form thin cappings to beds of breccia and sandstone, and are generally massive.

West Sandford–Sandford

An excavation [SS 8090 0283] at West Sandford Barton shows a lamprophyric lava flow resting on Knowle Sandstone. The lava is scoriaceous and vesicular towards the base, which is irregular, with lobes and tongues of lava in the underlying sandstone. The top 0.67 m of the Knowle Sandstone is red, coarse-grained and ashy sandstone, disturbed and convoluted at the top, and very pebbly towards the base. It overlies 0.05 m of hard, micaceous, brownish red, fine-grained sandstone, which overlies 0.34 m of red, sandy, fine-grained breccia resting on 0.03 m of sandstone.

At the southern end of Green Lane [SS 8107 0285], West Sandford, the Knowle Sandstone is exposed on the southern side of a fault. Approximately 2 m of crudely interbedded, reddish brown sandstone, breccia and rare mudstone are exposed (Figure 18). The beds of breccia are sharply or erosively based and are overlain by beds of sandstone with sharp or abruptly transitional bases. The breccia is generally matrix supported, with granules and small pebbles supported by a matrix of moderately sorted fine- to medium-grained sandstone. Grading is usually not developed, and clasts are subangular to subrounded, with a maximum particle size of approximately 0.05 m. Sandstone usually forms thin cappings, with occasional pebbles protruding from the underlying breccia into the sandstone. The sandstone is fine to coarse grained, and is poorly to moderately sorted, with occasional parallel or ripple cross-lamination. The sandstones are rarely overlain by thin (0.01 m), laterally impersistent beds of mudstone. Breccia clasts are of sandstone, slate, vein quartz, hornfels and igneous rocks including ignimbritic acid tuff and quartz-porphyry. In the lowest part of the section, the sandstone is ashy and tuffaceous.

At Ruxford Barton [SS 8167 0238], 0.7 km south-east of West Sandford, well-cemented, fine-grained, micaceous, reddish brown sandstone is interbedded with thin beds of fine-grained breccia. Beds dip at 15° to 170°. Further exposures of fine-grained, reddish brown sandstone are visible in the lane west of Town Barton, Sandford [SS 8253 0253 to 8266 0252].

A cutting [SS 8285 0237] in Sandford shows up to 6 m of reddish brown sandstone dipping at 22° to 160°, and forming thin (average bed thickness 0.2 m) poorly defined beds which are occasionally erosively based. Individual beds form fining-upwards

sequences of pebbly to granular sandstone, fine-grained sandstone and rare thin mudstone tops. Rare cross-lamination indicates that palaeocurrents flowed to the south.

Neopardy–Uton

A quarry [SX 8191 9845] west of Uton shows the following section:

	Thickness m
EXETER VOLCANIC ROCKS	
Basalt, purplish brown, weathered, with an irregular base	0.90
KNOWLE SANDSTONE	
Sandstone, reddish brown, with pale buff streaks, ashy in places, medium-grained	0.35
Sandstone, reddish brown, brecciated, with angular fragments of weathered basalt	0.30
Mudstone, lensoid, reddish brown	up to 0.02
Sandstone, reddish brown, medium-grained, weathered and friable, with irregular lenses of medium-grained breccia	(seen) 1.80

THORVERTON SANDSTONE

The Thorverton Sandstone was named from the village of Thorverton by Edwards (1984 a). The outcrop extends from near West Raddon [SS 896 024] through Thorverton to Silverton (Figure 6) and is repeated by faulting around Killerton Park [SS 970 010]. The Thorverton Sandstone is correlated with the Knowle Sandstone farther west on the basis of lithological and geochemical similarity and the presence of comparable interbedded volcanic rocks (Figure 7).

The formation is up to 120 m thick and consists predominantly of reddish brown, fine- to very fine-grained sandstone, with thin beds and partings of reddish brown clay in places. A lenticular bed of breccia occurs south of Raddon [about SS 915 016], and other breccia beds too thin to represent on the map occur within the formation elsewhere. Basaltic lavas (Chapter 7) occur at various levels within the formation, between Raddon and Thorverton, around Silverton, between Heazille and Killerton, and around Budlake (Figure 6). At Killerton, the prominent hill is composed of minette (biotite-lamprophyre), apparently forming a sheet-like body in the Thorverton Sandstone (Cornwell et al., 1990). The Thorverton Sandstone rests on the Cadbury Breccia, but the nature of the contact is uncertain. Regional relationships (Figure 8) suggest that the boundary may be unconformable.

The sandstones of the formation are generally reddish brown, with sporadic greyish green reduction bands and spots. The total range of mean size is from siltstone (4.6 ø) to fine-grained sandstone (2.2 ø), but most samples analysed are fine-grained sandstones in the range 2.2 to 3.0 ø. There is a wide range of sorting, from very well sorted (0.34 ø) to poorly sorted (1.84 ø), but the majority of samples are moderately sorted to moderately well sorted. The average sorting value is 0.91 ø (moderately sorted). The sandstones are composed predominantly of red-stained angular to subangular quartz with subordinate amounts of subrounded grains. Scattered mica flakes are commonly present. Cementation is generally weak, and the sandstone may weather to sand; well-cemented beds occur at some localities.

Few sedimentary structures are visible owing to poor exposures. Indistinct cross-bedding is present at a few localities, and some sandstones show parallel lamination. Intraformational breccias, consisting of angular clasts of reddish brown silty sand, set in a similar but somewhat less well-cemented matrix, occur near the top of the formation between Raddon [SS 910 015] and Berrysbridge [SS 923 012]. The Thorverton Sandstone at a locality [SS 9092 0255] near Raddon is interpreted as deposited within ephemeral fluvial (possibly braided) channels up to approximately 1 m deep. Flow within the channels was episodic, with periods of bedload transport marked by the formation and migration of small sinuous-crested dunes. These were commonly eroded by the next flood, producing clay rip-up clasts. Measurement of palaeo-currents indicates that flow within channels was approximately towards the south-west.

There is no palaeontological evidence of the age of the formation. Minette lavas apparently interbedded with the Thorverton Sandstone in the Killerton area have been dated at 290.8 ± 0.8 Ma (Early Permian), using the $^{40}Ar/^{39}Ar$ method (Chesley, personal communication, 1992).

Raddon–Thorverton

Exposures [SS 9066 0221] in Crownhill Lane near Raddon Court, close to the base of the Raddon lava, show 0.8 m of reddish brown, silty, fine-grained sandstone, irregularly bedded to wavy-bedded, with a few subspherical, greyish green reduction spots up to 30 mm across (especially in the upper 0.4 m), and with irregular, greyish green, reduced bands parallel to bedding. This bed overlies 1.0 of reddish brown, silty, fine-grained sandstone, more massive than the overlying bed, planar to wavy-bedded, parallel-laminated in places, and finely micaceous. The dip is 8° to 166°.

Thorverton Sandstone overlying the Raddon lava is exposed [SS 9078 0200] in the south-west corner of Raddon Quarry. The section shows 1.6 m of reddish brown, silty, fine-grained sandstone, with pods and lenses of friable, greyish green sandstone, harder in the basal 0.4 m. This bed overlies 0.3 m of hard, reddish brown, silty, fine-grained sandstone with clasts of basalt, resting on purple, amygdaloidal basalt.

Roadside sections [SS 9092 0255], about 300 m west of Chapel Farm, Raddon, show exposures near the base of the Thorverton Sandstone, as follows:

	Thickness m
THORVERTON SANDSTONE	
Sandstone, reddish brown, fine-grained, moderately sorted, weakly cemented, with a few partings and one (< 10 mm) bed of reddish brown clay; sandstone with some breccia about 0.6 m above the base	1.7
Breccia, reddish brown, lensoid, channelled into fine-grained sandstone; the proportion of breccia increases downwards	1.5

Small exposures [e.g. SS 9191 0121] near the top of the formation show intraformational breccia consisting of angular clasts

of reddish brown, very weakly cemented, silty, very-fine-grained sandstone set in a similar matrix.

A stream-bank [SS 9280 0184], south-east of Thorverton, shows the following section:

	Thickness m
THORVERTON SANDSTONE	
Sandstone, yellowish brown to reddish brown, weakly cemented	0.9
Sandstone, reddish brown, weakly cemented, with bed of purple clay	0.3
Sandstone, reddish brown, fine-grained, very well-sorted, weakly cemented, cross-bedded	0.2
Sandstone, reddish brown, silty, fine-grained, with blocky-fracturing band of very silty, reddish brown clay	0.5

The Yellowford Borehole [SS 9255 0089], about 1.3 km south of Thorverton, penetrated 16.01 m of Thorverton Sandstone between depths of 76.50 and 92.51 m (Figure 20). A borehole [about SS 939 015], near Latchmoor Green, penetrated Thorverton Sandstone to 46.33 m depth, beneath 7.62 m of river terrace deposits. The sequence was recorded as mainly fine-grained 'marly' sandstone (61 per cent), with units of 'marl' and sandy 'marl'.

Silverton

An exposure [SS 9567 0199] at Babylon Farm, about 1 km south of Silverton, shows the following sequence in weathered Thorverton Sandstone beneath 0.9 m of head:

	Thickness m
Sand, reddish brown, fine-grained, and sandy silt, with irregular thin beds (up to 0.1 m thick) of reddish brown and pale grey-green clay. The basal 0.4 m is fine- to medium-grained gravelly sand. Dip 14° to 123°	1.5
Breccia, reddish brown, fine-grained, and reddish brown gravelly sand. The breccia forms thin beds in hard gravelly sand. Clasts in the breccia are Culm sandstone and vein quartz to 5 cm maximum, most being < 1 cm	0.7

A section [SS 9705 0247] in a quarry near Combesatchfield, about 1.5 km east-south-east of Silverton, shows Thorverton Sandstone overlying basalt, as follows:

	Thickness m
THORVERTON SANDSTONE	
Sandstone, reddish brown, fine- to medium-grained, weakly to moderately well-cemented, finely laminated; dip 8° to 182°	0.3
Sand, gravelly, and fine breccia, lenticular	0 to 0.1
Sandstone, fine- to medium-grained, moderately sorted, with some breccia	0.1 to 0.2
Breccia, reddish brown, fine-grained	0.1
Sandstone, grey to reddish brown, rather rubbly, locally hard, with weathered derived igneous material	0.6
Sand, very clayey, and very sandy clay, patchily grey and reddish brown	0.3
Concealed	0.1 to 0.2
Basalt, pale red-purple, very vesicular	0.3

Old Heazille–Killerton

Reddish brown, moderately sorted, fine-grained sandstone at the base of the formation, underlying basalt, is exposed [SS 9573 0081] 100 m north of Old Heazille Farm. The dip is 16° to 128°.

Shallow quarries [SS 9690 0073] in Park Wood, Killerton Park, contain blocks of reddish brown, flaggy, moderately well-cemented, parallel-laminated fine-grained sandstone.

CREEDY PARK SANDSTONE

The Creedy Park Sandstone is named here from the type locality [SS 8386 0143] in Creedy Park, near Crediton, where it is most extensively developed. The outcrop extends as a narrow strip from the western edge of the district near Knowle [SS 783 015], eastwards to Sandford [SS 829 024] (Figure 6). South of Sandford, the low-lying ground of Creedy Park [SS 833 016] and its environs is underlain by a gentle anticline in the Creedy Park Sandstone, which forms an outcrop about 1.5 km in width from north to south. East of the Park, the outcrop is terminated by a fault along the valley of the River Creedy. A separate fault-bounded outcrop of the formation forms the low-lying ground that extends through Crediton from the west of the town [SS 820 005] to Lord's Meadow [SS 847 007] in the east (Figure 6).

The formation consists mainly of reddish brown, weakly cemented, clayey, fine-grained sandstone, weathering at outcrop to clayey sand. Interbeds of siltstone, claystone and breccia (weathering to silt, clay and gravel) are present throughout, the latter increasing in number towards the top of the formation. The thickness is about 10 m near Knowle, and about 80 m at Creedy Park and Crediton. Beds of breccia within the Creedy Park Sandstone contain clasts which are similar to those in the Crediton Breccia.

The Creedy Park Sandstone overlies the Knowle Sandstone unconformably. It rests on basalt in the west of the district, near Knowle; farther east, in the West Sandford area, there is angular unconformity between the Knowle Sandstone (dipping at up to 20°), and the Creedy Park Sandstone (dipping at less than 8°). East of Sandford, the Creedy Park Sandstone is faulted against Bow Breccia. East of the River Creedy, the Creedy Park Sandstone passes laterally into clay- and silt-rich breccia of the Crediton Breccia.

A sample from a section at Creedy Lakes [SS 8386 0143] (Figure 6:6) yielded the miospores *Crassispora kosankei*, *Lycospora* sp. and *Potonieisporites novicus*. *Lycospora* sp. and an indeterminate bisaccate pollen grain were recovered from material from a temporary exposure at St Saviours Way, Crediton [SS 8301 0016] (Figure 6:7). Samples from temporary exposures at Crediton Rugby Club ground [SS 8401 0061] (Figure 6:8) yielded the miospores *Crassispora kosankei*, *Florinites pumicosus*, ?*Lueckisporites virkkiae* (var. B of Clarke, 1965), and indeterminate taeniate and non-taeniate bisaccate pollen. The presence of ?*L. virkkiae* indicates a Late Permian age, and thus a correlation with the Whipton Formation (Figure 7).

Owing to poor exposure, few sedimentary features are visible. An exposure [SS 8401 0061] at Crediton Rugby Club is dominated by thinly bedded sandstones and fining-upward pebbly sandstones. The beds are sharp-based and variable in thickness, but typically average 10 cm. They are generally internally structureless, but are locally laminated. Rare thin mudstones cap these beds in places. The sandstones are interpreted as tractionally-deposited, formed from episodic sheetfloods. Individual beds represent deposition from single waning flows. The presence of parallel lamination probably represents upper phase plane beds, indicating deposition under conditions of high-flow velocity within the upper flow regime.

Knowle–West Sandford

An excavation [SS 8095 0212] near Aller Barton, about 0.8 km south of West Sandford, showed the following section; the beds dip 3° south:

	Thickness m
HEAD	
Loam, reddish brown, sandy and silty, with some gravel	up to 1.6
CREEDY PARK SANDSTONE	
Sand, reddish brown, clayey	0.10
Clay, dark reddish brown, silty and sandy	0.15
Sand, brownish red, fine-grained and extremely clayey	0.47
Clay, purplish brown, sandy	0.03
Sand, brownish red, with thin beds and lenses of purplish brown silty clay	0.61
Sand, reddish brown, with patches of black manganese-oxide staining	0.32
Clay, purplish brown, silty, with irregular pockets of grey-green clayey sand	0.38

Creedy Park

At Frogmire, an excavation [SS 8253 0149] showed the following section, with beds dipping about 3° south-west, beneath 1.4 m of made ground and head:

	Thickness m
CREEDY PARK SANDSTONE	
Sand, reddish brown, fine-grained, extremely clayey	0.59
Clay, purplish red, sandy	0.32
Clay, brownish red, with irregular thin beds of orange-red and greyish green, clay-rich sand	0.47
Sand, reddish brown, fine-grained, clayey, with patches of black Mn-oxide staining, and irregular beds of purplish red clay to 40 mm thick	(seen) 0.45

A channel [SS 8386 0143] at Creedy Lakes is the type section of the Creedy Park Sandstone. The western end shows 2.5 m of reddish brown and brownish red clayey sand, with sparse pebbles in the basal 0.8 m. About 20 m farther east, the section consists of 0.65 m of red and orange-red clayey very fine-grained sand, on 0.24 m of reddish purple silty clay, on 0.30 m of mottled, reddish brown and greyish green sandy clay.

Crediton

An embankment [SS 8219 0051] in the grounds of Crediton Hospital exposes, beneath 1.7 m of head, 0.8 m of fine breccia overlying 3 m of reddish brown sand with poorly defined beds of fine breccia up to 20 mm thick.

Excavations [SS 8226 0045] off Threshers showed 0.55 m of made ground and head resting on 1.15 m of reddish brown silty sand, interbedded with beds of fine breccia up to 30 mm thick. The junction of the Creedy Park Sandstone with the overlying Crediton Breccia was exposed [SS 8301 0016] near St Saviour's Way car park. The Crediton Breccia consists of breccia with clasts of sandstone, slate, vein quartz, hornfels and acid igneous rocks ranging up to 50 mm, in a red silty matrix. The Creedy Park Sandstone consists of 0.4 m of red, clayey, fine-grained sand. A cutting [SS 8302 0027] shows up to 2.0 m of weathered, friable, fine- to medium-grained sandstone, with thin beds of fine breccia.

At the western corner of Crediton Rugby Ground, an embankment [SS 8401 0061] shows up to 1.8 m of very poorly cemented sandstone, with scattered pebbles up to 30 mm throughout, thin, poorly defined breccia beds, and very thin interbeds of purplish brown mudstone. Samples from the mudstone yielded palynomorphs (p.67).

CREDITON BRECCIA

Hutchins (1963) divided the New Red Sandstone above his 'Bow Beds' into 'Crediton Beds' and overlying 'St Cyres Beds' (Table 6); in the present district, these units are called the Crediton Breccia and Newton St Cyres Breccia respectively (Scrivener, 1988). In the neighbouring Okehampton district (Edmonds et al., 1968), the Crediton Beds and St Cyres Beds of Hutchins were mapped as a single unit, the 'Crediton Conglomerates'.

The formation is between 100 m and 240 m thick, and consists predominantly of reddish brown, poorly to moderately cemented breccia with a poorly sorted silt, sand and clay matrix (Plate 6b). It weathers to gravelly silty sandy clay. Clasts do not generally exceed 40 mm, though some coarse channel-fills have been noted in places. Clasts in the lower part of the formation include Culm sandstone, slate, vein quartz, and igneous rock fragments including quartz-porphyry, tuff and rhyolite. The igneous clasts are untourmalinised, and range up to large cobble size; however, pebbles and small cobbles of tourmalinised breccia (possibly vent breccia) are present, though scarce. The clast population of the upper part of the formation is similar to the lower, except that the acid igneous debris is commonly tourmalinised, and the clasts of rhyolite, tuff and quartz-porphyry locally range up to boulder size.

The outcrop of the Crediton Breccia extends from the western boundary of the district, near Coombe House [SS 785 011], to Crediton. From Crediton, the outcrop extends eastwards through Shobrooke [SS 863 011] to Yendacott Copse [SS 906 009] (Figure 6), from where there is an apparent eastward passage into the Yellowford Formation.

The Crediton Breccia is unconformable upon older (Early Permian) formations, except around and to the west of Crediton where the unconformity is at the base of the Creedy Park Sandstone (Figure 8). Around Crediton, the Crediton Breccia overlies Creedy Park Sandstone, apparently conformably. East of the fault along the River Creedy, the Crediton Breccia rests unconformably on Bow Breccia and then, east of Great Gutton [SS 862 025], on Cadbury Breccia. East of West Raddon [SS 896 024], and thence to the Silverton area, the Crediton Breccia

and its correlative, the Yellowford Formation, overlie the Thorverton Sandstone (Figure 8).

The formation yielded a small pollen assemblage comprising *Perisaccus granulosus*, *?Lueckisporites* sp., *Lunatisporites* sp., *?Protohaploxypinus* sp., *?Klausipollenites schaubergeri* and indeterminate non-taeniate bisaccates from an exposure [SX 8406 9874] south of Fordton Cross (Figure 6:9). The presence of *?Lueckisporites* suggests a Late Permian age (p.31).

Sedimentary features of representative sections of the formation are shown graphically in Figure 19. The breccias at Forches Cross [SS 8318 0088], Crediton, show matrix-support, lack of grading and absence of sedimentary structures, indicating that deposition was from high-viscosity debris flows. Breccias at Chapel Downs [SS 8195 0065], Crediton (Figure 19), were also deposited from debris flows, but normal coarse-tail grading indicates deposition from low-viscosity flows. Sections [SS 8875 0185] near Raddon Cross, Uppincott (Figure 19) show better-bedded breccias with fine-grained sandstones and mudstones; the presence of sandstone lenses within individual beds, the presence of occasional pebbly lags, and the upward transition into sandstones which show parallel lamination and low-angle cross-bedding, suggest that this sequence was deposited during episodic, turbulent sheetflood events. Thin mudstones are interpreted as the final stage of sedimentation and represent the deposition of fine-grained sediment suspended in the flow.

Knowle–West Sandford

At Punch Bowl Tip [SS 792 008], about 1 km south of Spence Combe, the junction of the Crediton Breccia with the overlying Newton St Cyres Breccia is exposed [SS 7930 0073]; bedding dips at 10° to 170°. The Crediton Breccia consists of 3.3 m of red, poorly bedded breccia with a matrix of silty loam and clasts of slate, sandstone, hornfels, vein quartz and tourmalinite up to 40 mm; clasts of quartz-porphyry, acid tuff and rhyolite are also present, the last, commonly highly kaolinised, ranging in size up to 0.45 m. The overlying Newton St Cyres Breccia is distinguished from the Crediton Breccia by the presence of sparse cleavage fragments of murchisonite feldspar up to 15 mm.

The contact between the Crediton Breccia and underlying Creedy Park Sandstone is exposed in a track [SS 8066 0186] near Goldwell, about 1 km south-south-west of West Sandford. The Crediton Breccia is 4.6 m of brownish red silty breccia with subrounded and subangular clasts of sandstone, slate, hornfels, vein quartz, rhyolite and quartz-porphyry, ranging in size up to 70 mm. The underlying Creedy Park Sandstone is 1.0 m of reddish brown, medium-grained, well-cemented sandstone, with scattered thin beds of brown mudstone.

Crediton

At Chapel Downs on the west side of Crediton, cuttings [SS 8195 0065] show up to 4 m of reddish brown breccia, with scattered cobbles and small boulders of rhyolite and quartz-porphyry (Figure 19). The breccia consists of thin, poorly defined beds showing a marked normal coarse-tail grading, forming fining-upward sequences of pebbly breccia and more sand-rich horizons. Poorly sorted clasts are matrix-supported in mudstone and fine-grained sandstone. The clasts are angular and are composed generally of small pebble-grade material, with a maximum size of 0.12 m.

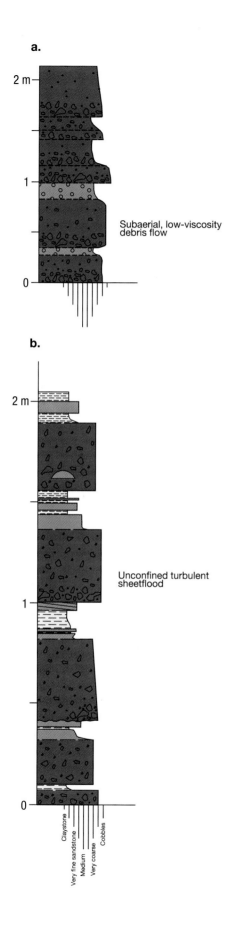

a.

2 m —

1 —

0 —

Subaerial, low-viscosity
debris flow

b.

2 m —

1 —

0 —

Claystone
Very fine sandstone
Medium
Very coarse
Cobbles

Unconfined turbulent
sheetflood

Road cuttings along the Sandford Road, extending north-west from Forches Cross [SS 8318 0088], show up to 3 m of reddish brown, poorly sorted, ungraded breccia (Plate 6b). Crude bedding is defined by variations in clast size. The breccia is matrix-supported, with clasts of small pebble-grade material in a matrix of mudstone. Clasts are subangular to angular, occasionally subrounded, and consist dominantly of sandstone, tourmalinised slate, quartz-porphyry, chert and lava. The average clast size is 0.02 m with a maximum of approximately 0.06 m. On the southeast of the exposure, two vertically stacked lens-shaped breccias, defined by a higher concentration of larger sized pebbles and clast-supported nature, are present. Each lens has diffuse boundaries, with marked convex-upwards bases and planar tops. The lowermost lens is 0.45 m thick and the upper is 0.80 m; the width is greater than 2.0 m. They are composed of very large angular pebbles and small cobbles, up to 0.10 m. The lenses are internally disorganised, with no internal stratification, and are poorly sorted and ungraded. Mudstone and sandstone form the matrix.

A cutting [SS 8323 0080] in Deep Lane exposes 2.2 m of poorly bedded, brownish red, silty breccia with scattered boulders, up to 0.6 m, of tourmalinised acid tuff and quartz-porphyry.

The cutting [SS 8425 0095] at Red Hill Cross shows up to 2.2 m of red, silty breccia, with clasts ranging in size up to large cobbles; thin, purplish brown mudstone beds (up to 5 mm) are present in places and are commonly underlain by a narrow (up to 10 mm) green selvedge. In the surrounding fields, the brash includes sandstone, slate, vein quartz, tourmaline breccia and untourmalinised acid igneous debris, mostly rhyolite and quartz-porphyry.

A quarry [SX 8404 9962] at Station Cross, Crediton, shows up to 7.6 m of weathered, reddish brown breccia in crudely parallel beds, up to 0.9 m thick, dipping at 4° to 160°. In places, the beds are coarse at their base, with cobble-size clasts of sandstone and vein quartz; a boulder of weathered acid tuff, 0.6 m in diameter, is present in the lowest breccia bed at the western end of the face.

East of Moorlake, a road cutting [SX 8138 9936 to 8160 9961] shows up to 5 m of red, silty breccia. Bedding is poorly developed in finer and coarser clast-size units, some of which have coarse channel fills at their bases, with clasts ranging in size up to small boulders. Clast lithologies include Culm sandstone, slate, vein quartz, hornfels, acid tuff, rhyolite, quartz-porphyry and tourmalinite.

A cutting [SX 8406 9874] south of Fordton Cross exposes 0.9 m of red-brown, moderately well-cemented breccia, with irregular green reduction bands up to 30 mm thick. A 10 mm-thick bed of micaceous blocky siltstone near the top of the section yielded pollen indicative of a Late Permian age (p.68).

Shobrooke

A cutting [SS 8546 0153] at Shobrooke House shows 2.1 m of breccia with a matrix of brownish red silty and sandy clay, and a poorly sorted clast population ranging in size from fine gravel to sparse cobbles up to 20 cm across. The clasts are mostly angular to subangular and include red- and purple-stained Culm

Figure 19 Representative graphical sedimentary logs and interpretation of the Crediton Breccia: (a) Chapel Downs [SS 8195 0065], Crediton; (b) near Raddon Cross [SS 8875 0185], Uppincott. See Figure 18 for key to graphical log symbols.

sandstone, slate, vein-quartz, acid to ?intermediate lava and tuff, quartz-porphyry and hornfels. Bedding is poorly developed and is marked by coarser clasts; the dip is about 4° to the south-south-east.

South from Shobrooke Cross, small exposures [SS 8604 0085 to 8614 0018] show reddish brown silty breccia with fine to medium gravel-size clasts. At the northern end of Church Lane, Shobrooke, a section [SS 8637 0134] shows up to 1.5 m of breccia with medium gravel-size clasts and a matrix of red, silty, fine-grained sand. Bedding, marked by poorly developed imbrication, is subhorizontal. At 150 m east, another lane-bank section shows 1.4 m of red silty breccia with poorly developed subhorizontal bedding marked by pale sand streaks. At 110 m farther east, the south side of the road cutting shows 1.2 m of breccia with a red silty and clayey matrix and angular to sub-rounded clasts of sandstone, slate, shale, quartz-porphyry, acid lava, acid tuff, hornfels and a range of tourmaline-bearing rocks.

A lane [SS 8699 0014] south of Shobrooke Mill Farm shows 1.4 m of reddish brown, silty sandy breccia with clasts up to 15 mm across (mostly less) of sandstone, slate, siltstone, acid lava and tuff, quartz-porphyry, vein quartz, microgranite, hornfels and tourmalinised rocks. The breccia shows poorly developed bedding, but thin, reddish brown mudstone or clay lenses up to 15 mm thick show dips of 4 or 5° to the south-east.

A stream-bank section [SS 8745 0215] south of Coombe Barton exposes (just above the Cadbury Breccia boundary) up to 1.9 m of red silty and clayey fine-grained breccia with scattered coarser clasts of Culm sandstone, quartz-porphyry and lava. Subhorizontal parallel bedding is poorly developed.

Uppincott–West Raddon

The stream and its tributaries draining southward from Uppincott show exposures in the Crediton Breccia [e.g. SS 8859 0189; 8865 0186; 8867 0188; 8866 0183 to 8869 0179] (Scrivener and Edwards, 1990). The best exposure is in a narrow gorge [SS 8875 0185] near Raddon Cross where about 7 m of thinly bedded, reddish brown breccia, fine-grained sandstone and mudstone are exposed (Figure 19). The breccias form beds up to about 0.4 m thick with sharp bases and abruptly-transitional tops. They are matrix-supported, with clasts of granule to small pebble size in a matrix of fine-grained sandstone and mudstone. The clasts and matrix are subangular to angular, poorly sorted, and lack imbrication. A few sandstone lenses up to 7 mm by 60 mm occur locally in the breccia, and have rare coarse pebbly lags at their bases. Clast types include sandstone, shale, slate and igneous fragments. The breccia commonly passes up abruptly into thin beds of sandstone, consisting of poorly to moderately sorted subangular grains of fine- to medium-grained sand with a common or abundant component of coarse and granular sand. The sandstone is typically pebbly at the base and commonly contains mudstone clasts; crude, low-angle cross-lamination and small-scale cross-bedding are present. The sandstone is repeatedly interbedded with thin impersistent beds of fine to coarse siltstone which are generally massive, although locally show rare dark and pale parallel lamination.

At West Raddon Farm, a tongue of sand within breccia is about 8 m thick. Exposures [SS 8932 0238] in a sunken track show 2 m of reddish brown fine- to very fine-grained sandstone with green reduced lenses and sandy clay lenses near the base. The dip is 8° to 196°.

Between West Raddon and Efford, the stream bed contains discontinuous exposures of reddish brown, plane-bedded, fine-grained breccia, commonly with thin beds of silty clay. The following exposures are typical: reddish brown, fine- to very fine-grained, shaly breccia with a few silty clay bands, dip 14° to 172° [SS 8916 0202]; reddish brown, plane-bedded breccia, predominantly of shale clasts, locally with porphyry clasts, sporadic 10 to 20 mm-thick beds of reddish brown silty clay, dip 14° to 188° [SS 8921 0190]; reddish brown shale and sandstone breccia with clasts mostly less than 10 mm, and a few centimetre-thick, very silty clay beds, dip 8° to 158° [SS 89360158]; reddish brown, fine-grained breccia with centimetre-thick silty clay beds, dip 18° to 218° [SS 8918 0129].

Exposures [SS 8903 0119] in West Efford Lane show up to 1.5 m of reddish brown breccia consisting predominantly of angular clasts of Culm sandstone and shale, mostly less than 10 mm across, with scattered porphyry clasts. The matrix is of reddish brown sandy clay; impersistent, pale greyish green layers pick out the bedding, which dips at 10° to 198°.

Shute–Yendacott Copse

The stream that flows from Carwithen Copse, about 1 km northeast of Yendacott Manor, southward through Yendacott Copse, contains discontinuous exposures [SS 9050 0132 to 9095 0055] of breccia. A typical exposure [SS 9050 0132] shows 1.0 m of reddish brown, fine-grained, clayey, sandy, friable breccia, composed predominantly of very fine-grained, angular, weathered shale fragments. A few pale greyish green irregular reduced layers are present. The dip is about 9° to 180°. Nearby [SS 9056 0114], reddish brown shaly fine-grained breccia, with a few centimetre-thick clay layers, dips at 6° to 176°. Exposures [SS 9047 0089] of breccia in the stream bed at Yendacott Copse show northerly dips (8° to 022°), and lie in the southern limb of the West Efford Syncline (p.129). The beds are reddish brown, fine-grained breccias composed mainly of Culm sandstone, shale and porphyry, with beds of gravelly sand and thin clay bands. Exposures [SS 9070 0083] of reddish brown breccia farther south also show northerly dips (8° to 346°), but southerly dips are again present [SS 9085 0075] on the southern limb of the Yendacott Anticline (p.129).

The Crediton Breccia gives rise to a moderately pronounced feature [SS 907 016 to 915 013] where it overlies Thorverton Sandstone south of Raddon; poor exposures [SS 9113 0135] of reddish brown slate- and shale-rich breccia, partly weathered to clay, are visible in the road cutting 250 m south of Poole Farm.

YELLOWFORD FORMATION

The Yellowford Formation (Edwards, 1987a), is a predominantly argillaceous sequence, laterally equivalent to the Crediton Breccia. The outcrop extends from about 0.7 km west of Yellowford [SS 924 009] (from where the formation is named), eastwards across the Exe valley to Ellerhayes [SS 976 020], near Silverton (Figure 6). Most of the outcrop is low-lying ground mantled by drift deposits.

The formation, which is between about 77 and 150 m thick, consists of reddish brown silty and sandy claystone, clayey sandy siltstone, and silty clayey fine-grained sandstone. Beds of clayey, fine-grained breccia are present locally. Exposures of the formation are few; near-surface weathering gives rise to clay, silt, fine-grained sand, and gravelly clay.

A thickness of about 76 m of the formation, recorded as mainly marl, with sandstone beds at some levels and a few beds of gravel (?breccia), was penetrated in the Yellowford Borehole [SS 9255 0089], where it rests on probable Thorverton Sandstone (Figure 20). A thicker sequence (at least 143 m) was recorded in the Rudway Barton Exploration Borehole [SS 9395 0057], Netherexe,

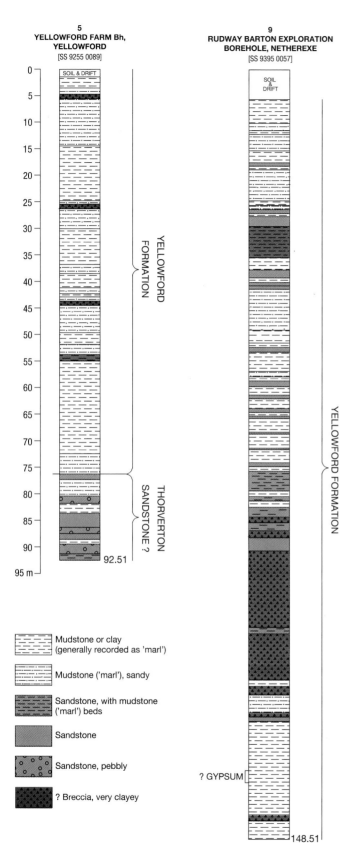

Figure 20 Graphic logs of boreholes in the Yellowford Formation. Borehole numbers are those of BGS records for sheet SS 90 SW. See Figure 15 for location map.

about 1.4 km east-south-east of Yellowford (Figure 20). The recorded sequence is 74 m of mainly marl with sandstone beds, on 35 m of marl and gravel with a few sandstone beds, on 29 m of mainly marl.

The Yellowford Formation rests on the Thorverton Sandstone, probably unconformably, but owing to extensive drift cover the junction is not exposed. In the Dunsmore area [SS 957 015], near Silverton, basalt at the top of the Thorverton Sandstone is overlain by clay of the basal Yellowford Formation, but east of Poundsland [SS 961 013], the distinction between the Thorverton Sandstone and Yellowford Formation is less clear owing to the greater proportion of sand in the latter.

There is no palaeontological evidence of age; however, the field evidence indicates that the formation is laterally equivalent to the Crediton Breccia, of Late Permian age.

The formation is too poorly exposed for detailed sedimentological study; its stratigraphical position indicates that it was probably deposited on the distal part of the Crediton Breccia alluvial fan.

Yellowford

The Yellowford Formation of the Yellowford Borehole [SS 9255 0089] is shown on Figure 20. Between Yellowford and the River Exe, augering proves reddish brown variably silty clay, silt, and very clayey, very fine-grained sand; a few exposures [e.g. SS 9306 0113] of reddish brown, clayey, fine-grained breccia are present in Hulk Lane.

Exposures [SS 9294 0053 to 9292 0045] in the west bank of the River Exe near the top of the formation consist predominantly of reddish brown, very fine-grained, clayey breccia, mainly of shale clasts less than 20 mm and scattered porphyry clasts. The breccias are interbedded with reddish brown, silty, very fine-grained sandstone, probably in transition to the overlying Shute Sandstone.

The Rudway Barton Exploration Borehole [SS 9395 0057] near Netherexe penetrated Yellowford Formation to 148.51 m, without bottoming the formation, indicating a considerable thickness increase over that in the Yellowford Borehole (see above). A graphic log is given in Figure 20.

Silverton

Auger holes in the partly drift-mantled low ground between the western edge of the district [SS 950 010] near Yonder Down, and near Poundsland [SS 960 012], about 1.7 km south of Silverton, prove predominantly reddish brown, variably silty and sandy clay, with lesser proportions of clayey, sandy silt and clayey, silty, very fine-grained sand. A borehole [about SS 961 022] near Waterleat House penetrated 43.28 m of 'sandy marl' beneath 2.44 m of soil and river terrace deposits.

Between Poundsland and the Ellerhayes area, the formation is mainly concealed beneath drift; auger holes reveal a variety of lithologies: for example, clayey fine-grained breccia [SS 9641 0213], very silty fine- to very fine-grained micaceous sand [SS 9660 0196], and firm clay 0.4 m on clayey sand [SS 9669 0169]. Temporary exposures [SS 9647 0178] near Hayne House showed up to 0.8 m of moderately well-sorted fine- to very fine-grained sand. Between the Heal-eye Stream and the Ellerhayes area, the range of lithologies within the Yellowford Formation is revealed in selected auger holes: very clayey, silty, fine-grained sand passing to extremely sandy, silty clay [SS 9667 0232]; very clayey, sandy silt [SS 9742 0204]; extremely clayey silt and silty clay [SS 9819 0177].

NEWTON ST CYRES BRECCIA

The term Newton St Cyres Breccia was introduced by Scrivener (1988) to include rocks formerly named the St Cyres Beds (Hutchins, 1963) and the St Cyres Breccias (Laming, *in* Durrance and Laming, 1982). In the adjacent Okehampton district (Edmonds et al., 1968), the St Cyres Beds and Crediton Beds of Hutchins were mapped as a single unit, the Crediton Conglomerates (Table 6). The name is derived from the type locality [SX 8799 9805] in the village of Newton St Cyres. The Newton St Cyres Breccia is distinguished from the underlying Crediton Breccia by its more sandy matrix and greater degree of cementation. In addition, it contains, in increasing abundance towards the top of the formation, fragments of the potassium feldspar 'murchisonite' (Levy, 1827). The murchisonite links the Newton St Cyres Breccia with the Heavitree Breccia of the Exeter area (p.47), with which it is correlated (Figure 7).

Typical Newton St Cyres Breccia is reddish brown, weakly to well-cemented breccia with a matrix that varies from predominantly sandy to a mixture of mudstone and fine-grained sandstone. It weathers to gravelly clayey sand or gravelly sandy clay. The clasts are subrounded to angular and are generally granule to medium pebble sized; they seldom exceed 50 mm. They include shale, slate, pelitic hornfels, chert, potassium feldspar (murchisonite), vein quartz, quartz-porphyry, acid lava and tuff, microgranite and tourmalinised igneous rocks. Dangerfield and Hawkes (1969) recorded Dartmoor Granite material at Newton St Cyres.

Bodies of sandstone are present throughout the sequence, and range from mappable units to centimetre-scale beds and lenses. Much of the sandstone is very weakly cemented, medium- to fine-grained, silty and clayey. Irregular thin beds or partings of coarse-grained, relatively clean 'millet seed' sand may be present towards the top of breccia beds. The sand content of the Newton St Cyres Breccia increases towards the top of the sequence and the junction with the overlying Shute Sandstone is gradational, the frequency of sand interbeds increasing at the expense of breccia, which also decreases in clast size.

The outcrop extends from the western margin of the district north of Neopardy [SX 790 990], to just west of Crediton. A separate outcrop is present along the southern side of the trough between Trobridge House [SX 835 979] in the west and Newton St Cyres in the east; the outcrop continues around a synclinal core of Shute Sandstone through Wyke [SX 873 999] to near Shute [SS 891 002], where there is a probable lateral passage into Shute Sandstone (Figure 6).

The greatest thickness is an estimated 180 m south-west of Newton St Cyres village. The thickness is greatly reduced, to about 10 m, near Creedy Barton [SX 865 995].

At 'Cromwell's Cutting', Pitt Hill [SS 8165 0020], near Crediton, murchisonite clasts may, in rare cases, be found with their igneous matrix adhering. One of these, in thin section (E 66093), shows the matrix to be brecciated, coarsely crystalline, biotite-granite with much secondary

blue tourmaline and quartz. The murchisonite feldspar appears to be microcline, rather than the sanidine suggested by Scrivenor (1948).

Samples from an exposure of the formation [SX 8406 9861] south of Fordton Cross (Figure 6:10) yielded solitary specimens of indeterminate taeniate and non-taeniate bisaccate pollen and solitary specimens of *Lycospora* sp., the last considered to be reworked from Carboniferous rocks. The stratigraphical position of the Newton St Cyres Breccia suggests a Late Permian age.

Sedimentary features of representative sections of the formation are shown graphically in Figure 21. At Newton St Cyres [SX 8799 9805], bedding is moderately developed, with individual erosively based beds varying in thickness from 0.5 to 1.0 m. The breccia is predominantly matrix-supported. Rare normal grading is present. The clast size population is bimodal (granules and medium-sized pebbles, with a maximum size of 0.06 m). Trough cross-bedding and ripple cross-lamination occur rarely, and form sets up to 0.18 m in height. Palaeocurrents, though variable, flowed from west to east, with a vector mean from 286°. Within the breccia, rare sandstone lenses up to 2.0 m long and 0.07 m thick occur; generally, they are horizontal with rare lenses inclined along cross-bedding foresets. The lenses are internally massive, parallel laminated or cross-bedded. A distinctive feature is the presence of vertical pipes filled with structureless, moderately sorted, fine-grained sandstone. They occur within individual beds, pass through more than one bed or are truncated at the base of an overlying bed. The pipes are bulbous at their base, taper upwards and attain widths up to 0.3 m; their vertical extent is approximately 1.2 m. Undisturbed horizontal beds of sandstone can be traced laterally and deflect upwards into sandstone-filled pipes.

The thickly bedded and erosively based nature of the breccia at Newton St Cyres may indicate deposition in unconfined proximal sheetfloods or confined channel systems; both are capable of scouring at their bases. However, the basal erosion surfaces, together with a relatively complex fill of cross-bedded breccia and sandstone lenses, are more indicative of sedimentation within channels. The sand-filled pipes are considered to have formed by catastrophic liquefaction and fluidisation. Dewatering commonly led to the complete destruction of beds or lenses of sandstone. Incomplete dewatering can be seen by the passage from vertical pipes into horizontal beds of sandstone.

A section at 'Cromwell's Cutting', Pitt Hill [SS 8165 0020], near Crediton, close to the base of the formation, shows up to 10 m of weak, unconsolidated breccia, the lower part of which consists of stacked sets of planar cross-bedding (Figure 21). The planar cross-bedded breccias are interpreted as transverse barforms formed within a fluvial channel. The sharp-based and sharp-topped nature of breccia foresets suggests discrete, episodic depositional events.

Knowle

The base of the Newton St Cyres Breccia is exposed at Punch Bowl Tip [SS 792 008], where it overlies the Crediton Breccia (p.68).

a.

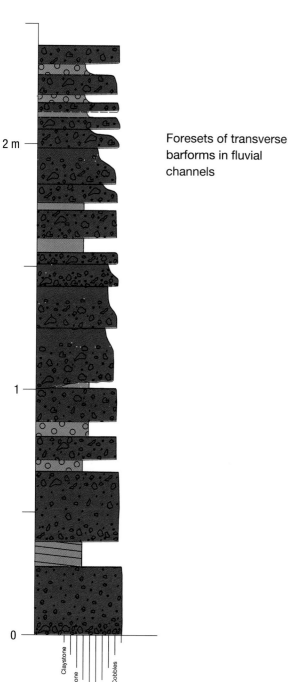

Foresets of transverse barforms in fluvial channels

b.

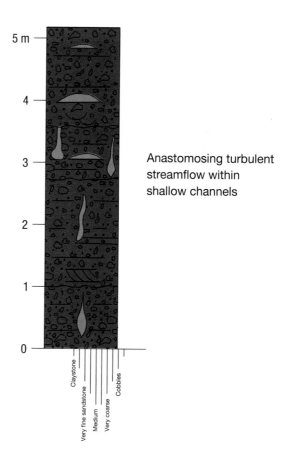

Anastomosing turbulent streamflow within shallow channels

Figure 21 Representative graphical sedimentary logs and interpretation of the Newton St Cyres Breccia: (a) 'Cromwell's Cutting', Pitt Hill [SS 8165 0020], near Crediton; (b) Newton St Cyres [SX 8799 9805]. See Figure 18 for key to graphical log symbols.

Neopardy

An excavation [SX 8012 9890] at Stockeydown Farm shows 1.9 m of red sandy breccia in well-defined parallel beds dipping at 15° to 150°. At 800 m north-east, a field entrance [SX 8079 9922] exposes up to 3 m of red sandy and silty breccia in crudely cross-bedded units, the bases of which are marked by coarse gravel lags. The cross-bedding indicates transport from the south-west.

Crediton

'Cromwell's Cutting' [SS 8165 0020], at Pitt Hill, near Crediton, exposes up to 10 m of breccia. The sedimentary features are described on p.72; a graphic log of the section is given in Figure 21. Clasts include sandstone, slate, hornfels, vein quartz, murchisonite, acid tuff, rhyolite, rare microgranite and tourmalinite; the igneous clasts are commonly tourmalinised. The murchisonite fragments are commonly very fresh and up to 20 mm in length.

A road cutting [SX 8406 9861] 870 m south of Fordton Cross shows the following section:

	Thickness m
Breccia, reddish brown, with a sandy and silty matrix; clasts up to cobble size, mostly of Culm sandstone, showing marked imbrication	1.10
Siltstone, reddish brown, hard, with some coarse-grained sand, passing down into	0.07
Breccia, red, with sandy matrix, clasts ranging up to 30 mm, including sparse murchisonite fragments	(seen) 1.80

Eastacott–Newton St Cyres

At Eastacott, an excavation [SX 8447 9761] exposes up to 5 m of red sandy breccia, in crudely bedded parallel units up to 0.9 m thick. The breccia is matrix-supported, with poorly sorted clasts ranging in size to 0.2 m in length. Clasts are subangular to subrounded and include sandstone, slate, hornfels, vein quartz, and acid igneous debris, notably rhyolite, tuff, quartz-porphyry, microgranite and cleavage fragments of murchisonite. A particular feature at this locality is the extent to which the igneous fragments have been tourmalinised.

At Creedy Barton, a section [SX 8652 9954] beneath a farm building, with beds dipping at 12° to 150° shows:

	Thickness m
SHUTE SANDSTONE	
Sandstone, red, friable, fine- to medium-grained; thin parallel-bedded, but with traces of small-scale channel cross-bedding	1.05
NEWTON ST CYRES BRECCIA	
Breccia, red, well-cemented, with a red sandy matrix and clasts of quartz-porphyry, hornfels, sandstone, chert and murchisonite. Sand lenses and thin beds are present throughout	0.95

A cutting [SX 8577 9908] east-south-east of Downes Mill Farm, and near the base of the formation, shows 2.1 m of friable, red, silty breccia, with medium to coarse clasts of quartz-porphyry, sandstone, hornfels, basalt and sparse murchisonite. At 150 m to the east, the cutting [SX 8592 9907] of the Exeter–Barnstaple railway shows up to 7.0 m of well-cemented red breccia, in parallel beds up to 0.75 m thick dipping 4 to 6° due east.

Near the southern boundary fault of the Crediton Trough, a unit of sand has been mapped within the Newton St Cyres Breccia. A section [SX 8560 9757] adjacent to the Small Brook shows 1.8 m of red, clayey, fine-grained sand with indistinct parallel-bedding dipping at 1 or 2° south.

Cuttings [SX 8779 9812 to 8799 9805] for the A377 road near Newton House, Newton St Cyres, are the type sections of the Newton St Cyres Breccia. The sedimentary features are described above and shown in Figure 21. The subrounded to subangular clasts consist of sandstone, slate, chert, hornfels, volcanic rocks, quartz-porphyry, granite, vein quartz, tourmalinite and abundant murchisonite feldspar. In places the matrix is rich in flakes of biotite.

Brampford Speke–Upton Pyne

The partially cored Starved Oak Cross No.2 Borehole [SX 9130 9879] (Figure 15) proved 38.6 m of pebbly sandstones and breccias beneath 52.4 m of Dawlish Sandstone (p.54); this sequence can be classified mainly as Newton St Cyres Breccia, but might also include some Shute Sandstone. The breccias typically consist of brownish red sandstone or silty sandstone with bands of fine pebbles of Culm sandstone, hornfels, igneous rocks and feldspar fragments. Fresh cleavage fragments of flesh-coloured murchisonite feldspar are very abundant in the coarser breccia beds.

SHUTE SANDSTONE

The Shute Sandstone (Edwards, 1984 a) is named from the hamlet of Shute [SS 891 002], about 2.5 km north-north-east of Newton St Cyres. The outcrop extends around the limbs of a synclinal structure from Upton Pyne to Sweetham [SX 880 990] and thence to near Yellowford [SS 924 009] (Figure 6). The formation gives rise to an area of subdued topography, largely mantled in its western part by alluvium and river terrace deposits.

The formation is up to 80 m thick, and consists predominantly of reddish brown silty sandstone and sandy siltstone, both commonly weathering to silty sand and sandy silt; lenses of breccia occur at a few localities. Augering indicates the presence of very sandy silty clay at some levels in the formation. Grain-size analyses of eight samples show a range of mean grain size between 3.53 and 4.28 ø, with an average value of 3.9 ø, close to the borderline between very fine-grained sandstone and coarse-grained siltstone. All samples analysed are poorly sorted, except one moderately sorted sample (range of sorting values 0.84 to 1.97 ø).

The sandstones consist predominantly of red-stained quartz grains, with scattered flakes of mica commonly present. Beds of intraformational breccia are widespread and typically consist of reddish brown, angular to platy fragments of finely micaceous, weakly cemented, silty, very fine-grained sandstone in a similar, but somewhat softer matrix. In beds without intraformational breccia, bedding is poorly differentiated by indistinct lenses and pods of coarser-grained sandstone. A few subspherical pods and streaks of greyish green sandstone occur at some localities.

The base of the formation is fairly sharply defined on the Crediton Breccia or Yellowford Formation, but in the River Exe exposures [SS 9292 0044], there is evidence of interdigitation of silty sandstone with breccia at the top of the Yellowford Formation.

The probable lateral equivalence of the formation to the Newton St Cyres Breccia indicates a probable Late Permian age.

The moderately well-sorted nature of the sandstone, together with the presence of cross-bedding, is interpreted as reflecting deposition from unconfined tractional sheetfloods, possibly sourced from the west.

Wyke

At the western limit of the Shute Sandstone outcrop, a section [SX 8652 9950] near Creedy Barton shows the formation resting on Newton St Cyres Breccia. Up to 1.05 m of friable red sandstone, mostly parallel-bedded but with some small-scale channel cross-bedding, dip at 12° to 150°. Vertical and near-vertical joints are present throughout.

Between Wyke [SX 8726 9971] and Rewe [SX 8819 9970], the boundary between the Shute Sandstone and the Newton St Cyres

Breccia is marked by a slight feature. Scattered exposures of red clayey and silty sand and, in places, sandy and silty clay are present in the lane between Wyke Cross [SX 8731 9942] and Rewe Cross [SX 8848 9969].

Shute

A cutting [SX 8930 9998] near Shute Cross is the type section of the formation. The locality shows Shute Sandstone, and a breccia which is either the underlying Newton St Cyres Breccia or a breccia lens within the Shute Sandstone. The Shute Sandstone consists of 2.1 m of reddish brown, moderately sorted, very fine-grained sand, with scattered beds of fine breccia up to 60 mm thick; the sandstone is parallel-bedded and dips 3 to 4° south. The northern part of the exposure is predominantly breccia, which contains, in addition to Culm and igneous rock types, fragments of murchisonite feldspar.

A roadside section [SX 8870 9977] east of Rewe Cross shows 0.5 m of red fine-grained sandstone weathered in thin parallel beds dipping at 5° to 225°.

Exposures [SS 8944 0002] in Rivenford Lane, about 300 m south-east of Shute, show 1.5 m of reddish brown, very fine-grained, very silty, poorly sorted, weakly cemented, roughly bedded, finely micaceous sandstone, with black-stained fractures; the dip is 12° to 180° Small exposures [SS 8965 0021 to 8965 0015] in Yendacott Lane show reddish brown sand-breccia, consisting of angular clasts of silty, very fine-grained sand in a similar, but less well-cemented matrix.

Yendacott

Shute Sandstone preserved in the core of the West Efford Syncline is exposed beneath up to 1.2 m of valley head in a stream [SS 9039 0085] west of Yendacott Copse. The section shows 1.0 m of reddish brown, weakly cemented, very silty, very fine-grained sandstone. Weathered surfaces show brecciated structure, consisting of angular clasts normally less than 10 mm across of very fine-grained sandstone or siltstone set in a similar, but less well-cemented matrix. An 0.6 m-thick lens of reddish brown breccia, consisting of Culm sandstone and black ?tourmalinised hornfels, in a silty sand matrix, overlies the sand on the east side of the section.

A few breccia lenses have been mapped on the basis of soil brash; a ditch [SS 9051 0011] showed spoil of breccia with clasts less than 50 mm across mainly of Culm sandstone, vein quartz, ?chert and ?hornfels.

Exposures [SS 9161 0064] about 300 m north-north-west of Heathfield show up to 0.2 m of reddish brown, brecciated, poorly sorted, very silty, very fine-grained sandstone, consisting of angular clasts of very weakly cemented sandstone in a similar, but softer matrix.

Yellowford

Excavations [SS 9243 0085 to 9248 0083] at Yellowford showed up to 2 m of very weakly cemented reddish brown, very silty, very fine-grained sandstone, partly brecciated, with platy clasts of micaceous silty sandstone in a similar matrix. River-cliff exposures [SS 9292 0063] about 600 m south-east of Yellowford show up to 3 m of reddish brown, poorly sorted, very fine-grained silty sandstone, weakly cemented, indistinctly bedded, with pods and lenses of coarser-grained sand. A few beds have scattered coarse to very coarse grains; others are siltier and clayier, with concentrations of mica flakes. A few subspherical pale greyish green reduction spots are present. The beds dip at 12° to 206°.

Exposures [SS 9290 0039 to 9290 0024] in the west bank of the River Exe show weakly to well-cemented, poorly sorted, reddish brown, very silty, very fine-grained sandstone. Dips [at SS 9290 0031] are 16° to 171°.

SUMMARY OF THE SEDIMENTARY HISTORY OF THE EXETER GROUP

A thick, coarse, clastic alluvial fan sequence — the Cadbury Breccia — formed on the northern margin of the Crediton Trough, with sedimentation concentrated in the east of the basin. The predominance of Bude Formation clasts, and clasts with Devonian fossils from north Devon, show that sediment supply was from the north. Debris-flow deposits in the Cadbury Breccia indicate sedimentation on the upper or mid-fan area. The lack of sediment from the south suggests that an area of low relief lay south of the Crediton Trough at that time.

As the Crediton Trough evolved and extension continued, sedimentation on the Cadbury Breccia alluvial fan decreased in importance, and coarse-sediment infill switched to the west of the trough with the onset of alluvial-fan sedimentation forming the Bow Breccia. A change in the direction of sediment supply is also indicated, on the basis of marked thickness changes in the Bow Breccia from about 430 m in the west to about 40 m in the east of the trough, and by changes in clast types. The Middle Devonian limestone pebbles in the Bow Breccia indicate a likely source from the west or southwest. The breccia probably prograded axially into the basin from west to east, controlled by topographical lows within the graben. Sedimentation was by density-modified grain flows and debris flows with fluvially reworked tops, representing deposition on the upper or mid-fan part of an alluvial fan.

The Knowle Sandstone contains clasts of similar type to those of the Bow Breccia, and represents the upper, outer and finer-grained part of the Bow Breccia alluvial fan. It shows a west to east thickening from about 18 m to about 150 m. The fining-outwards of breccia to sandstone represents the transition from the proximal to distal part of the Bow Breccia alluvial fan. Sedimentation was characterised by unconfined tractional sheetfloods, typical of sedimentation on a distal fan.

A prolonged phase of active extension and uplift occurred before deposition of the Creedy Park Sandstone, and was probably responsible for the abandonment of the Bow Breccia alluvial fan. The Creedy Park Sandstone was deposited disconformably above the Knowle Sandstone. The overlying Crediton Breccia passes laterally into finer-grained breccias, which in turn pass eastwards into the Yellowford Formation. The rhyolite and quartz-porphyry fragments of the Crediton Breccia have rare earth and isotopic signatures that suggest a Dartmoor source, and the presence of tourmalinised slate clasts also indicates erosion of the granite aureole, although the granite was probably not unroofed at this time. The direction of fill of the basin therefore changed from west to east, to southwest to north-east. The Crediton Breccia probably represents the proximal part of the alluvial fan, with the distal part represented by the Yellowford Formation. Crediton Breccia sedimentation was dominated by low- and high-viscosity debris flows in the west of the district and by tractional sheetfloods in the eastern part of the outcrop. The decrease in clast size coupled with a change from disorganised debris-flow sedimentation to more organised fluvial sedimentation is characteristic of proximal to distal variations on an alluvial fan.

At the same time as the Crediton Breccia fan was extending axially along the Crediton Trough, the Alphington Breccia fan, which has a similar clast content and thus probably the same source area, extended over much of the remainder of the southern part of the district. It is not clear whether the Newton St Cyres Breccia and the Heavitree Breccia, which have similar clast contents, are the product of deposition from a single alluvial fan. The presence in the area north of Exeter of Dawlish Sandstone resting directly on Crackington Formation suggests that there was a fault-bounded block in this area against which the Alphington Breccia and Heavitree Breccia were banked, and which was not overtopped until Dawlish Sandstone times. It is uncertain to what extent this topographical high persisted westwards and served as a divide between the Late Permian alluvial fans.

The Newton St Cyres Breccia and Heavitree Breccia show greater organisation than the older breccias, with deposits of fluvial channels dominated by transverse bars and small dunes, and also the deposits of periodic sheetfloods. Although channels are common features on upper-fan areas, the degree of complexity in the Newton St Cyres Breccia channels is atypical of such an environment. The type of sedimentation represented here is more likely to indicate deposition on the lower parts of an alluvial fan, transitional to a proximal braidplain. The common development of dewatering pipes within the formation may indicate seismic activity along faults.

The Newton St Cyres Breccia shows a distal thinning and fining into the Shute Sandstone, and the presence of unconfined sheetflood deposits within the Shute Sandstone is indicative of deposition in a more distal setting on the fan.

Active extension is likely to have ceased towards the end of the Permian, and was marked by a distinct change in sedimentary facies within the basin and a change in the source direction. This change is marked by the Dawlish Sandstone, an aeolian deposit formed by the migration of small dunes into the east of the basin under the influence of a northerly directed wind. Coarse-grained interdune deposits of fluvial origin are similar in composition to the Newton St Cyres Breccia, although a source from the east is suggested. However, some channelling along interdune lows may have obscured the original source direction.

SIX

Triassic rocks: Aylesbeare Mudstone Group and Sherwood Sandstone Group

AYLESBEARE MUDSTONE GROUP

An area of about 100 km² in the district is underlain by the Aylesbeare Mudstone Group, a sequence of up to 400 m of reddish brown, weakly calcareous siltstone and silty claystone, with lesser proportions of sandy siltstone, sandstone, claystone and silty sandstone.

The outcrop of the Aylesbeare Mudstone is 5 to 6 km wide between Woodbury and Whimple, narrowing to about 3 to 4 km wide north of Whimple (Figure 22). It gives rise to gently undulating lowland, mainly 15 m to 75 m above OD, but rising eastwards to over 120 m above OD near the scarp of the overlying Budleigh Salterton Pebble Beds. The Aylesbeare Mudstone outcrop is mantled in many places by a thin spread of rounded pebbles and cobbles derived from the Budleigh Salterton Pebble Beds. Exposures are generally poor, the claystones and siltstones being readily degraded to clay and silt by weathering; natural exposures occur mainly in the banks and beds of streams beneath alluvium or head. Lenticular beds of sandstone are more resistant to weathering, and steep-sided topographical features are commonly related to a sandstone bed or beds. The sandstones have been quarried locally.

The sequence was called Lower Marls by Ussher (1902), and renamed Aylesbeare Group by Smith et al. (1974). The evolution of the nomenclature is summarised in Table 7 (p.30). The type area was named by Smith et al. (1974) as the village of Aylesbeare [SY 040 920], but inland exposures are few; the best section is the sea-cliff between Exmouth [SY 003 805] and Budleigh Salterton [SY 059 815], south of the present district.

The Aylesbeare Mudstone of the adjacent Newton Abbot district was divided by Selwood et al. (1984) into two formations corresponding to the 'marls without sandstones upon marls with sandstones' of Ussher (1902, 1913), and this subdivision has been extended into the Exeter district as far north as Aylesbeare (Figure 22). The subdivision depends on the presence of a lenticular but generally persistent sandstone (correlated with the Straight Point Sandstone of the coastal section (Selwood et al., 1984)), the top of which defines the boundary between the two formations — Exmouth Mudstone and Sandstone below and Littleham Mudstone above. North of Aylesbeare, although sandstones are present in the Aylesbeare Mudstone (for example south of Whimple, Figure 22), they are fewer, thinner, and more impersistent than those south of Aylesbeare and it is uncertain whether they occur at the same horizon as the Straight Point Sandstone. For this reason, the subdivision of the Aylesbeare Mudstone has not been extended north of Aylesbeare.

North of Westwood [SY 020 990], a basal unit of sandstone and siltstone, the Clyst St Lawrence Formation, is recognised (Edwards, 1988). Between the southern boundary of the district, near Exton [SX 980 860], and as far north as Burrow [SX 999 980], near Broadclyst, the base of the Aylesbeare Mudstone is taken at a sharp lithological change, to mudstone, from the sandstone of the underlying Dawlish Sandstone. The contact is probably unconformable. The junction is nowhere well exposed, but can readily be traced inland by features and augering. Locally, near Topsham, a lens of breccia is present at the junction, and farther south, in the Exmouth area, the Exe Breccia intervenes between the Dawlish Sandstone and the Exmouth Mudstone and Sandstone (Selwood et al., 1984). North of the southern boundary fault of the Crediton Trough near Westwood, the Clyst St Lawrence Formation rests unconformably on the Crackington Formation.

There is no direct evidence for the age of the Aylesbeare Mudstone Group. The Whipton Formation at Exeter is Late Permian in age (Warrington and Scrivener, 1988, 1990), as, probably, is the Dawlish Sandstone, while the Otter Sandstone is Mid Triassic. The Permian–Triassic boundary has therefore been taken at the base of the Aylesbeare Mudstone Group.

Lithology and sedimentary features

Siltstones and claystones, the predominant lithologies of the Aylesbeare Mudstone, are generally massive and blocky-weathering. They are normally moderate reddish brown in colour, but commonly contain pale greenish grey spherical spots, irregular patches, and bands (p.79). The reddish brown colour of the mudstones is due to ferric oxide, usually haematite, which occurs as a coating to grains and as finely divided material in the clay-mineral matrix. The mudstones consist predominantly of silt- and very fine sand-sized grains of angular to subangular quartz (with a few feldspar grains) set in a haematite-stained clay matrix. Two mudstone samples contained 3.9 per cent and 12 per cent calcium carbonate. The clay mineralogy of the Aylesbeare Mudstone of the coastal sequence between Exmouth and Budleigh Salterton is dominated by illite, with subordinate amounts of mixed-layer illite, kaolinite, chlorite, and swelling chlorite (Henson, 1973).

Sandstones in the Aylesbeare Mudstone are generally reddish brown, medium to fine grained, and weakly cemented where weathered at outcrop. The sandstones are composed predominantly of subangular to subrounded quartz and metaquartzite grains, with an estimated 15 to 25 per cent of the total sample weight consisting of feldspar; they are subarkoses and arkoses in the classification of Folk (1968). Grain-size analysis of 20 samples of sandstone from the Exmouth Mudstone and Sandstone and from the undivided Aylesbeare Group shows that most are fine to medium grained. The size

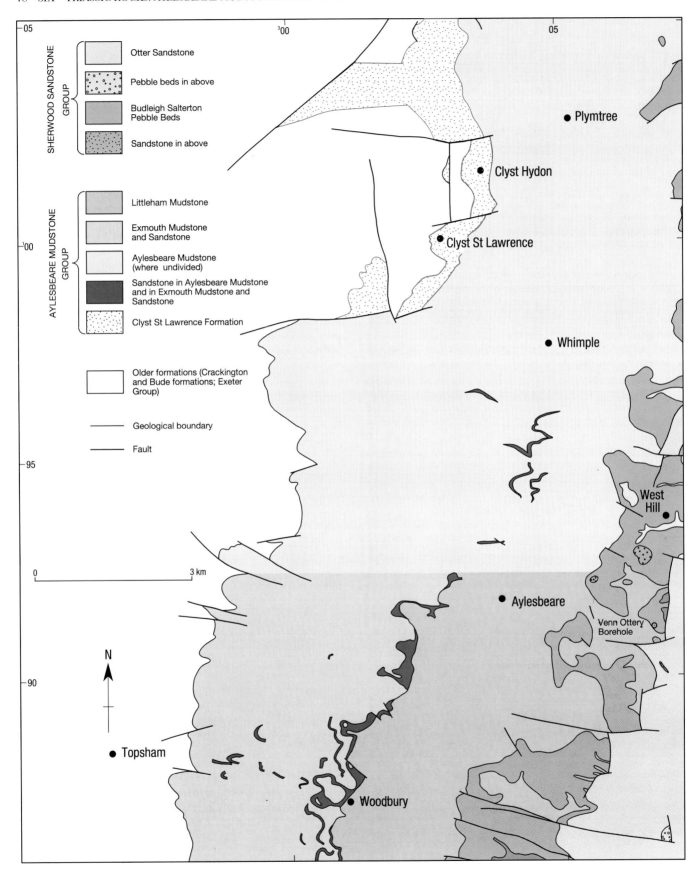

Figure 22   Distribution of the Aylesbeare Mudstone Group and the Sherwood Sandstone Group in the district.

range is between 1.60 and 2.68 ø, and the average value is 2.2 ø (fine-grained). Most samples are moderately well to moderately sorted, the range being from 0.30 to 1.14 ø (very well to poorly sorted).

DEPOSITIONAL ENVIRONMENTS

Henson (1970, 1971) considered that the Aylesbeare Mudstone Group formed on a river floodplain, and recognised the deposits of fluvial channels, crevasse splays and levees. However, the results of studies carried out during the resurvey of the district (Jones, 1992 b) indicate that the sequence was deposited in a lake that periodically dried out and became evaporitic; this type of lake is termed a sabkha-playa. The Aylesbeare Mudstone consists of thick sequences of mudstone deposited largely from suspension. Emergent conditions are indicated by the common occurrence of desiccation-related features, and at times the environment became evaporitic. A floodplain origin is unlikely since channels are absent, and it is improbable that about 400 m of mudstone would be deposited on a floodplain devoid of channels.

The thicker, fine- to medium-grained sandstones of the sequence, including the Straight Point Sandstone, interpreted as fluviatile by Henson (1971) and Selwood et al. (1984), are now considered to be largely aeolian in origin, formed by the migration of small dunes across a dry playa surface. The playa mudstones commonly contain scattered grains of well-rounded coarse-grained sand which are of probable wind-blown origin.

Minor coarsening-upwards sequences of very fine- to fine-grained sand with interlaminated silt are the deposits of fluvial sheetfloods. The source of sandstone lenses in the widespread mudstone with sand patch facies was probably also fluvial sheetfloods, with minor amounts of wind-blown sediment. Thick successions of mudstones with sand patches are likely to form only when the level of the water table keeps pace with sedimentation; this suggests the existence of fairly permanent sabkha conditions (Glennie, 1972).

The sandstones at outcrop are commonly cross-bedded, but exposures are too few for many measurements of cross-bedding to have been made, or for the vertical sequence of sedimentary structures to be established. At Windmill Hill [SY 0162 9009], Higher Greendale, foresets dip to the east-south-east and south-south-east, giving a suggestion of transport from the north-west.

Owing to the poor exposure, most of the sedimentary features of the Aylesbeare Mudstone described below are based on examination of cores from the Venn Ottery Borehole [SY 0659 9111], in which the upper 306 m of the group were proved (Jones, 1992 b). A graphic lithological log is given in Figure 23. The mudstones belong to a playa mudflat facies association (Jones, 1992 b), which comprises i) ephemeral lacustrine mudstones ii) desiccated playa mudstones iii) gypsiferous mudstones, and iv) mudstones with sand patches. The ephemeral lacustrine mudstone facies consists of reddish brown, interbedded claystone, silty claystone and siltstone, with rare sandstone, in thin sequences generally 2 m thick but up to 6 m; the facies is typically interbedded with other playa facies, in particular desiccated mudstones. The desiccated playa-mudstone facies comprises combinations of claystones, siltstones and rare sandstones characterised by a disrupted brecciated texture composed of clasts of broken mudstone, typically associated with desiccation cracks. Also present are mudstones composed of 2 mm-wide flattened cuboids, commonly surrounded by a compactionally induced listric surface. Gypsiferous mudstones are common in the Venn Ottery Borehole. The gypsum occurs as greyish white, irregularly shaped patches, and as thin veins of satin spar. The mudstones with sand patches facies is characterised by claystone and siltstone with common fine- to coarse-grained, irregularly shaped, sandstone wisps.

Very fine- to fine-grained sandstone with minor interlaminated siltstone, in coarsening-upward sequences, is interpreted as a fluvial facies association (Jones, 1992 b). Sedimentary structures include parallel lamination, unidirectional ripple cross-lamination, climbing ripples and small scours. In parts of the sequence dominated by subaqueous deposits, sandstone occurs in undisturbed interlaminated sequences of silts and sands with well-defined ripple cross-lamination. Desiccation cracks are locally present.

An aeolian facies association in the Venn Ottery Borehole generally forms thinly bedded units of sandstone, up to 1.0 m thick, in sequences up to 3.5 m thick (Figure 23). Within this association three main types can be recognised: i) small aeolian dunes ii) wind-rippled sands, and iii) scattered aeolian grains (Jones, 1992 b).

NODULES AND REDUCTION FEATURES

Reduction features in the Aylesbeare Mudstone are of four main types: (a) bands and discontinuous beds and zones; (b) green spherical spots with no apparent nucleus; (c) grey and black nodules, and smaller varieties ('fisheyes') that are surrounded by a halo of pale green reduced material; and (d) diffuse patches of reduced material (Bateson and Johnson, 1992).

The reduction features are locally associated with dark grey radioactive nodules that contain high proportions of metallic elements including vanadium, uranium, copper and nickel (Carter, 1931; Perutz, 1939; Harrison, 1975; Durrance and George, 1976; Durrance et al., 1980). The nodules occur mainly within the Littleham Mudstone, and are apparently most abundant near the base of the formation; the best-known locality is at Littleham Cove [SY 040 803], south of the present district. Tandy (1973, 1974) demonstrated the widespread occurrence of dissolved uranium salts over the Aylesbeare Mudstone outcrop, and showed that the nodules were more widespread than previously known, extending at least 16 km north of the coast; he mapped the distribution of radioactive reduction features between Littleham Cove and Aylesbeare.

The chemistry of 13 nodules was examined by Nancarrow (1985). Type 1 nodules, all from the Littleham Cove locality, show marked enrichment in U, As, Se, Cu and Co relative to Type 2 nodules, which occur at diverse localities along the Littleham Mudstone outcrop. Detailed chemical data for nodules are given by Nancarrow (1985), and by Bateson and Johnson (1992).

a. 3.25 m to 108.00 m

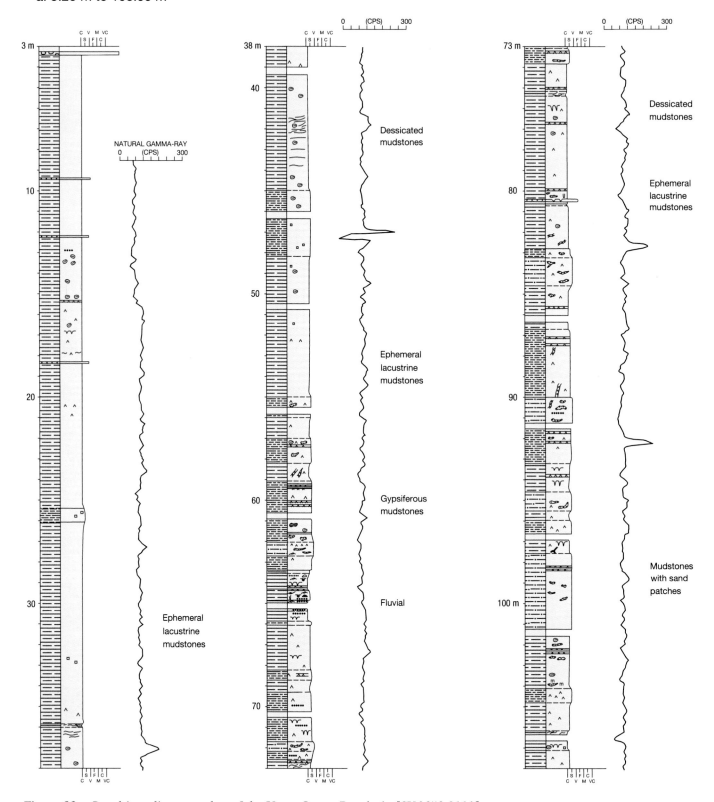

Figure 23   Graphic sedimentary log of the Venn Ottery Borehole [SY 0659 9111].

b. 108.00 m to 213.00 m

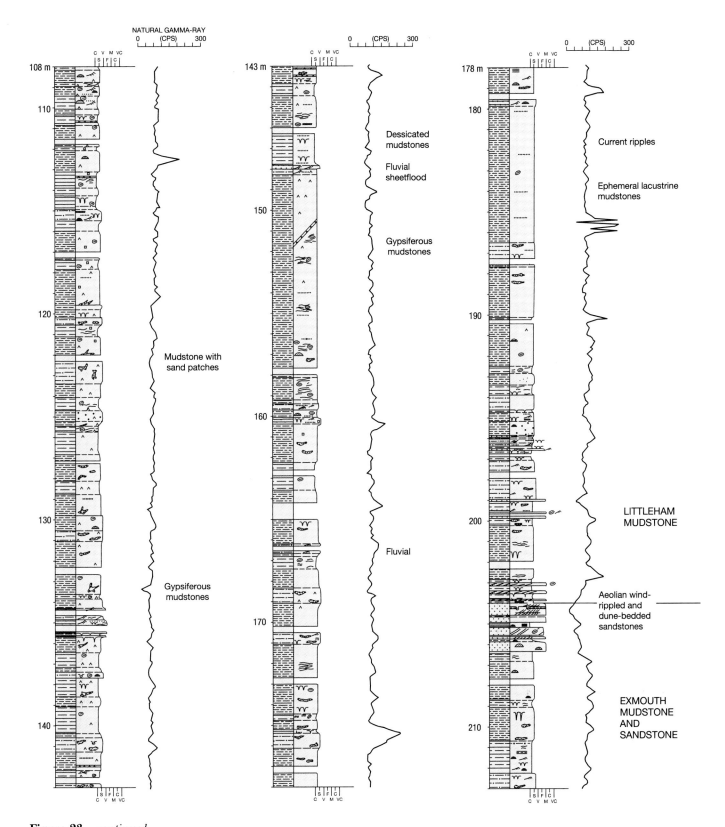

Figure 23   *continued*

c. 213.00 m to 309.70 m

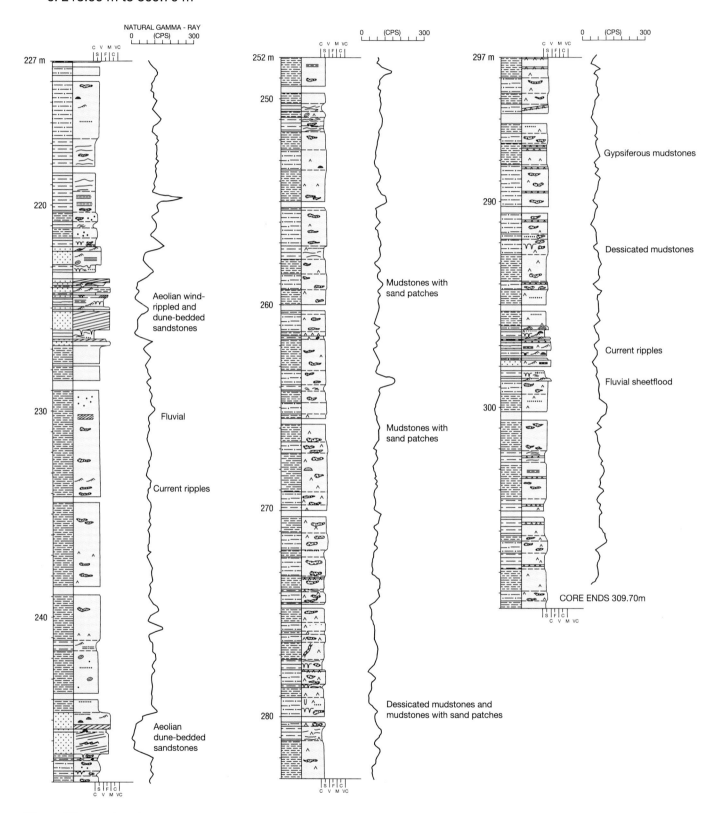

Figure 23   *continued*

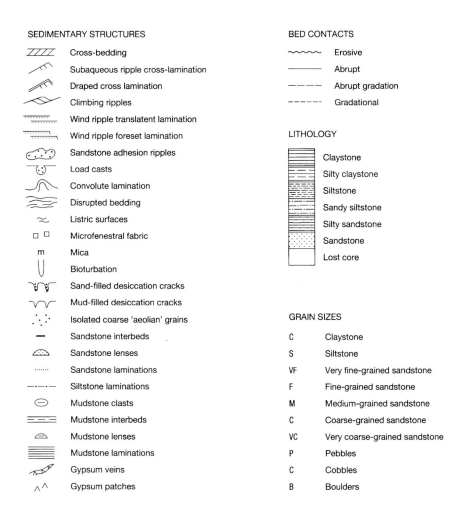

SEDIMENTARY STRUCTURES

/////	Cross-bedding
	Subaqueous ripple cross-lamination
	Draped cross lamination
	Climbing ripples
	Wind ripple translatent lamination
	Wind ripple foreset lamination
	Sandstone adhesion ripples
	Load casts
	Convolute lamination
	Disrupted bedding
	Listric surfaces
□ □	Microfenestral fabric
m	Mica
	Bioturbation
	Sand-filled desiccation cracks
	Mud-filled desiccation cracks
	Isolated coarse 'aeolian' grains
—	Sandstone interbeds
	Sandstone lenses
....	Sandstone laminations
–·–·–	Siltstone laminations
	Mudstone clasts
===	Mudstone interbeds
	Mudstone lenses
	Mudstone laminations
	Gypsum veins
^ ^	Gypsum patches

BED CONTACTS

~~	Erosive
——	Abrupt
– – –	Abrupt gradation
- - - -	Gradational

LITHOLOGY

	Claystone
	Silty claystone
	Siltstone
	Sandy siltstone
	Silty sandstone
	Sandstone
	Lost core

GRAIN SIZES

C	Claystone
S	Siltstone
VF	Very fine-grained sandstone
F	Fine-grained sandstone
M	Medium-grained sandstone
C	Coarse-grained sandstone
VC	Very coarse-grained sandstone
P	Pebbles
C	Cobbles
B	Boulders

Harrison (1975) considered that the nodules originated by precipitation of mineralised solutions at permeable/impermeable interfaces, the ultimate source of the solutions being hot springs associated with waning phases of the metalliferous mineralisation of south-west England. However, Durrance and George (1976) suggested that precipitation took place around fragments of contemporaneous plant material; Bateson and Johnson (1992) disagreed in view of the lack of plant material in the Littleham Mudstone. Hofmann (1991) concluded that the reduction spheroids developed as a result of bacterial activity, using dissolved reductants such as methane or low molecular weight organic acid anions, or inorganic compounds such as ammonia or hydrogen. Durrance et al. (1978) proposed that the 'reduction' spots of the Aylesbeare Mudstone were caused by the inhibition of oxidation in sediments buried beneath about 1000 m of overburden.

GEOCHEMISTRY OF MUDSTONES

Chemical analyses of 200 samples from the Aylesbeare Mudstone cored in the Venn Ottery, Blackhill, and Withycombe Raleigh Common boreholes are given by Bateson and Johnson (1992, table 1). Many of the elements have values close to those of 'average' shale, except that Mn and Al are somewhat reduced in the Venn Ottery Borehole samples, and Cu and Sr are somewhat enriched in the same samples. Samples from the three boreholes contain on average 146 parts per million (ppm) V, close to the mean value of 120–130 ppm for shales (Levinson, 1974). There is no enrichment in uranium in the host-rock samples; 72 per cent of the samples analysed gave values less than 5 ppm (the detection limit of the analytical technique used), similar to the average shale value of 4 ppm (Levinson, 1974).

Southern area

South of Aylesbeare, the Aylesbeare Mudstone is divided into roughly equal thicknesses of Exmouth Mudstone and Sandstone, below, and Littleham Mudstone, above.

EXMOUTH MUDSTONE AND SANDSTONE

The Exmouth Mudstone and Sandstone consists of reddish brown, weakly calcareous siltstone, sandy siltstone and silty claystone, with lesser proportions of sandstone, claystone, and silty sandstone. The sandstone beds are lenticular, and locally up to 30 m thick. The outcrop of the formation lies west of a line between Farringdon, Woodbury Salterton and Woodbury; it is not mapped north of a line extending westwards from Rill Farm [SY 030 923], Aylesbeare (Figure 22).

The thickness in the district is estimated to be about 200 m. The thickness in the coastal exposure was estimated by Selwood et al. (1984) to be 255 m. Analysis (from data in Jones, 1992 b) of the proportions of different lithologies within the topmost 105 m of the formation cored in the Venn Ottery Borehole [SY 0659 9111] indicates that they are present in the following percentages: claystone 2 per cent, silty claystone 15 per cent, siltstone 50 per cent, sandy siltstone 25 per cent, silty sandstone 1 per cent, and sandstone 7 per cent.

The Blackhill Borehole [SY 0319 8545], sited about 300 m south of the district, proved the topmost 56 m of the formation to consist of 62 per cent sandstone; the Withycombe Raleigh Common Borehole [SY 0330 8407], 1.4 km south of the Blackhill Borehole, proved the cored part of the formation to be 60 per cent sandstone. These sandstone beds probably correlate with the Straight Point Sandstone of the coastal exposures (Selwood et al., 1984, fig. 17). When traced inland, the Straight Point Sandstone is lenticular and impersistent, and the topmost mapped sandstone of the formation in the district may not everywhere be at the same stratigraphical level. A moderately persistent sandstone has been traced from near the southern margin of the district to near Aylesbeare (Figure 22). North of there, the sandstone beds are fewer and more impersistent, and cannot be satisfactorily correlated.

Beds of gypsiferous mudstone are present in the Venn Ottery Borehole; the gypsum occurs as greyish white irregularly shaped patches and as thin veins of satin spar. Gypsum was not recorded at outcrop, nor in the Blackhill and Withycombe Raleigh Common boreholes and the coastal section.

Perkin's Village–Farringdon

In Perkin's Village (near Aylesbeare), the junction with the overlying Littleham Mudstone is exposed in a stream section [SY 0262 9132] south of the village. Other stream-bed exposures [SY 0250 9140] west of the village show up to 1.5 m of reddish brown, flaggy, poorly sorted, fine- to medium-grained sandstone. Exposures [SY 0238 9189] in the stream north of Perkins's Village are of reddish brown mudstone and clayey siltstone with bands of calcareous, greenish grey, silty, fine- to very fine-grained sandstone.

A sandpit [SY 021 905] south of Upham Farm, Farringdon, contains small exposures of reddish brown, moderately well-sorted, medium-grained sand.

An exposure [SY 0098 9253] beneath alluvium in the stream north of Farringdon Wood shows about 0.2 m of reddish brown mudstone with patches and streaks of greyish green, sandy silt.

Woodbury Salterton

South of the A3052 road, in the Higher Greendale area north of Woodbury Salterton, the sandstone at the top of the Exmouth Mudstone and Sandstone thickens to a maximum of about 30 m, and bifurcates to form two scarp features, those of Windmill Hill [SY 0162 9009] and the unnamed hill [SY 0187 9013] 250 m farther east. Exposures [SY 0166 9001] on the south side of Windmill Hill show about 1 m of cross-bedded, reddish brown, moderately sorted, medium-grained sandstone with cross-bedding dips of 26° to 100°. Cross-bedding in a nearby sandstone exposure [SY 0180 9004] dips at 13° to 148°.

South-west of Woodbury Salterton, a weakly cemented, moderately sorted, fine-grained sandstone bed about 5 m thick, lower in the sequence than the Straight Point Sandstone, is exposed [SY 0077 8867].

Woodbury

North-west of Woodbury, a borehole [SY 0082 8748] penetrated about 56 m of the Exmouth Mudstone and Sandstone, including 7.31 m of Straight Point Sandstone at the top. The driller's record is as follows:

	Thickness m	Depth m
Soil and drift	4.8	4.88
EXMOUTH MUDSTONE AND SANDSTONE		
Sandstone	7.31	12.19
Sandstone and red clay beds	9.15	21.34
Clay, red	12.19	33.53
Sandstone	0.61	34.14
Clay, red	0.91	35.05
Clay, red, and beds of sandstone	21.64	56.69
Clay, red and grey	4.27	60.96

LITTLEHAM MUDSTONE

The Littleham Mudstone consists of reddish brown, weakly calcareous silty claystone and siltstone, with minor proportions of sandy siltstone, claystone, silty sandstone and sandstone. Analysis of the lithologies in the formation in the Venn Ottery Borehole (Jones, 1992 b) showed silty claystone 46 per cent, siltstone 40 per cent, sandy siltstone 10 per cent, claystone 2 per cent, silty sandstone 1 per cent and sandstone 1 per cent. The base of the Littleham Mudstone is taken at the top of the uppermost thick sandstone of the Exmouth Mudstone and Sandstone. The formation was fully cored in the Venn Ottery Borehole [SY 0659 9111], where it was 201 m thick; in the Blackhill Borehole it was 223 m thick; in the Withycombe Raleigh Common Borehole it was 215 m thick; and in the coastal section, 275 m thick (Selwood et al., 1984). Gypsiferous mudstones are present throughout the Littleham Mudstone of the Venn Ottery Borehole, occurring as greyish white irregularly shaped patches and as thin veins of satin spar. No gypsum was recorded in the formation in the Blackhill Borehole, or in the coastal section.

Aylesbeare–Higher Greendale

The stream that flows through Aylesbeare contains small exposures [e.g. SY 0430 9181; 0410 9181; 0371 9186; 0348 9184] of up to 2 m of reddish brown silty clay and silty claystone. One exposure [SY 0348 9184] shows 1.5 m of reddish brown silty claystone with 10 cm-thick bands of greyish green clayey siltstone.

A stream-bed exposure [SY 0295 9158] near the base of the formation shows 2.5 m of reddish brown siltstone with irregular bands of pale greyish green, clayey silt and silty, very fine-grained sand. One 0.15 m-thick bed of pale greyish green sandstone forms a ledge protruding from the mudstone outcrop. About 300 m east-south-east, exposures [SY 0331 9154 to 0335 9154] of mudstone contain black to dark grey nodules; two of the nodules, 30 and 50 mm in diameter, occur at the centre of reduction spots that are 80 and 90 mm in diameter respectively.

Exposures in the Littleham Mudstone south of Aylesbeare occur beneath head or alluvium in most stream beds. The stream east of Withen Copse [SY 0282 9098 to 0287 9086] contains exposures of blocky-fracturing reddish brown mudstone with abundant greyish green spots. The upper reaches of the Grindle Brook expose [SY 0361 9027] reddish brown mudstone with a 20 to 30 mm-thick band of very clayey, silty, fine-grained sandstone. A nearby exposure [SY 0405 9029], at a higher stratigraphical level, shows reddish brown mudstone with green spots and impersistent greyish green clayey silt bands. Exposures [SY 0251 8977] in the bank and bed of the Grindle Brook show about 0.4 m of fissured, blocky-fracturing, reddish brown mudstone, soft in the top 0.1 to 0.2 m. A sample from this locality contains 7.4 per cent by weight calcium carbonate. A nearby stream section [SY 0268 8955] exposes up to 3.0 m of reddish brown silty clay, soft in the top 2.0 m, blocky-fracturing at the base, with a few thin bands (up to 10 mm) of friable, greyish green, clayey, silty fine-grained sand and weakly cemented sandstone.

A stream section [SY 0333 8978] shows 0.7 m of soft reddish brown, extremely clayey silt and very silty clay on 0.3 m of fairly hard, blocky-fracturing, extremely clayey siltstone. Upstream [SY 0410 8945], about 0.3 m of fairly hard, reddish brown, very clayey siltstone with abundant greyish green spots and bands is exposed. A sample from this locality has a calcium carbonate content of 12.6 per cent by weight.

Venn Ottery Borehole

In the borehole [SY 0659 9111] near Venn Ottery Common, Aylesbeare Mudstone Group was cored between 3.60 and 309.7 m depth. A graphical log of the sequence is given in Figure 23, and a description of the lithologies and sedimentary features is given on p.79.

Woodbury Salterton–Woodbury

Excavations [SY 0305 8879] for a lake near Lyndhane showed up to 3.0 m of moderately hard, blocky-fracturing, reddish brown mudstone, with scattered greyish green reduction spots and patches up to a few centimetres across. Reduction spots were abundant in the lowest part of the section, with a 10 mm-thick band of greyish green, clayey silt. Indistinct traces of lamination were locally present in the mudstone.

Northern area

North of Aylesbeare, the group is undivided, except where the basal Clyst St Lawrence Formation (Figure 22) is distinguished.

CLYST ST LAWRENCE FORMATION

The outcrop of the Clyst St Lawrence Formation extends from Westwood [SY 020 990] to the district boundary at Mutterton [ST 030 050] (Figure 22). In the Exeter district, the formation rests unconformably on the Crackington Formation; just north of the district, it rests unconformably on Cadbury Breccia and Bude Formation. West of Langford, the outcrop is fault-bounded and extends westwards to the valley of the River Culm. The upper boundary is transitional to the overlying undivided part of the Aylesbeare Group, and thus there is, at some localities, difficulty in the classification of strata near the boundary.

The formation consists of 30 to 50 m of reddish brown silt and sand, ranging from sand, through silty sand, silty clayey sand, sandy silt, clayey silt, and sandy clayey silt. Grain-size analysis of 24 samples shows that most (17) have mean sizes in the range 4.10 to 7.57 ø (silt); the range of all samples is 1.90 to 7.57 ø (medium-grained sand to very

fine silt); the average mean size value for the 24 samples analysed is 4.9 ø (coarse silt). Most samples are poorly to very poorly sorted, the range being 0.79 to 2.48 ø (average of 24 samples is 1.8 ø). Skewness ranges from strongly negatively skewed to strongly positively skewed, with over half (14) having positively to strongly positively skewed distributions, reflecting the presence of substantial 'tails' of fine-grained material; however, eight samples have near-symmetrical distributions.

Mutterton–M5 Motorway

At Middle Bolealler, Mutterton, a section [ST 0270 0428] shows up to 2 m of reddish brown very poorly sorted clayey silt.

Boreholes [ST 0146 0406; 0129 0380; 0111 0354] drilled for the M5 Motorway penetrated up to 1.8 m of reddish brown, slightly sandy, clayey silt beneath up to 4 m of drift. Farther south-west, a borehole [ST 0040 0285] near Kensham Lodge proved, beneath 4.42 m of made ground, 3.20 m of reddish brown, slightly clayey, fine- and medium-grained sand, with some fine gravel at 6.5 m, resting on 3.28 m of reddish brown, very weakly cemented, clayey, fine-grained sandstone.

Langford

At Tye Farm, a section in a lane [ST 0305 0362], near the top of the formation, shows 1.5 m of reddish brown, extremely silty, sandy, brecciated clay (locally extremely clayey, sandy silt), with some pale greyish green streaks, on 0.3 m of reddish brown, very poorly sorted, clayey silt.

Near Langford Court, a stream section [ST 0277 0240] shows about 1 m of head on 0.4 m of reddish brown, poorly sorted, clayey, very fine-grained sand; cross-bedding forests dip at 16° to 064°. A nearby stream section [ST 0299 0224] shows about 1 m of reddish brown, very poorly sorted, sandy, clayey silt, locally with patches and thin beds of clayey, extremely sandy silt and silty fine-grained sand.

Clyst Hydon–Clyst St Lawrence

A faulted outlier of Clyst St Lawrence Formation is present about 0.6 km west of Clyst Hydon. A sample [ST 0293 0155] from a small ditch exposure near One Fir is reddish brown, very poorly sorted, silty, clayey sand.

The base of the formation was exposed in a small outlier [ST 0257 0002] west of Clyst St Lawrence, which showed 0.15 m of reddish brown very fine breccia, on 0.30 m of reddish brown poorly sorted, fine- to medium-grained sand, on 0.40 m of reddish brown, clayey, fine-grained sand, on 1.2 m of purple silty shale (Crackington Formation).

Excavation near Town Farm, Clyst St Lawrence, revealed [ST 0281 0002] reddish brown, poorly sorted, fine- to very fine-grained sand in the base of a bank, with clayier sand above. Nearby [ST 0283 0003] are small exposures of reddish brown, poorly sorted, clayey silt. On the south side of the lane, excavation [ST 0285 0004] revealed reddish brown, very poorly sorted, silty, clayey sand. Farther east, excavations [ST 0292 0007; 0299 0012] showed reddish brown, very poorly sorted, sandy, clayey silt.

Westwood

About 600 m east-north-east of Ashclyst Farm, 0.5 m of red-stained fine- to medium-grained sand is exposed in the river bank [SY 0164 9838] beneath 1 m of alluvial gravel. About 0.5 m of fine-grained sand is exposed in the river bank [SY 0273 9903] beneath gravel east of Westwood.

A pit [SY 0144 9915] 150 m north-west of Channons Farm exposes 1.2 m of thinly bedded, moderately well-sorted, clayey, fine-grained sandstone beneath 0.9 m of rounded pebble gravel.

AYLESBEARE MUDSTONE GROUP (WHERE UNDIVIDED)

Between Aylesbeare and Westwood, the Aylesbeare Mudstone contains a few scattered laterally impersistent sandstones, and between Clyst Honiton and Burrow [SX 993 976], near Broadclyst, it rests with sharp lithological contrast on the Dawlish Sandstone. North of Westwood, it rests on the Clyst St Lawrence Formation, the boundary between the two being gradational. A few impersistent beds of fine- to medium-grained sandstone are present between Whimple and Marsh Green (Figure 22), and form small features. Other topographical features may be formed by sandy siltstone, possibly in combination with thin beds of sandstone.

The Aylesbeare Mudstone Group is about 400 m thick in the Whimple area. North of the bounding fault of the Crediton Trough, the outcrop of the group narrows from about 6 km south of the fault to about 4 km north of the fault. In this northern outcrop (Talaton–Plymtree area), the thickness of the Aylesbeare Mudstone (including the Clyst St Lawrence Formation) is estimated to be 250–350 m. This apparent thickness change is coincident with the incoming of the Clyst St Lawrence Formation north of the fault, and may be related to fault movements which caused the Aylesbeare Mudstone to thin against an upstanding fault-controlled block of folded Carboniferous rocks extending eastwards from the Ashclyst Forest area.

Plymtree

A sample from a gully [ST 0660 0340] near Clyst William consists of reddish brown clayey siltstone with small green reduction spots.

Grain-size analyses of samples [from ST 0515 0295; 0516 0291; 0517 0288; 0529 0284] in and around Plymtree indicate that the first three are clayey silt, and the fourth is silt.

Moderately well-defined features [ST 052 019 to 059 024] east of Peradon Farm may be partly related to beds of siltstone. A sample from a disused pit [ST 0550 0198] is clayey silt.

Talaton–Larkbeare

A sample from a pit [SY 0628 9835] south of Newtown is reddish brown clayey silt. Pits farther east along the same ridge contain small exposures [e.g. SY 0659 9835] of reddish brown silty clay and siltstone with small green spots.

A 2 to 3 m-deep ditch [SY 0655 9781] near Larkbeare shows reddish brown silt and siltstone with a few small green reduction spots.

Several boreholes and trial pits penetrated the Aylesbeare Mudstone between Whimple Wood and Larkbeare Avenue [SY 0502 9616 to 0735 9726]. One borehole [SY 0658 9704] proved the following sequence:

	Thickness m	Depth m
Soil, peaty	0.25	0.25
DRIFT		
Clay, soft, grey, sandy, with gravel	2.00	1.75

AYLESBEARE MUDSTONE

Clay, reddish brown, very stiff, silty; gradational base	8.00	6.00
Clay, reddish brown, hard, friable, silty, with pockets of greyish green silty clay; gradational base	14.55	6.55
Mudstone, reddish brown, weakly cemented, with pockets of greyish green silty clay	18.50	3.95

Whimple

A 304 m-deep borehole [SY 0453 9745] at Whimple is recorded by the driller as follows:

	Thickness m	Depth m
Clay, red	1.52	1.52
Marl, brown, red	2.14	3.66
Marl, red, hard	8.63	12.29
Marlrock, red, hard	235.21	247.50
Marl, hard, and sandstone	4.26	251.76
Marlrock, red, hard	2.75	254.51
Marl, hard, and sandstone	28.95	283.46
Shellit, hard	9.15	292.61
Breccia, hard	11.73	304.34

The beds down to 254.51 m can be classified with reasonable certainty as Aylesbeare Mudstone; it is uncertain whether the Dawlish Sandstone is present, although it might be represented in the 28.95 m of beds recorded as 'hard marl and sandstone', present to a depth of 283.46 m. Samples of core from 298.86 m are hard greyish red siltstone, locally finely micaceous, with some plant material on bedding surfaces; bedding dips at 22° to the core axis. These samples indicate that the beds recorded by the driller as 'shellit' and 'breccia' below 283.46 m are probably Crackington Formation.

A road bank [SY 0382 9704] west of Whimple exposes 2.5 m of red-stained, weakly calcareous mudstone with impersistent thin (1 to 2 cm) medium-grained, locally greenish grey, sands in the upper part. In the lower part, a 0.3 m-thick, dominantly fine-grained, moderately well-sorted sand is present.

Rockbeare–Strete Ralegh

An exposure [SY 0410 9527] at Strete Farm revealed 0.5 m of gravel above 1.5 m of red, very well-sorted, fine-grained sand. The most westerly occurrence of the sand is in a borehole [SY 0386 9524] 300 m west of the farm, where 0.7 m of dense, greyish green, clayey, silty, fine-grained sand was proved between depths of 5.2 and 5.9 m. At Strete Ralegh Farm, medium-grained sand augered [SY 0465 9554] beneath red silty clay may be the same bed as that at Strete Farm. Similarly, 1.55 m of dense, reddish brown, silty sand and 1.95 m of red sand encountered in two boreholes [SY 0454 9554 and 0490 9599] at Hand and Pen may also be at the same stratigraphical level.

In a deeply incised stream course [SY 0527 9610] near Northcotts Farm, beneath about 2 m of head, up to 3 m of reddish brown clay with angular fragments of mudstone overlie 0.2 m of reddish brown, sandy, clayey siltstone with green reduction spots.

A ditch [SY 0546 9528] between Brickyard Copse and Bob's Close Copse shows reddish brown and greyish green, very silty to extremely silty clay, and clayey silt, with a thin bed of yellowish brown and greyish orange, poorly sorted, silty, fine- to very fine-grained sand.

Rockbeare–Marsh Green–Aylesbeare

A borehole [SY 0086 9467] north of Treasbeare Farm, Rockbeare, penetrated (beneath 3.55 m of older head) very stiff to hard, reddish brown, silty clay to 12.00 m depth, on reddish brown mudstone with a few beds up to 0.55 m thick of weakly cemented, greyish green, silty, fine-grained sandstone, penetrated to 19.60 m depth.

A borehole [SY 0020 9305], south of Exeter Airport, penetrated Aylesbeare Mudstone (recorded as alternations of 'red rock' and 'red marl') to 47.25 m depth, on Dawlish Sandstone.

A borehole [SY 0441 9487] near Allercombe penetrated 10.3 m of stiff to hard, red, silty clay, with two beds of weakly cemented, reddish brown sandstone at 7.0 to 8.9 m and 10.05 to 10.3 m depth. The higher sandstone crops out on the hillside to the west, and the same horizon is exposed in stream bank exposures [SY 0440 9468 to 0444 9470] 100 m west of Turkey Lane, which show (beneath 1.9 m of alluvium) 0.15 m of reddish brown sand on reddish brown, well-laminated, possibly cross-bedded, weakly cemented, moderately sorted, fine- to medium-grained sandstone.

Exposures [SY 0442 9384] in a ditch along the north side of Marsh Green Hill show about 1.2 m of well-bedded, reddish brown mudstone, with five beds of pale olive-grey, fine- to medium-grained, calcareous to very calcareous, moderately sorted, fine- to medium-grained sandstone from 25 to 50 mm thick. Two of the sandstone beds are joined by a number of vertical sandstone pipes which transect the intervening mudstone.

Rockbeare Hill–Manor Farm

A borehole [SY 0552 9420] near Melton Court, Rockbeare Hill, is recorded as having penetrated 13.11 m of red sandstone beneath 28.65 m of 'soft red marl with occasional sandstone'. The thick sandstone recorded is estimated to be about 100 m below the top of the Aylesbeare Mudstone.

Exposures of Aylesbeare Mudstone are locally present in stream banks near Little Houndbeare Farm; for example, exposures [SY 0508 9338] in Furzy Copse show 0.4 m of reddish brown, silty clay beneath 1.2 m of drift. Similar clay is present beneath 1.4 m of drift in the stream bed [SY 0546 9364], and in the goyle [SY 0536 9308] about 150 m west of Little Houndbeare Farm.

SHERWOOD SANDSTONE GROUP

In the Exeter district, the Sherwood Sandstone comprises the Budleigh Salterton Pebble Beds, up to 30 m thick, overlain by the Otter Sandstone, up to 160 m thick. The group is of economic importance because it contains large resources of sand and gravel, and also forms the principal aquifer in east Devon (Chapter 11).

Budleigh Salterton Pebble Beds

The Budleigh Salterton Pebble Beds were first recognised as a distinct formation, traceable from Minehead in Somerset to Budleigh Salterton in Devon, by Ussher (1875). Disconnected outcrops of the formation extend along the eastern border of the district from Clyst William (near Plymtree), through West Hill, to Bicton Common (Figure 22). Along most of its outcrop, the formation forms a well-defined, westward-facing scarp up to 25 m high. The contrast between the Pebble Beds and the underlying

mudstone is accentuated by differences in land use; the mudstones are largely in agricultural use, while much of the Pebble Beds outcrop south of Aylesbeare Common [SY 050 910] is occupied by heathland.

In the district, the formation consists predominantly of brown, horizontally bedded gravel with subordinate lenticular beds of trough cross-bedded pebbly sand and sand. Sand beds of mappable thickness are shown on the published 1:50 000 scale map. The gravel is composed of well-rounded pebbles, cobbles and boulders in a coarse to fine gravel and silty sand matrix (Plate 7a). The clasts are mainly (up to about 90 per cent) of metaquartzite, with a few per cent each of porphyry, vein quartz, tourmalinite and feldspathic conglomerate (Table 12).

The Pebble Beds rest with marked lithological contrast on the Aylesbeare Mudstone Group. Resistivity surveys (Henson, 1971, fig. 6.6; Sherrell, 1970, fig. 7) have shown that the base of the formation is irregular, with ridges and troughs along east–west and north-west–south-east axes. Thickness variations in the formation may represent erosional relief on this surface.

The formation is between 20 and 30 m thick in the district. It is typically about 21 m thick around West Hill [SY 060 940], and nearly 30 m thick on Bicton Common [SY 030 860]. About 2 km east of the district, boreholes [SY 0915 9102 to 0932 9041] in the Harpford area show it to thicken southwards from 11.6 m to 19.5 m over a distance of only 0.6 km (Edwards, 1990).

There is no palaeontological evidence of age for the Pebble Beds, but from their stratigraphical position they are considered to be Early Triassic (Scythian) in age (Warrington et al., 1980). The metaquartzite pebbles contain brachiopods, bivalves, and trilobites. They have been described by Salter (in Vicary and Salter, 1864), Davidson (1866–71, 1870, 1880, 1881), Cocks and Lockley (1981) and Cocks (1989, 1993). The last named have shown that the brachiopods are indicative of the Ordovician Arenig and Llandeilo stages, and the Devonian Gedinnian/Siegenian and Frasnian stages. Cocks and Lockley (1981) noted that the Ordovician brachiopods are similar to those found in situ in the Grès Armoricain and Grès de May formations in Britanny and Normandy, and to those from the Gorran Quartzites of south Cornwall (Sadler, 1974). The trilobites are also similar to the Llandeilo trilobites in the Gorran Quartzites (Sadler, 1974). The Gedinnian/Siegenian faunas match forms in the Landevennec and Gahard formations in Brittany, but the Frasnian faunas do not match with any outcrop (Cocks, 1989).

The sedimentary features of the Pebble Beds were described by Henson (1971), Mader (1985 b), Smith (1989, 1990), and Smith and Edwards (1991). The sequence in the Exeter district consists of couplets of horizontally bedded gravel and trough cross-bedded sand. The gravels are sandy, poorly sorted, and clast-supported with a closed-framework fabric. The clasts are mainly small cobbles and pebbles with high sphericities; imbrication is poorly developed. Bedding is defined by a crude segregation of clast sizes and sand content, by thin (less than 20 cm) lenses and sheets of sand and, locally, by lenses of fine-grained gravel with open framework. The trough cross-bedded units consist of medium- to coarse-grained sand,

with pebbles occurring as discontinuous lenses and as scattered clasts in the foresets. Counter-flow ripples and drapes of red mudstone are locally present.

Plate 7a   Typical horizontally bedded gravel facies in the Budleigh Salterton Pebble Beds, consisting of well-rounded pebbles, cobbles and boulders, mainly of quartzite, in a coarse to fine gravel and silty sand matrix. The gravels are sandy, poorly sorted, and clast-supported, with a closed-framework fabric. Beggars Roost Quarry [SY 062 941] (A 15287).

These lithologies typically form 4 to 5 m-thick couplets consisting of a lower unit of horizontally bedded gravel and an upper unit of trough cross-bedded sand. The gravel units are 1.5 to 4 m thick, and have erosive bases with up to 2 m of relief. The troughs of the sand units are 8 to 10 m wide, 0.25 to 0.5 m thick, and locally show mudstone drapes. The couplets are stacked both laterally and vertically, forming multistorey bodies. There is a change in the style of stratification towards the upper parts of the sequence. The couplets become progressively smaller, and the uppermost part of the sequence shows small, gravel-filled, ribbon-like channels, 1 to 2 m deep, which are flanked by laterally continuous sheets of gravel. Some of these ribbon-like units contain trough cross-bedded gravels. Palaeocurrent measurements indicate that flow was towards the north and north-east (Figure 24). The mean maximum clast size of the formation decreases northwards, from about 16 cm at the coast to about 10 cm in the district over a distance of about 8 km (Table 12).

At the top of the Pebble Beds is a bed of reddish brown silty clay up to 0.3 m thick containing polished, pitted and (less commonly) faceted pebbles (Plate 7b). These are interpreted as wind-faceted clasts (ventifacts), and the bed has been interpreted as a 'reg' type palaeosol by Wright et al. (1991). The ventifact bed has been recognised at Budleigh Salterton (Henson, 1970; Wright et al., 1991) and as far north as Hillhead Quarry [ST 065 135], Uffculme (Smith and Edwards, 1991). It represents a hiatus of

Plate 7b   Ventifact-bearing bed at the top of the Budleigh Salterton Pebble Beds, consisting of reddish brown silty clay up to 0.3 m thick, containing polished, pitted and faceted pebbles; the bed is interpreted as a 'reg' type palaeosol horizon. Beggars Roost Quarry [SY 0606 9414]. The hammer is 0.3 m long (A 15285).

Plate 7c   Section in the east face of Beggars Roost Quarry [SY 061 941], showing Otter Sandstone (OS) overlying Budleigh Salterton Pebble Beds (BSPB). The boundary is marked by a ventifact bed, indicated by Vs. An upper ventifact bed within the Otter Sandstone, indicated by Xs, is about 6 m higher (A 15283).

regional significance, during which the area was a windswept stony desert. Ventifacts at Woodbury Common [SY 0425 8675 to 0410 8644] were also considered by Leonard et al.(1982) to be at the top of the Pebble Beds, but Wright et al.(1991) considered that they formed on a Pleistocene or early Holocene surface.

An account of the petrography of the formation was given by Henson (1971); a selection of pebbles of igneous rocks was described by Campbell-Smith (1963). Other pebbles have been described by Strong and Smith (1989). The heavy-mineral assemblages have been described by Thomas (1902, 1909). The clast types identified in hand specimens are given in Table 12. Metaquartzites are mainly grey, locally with reddish brown mottling. 'Liver-coloured' quartzites form less than 5 per cent of the total. Henson (1970) identified the metaquartzites petrographically as medium-grained quartzarenites, subarkoses, and arkoses. Fine-grained conglomerate and grit also occur. Quartz and feldspar porphyries form less than 5 per cent of the clast types; they are commonly altered in situ to quartz and kaolinite. All the specimens examined by Campbell-Smith (1963) were either fine-grained quartz-

Table 12   Composition of clasts in the Budleigh Salterton Pebble Beds of the Exeter and adjacent districts.

Locality and SY grid reference	Mean max. clast size (cm)	Meta-quartzite %	Porphyry %	Tourm-alinite %	Vein quartz %	Conglomerate/ grit %
Beggars Roost [062 941]	9.4	86	1	4	5	4
Foxenholes [078 948]						
near top		90	2	2	3	3
	9.4					
near base		91	—	3	3	3
Venn Ottery [066 910]						
near top		84	—	1	7	8
	10.5					
near base		84	—	4	7	5
Blackhill [030 857]						
near top		86	1	8	3	2
	13					
near base		81	5	5	5	4

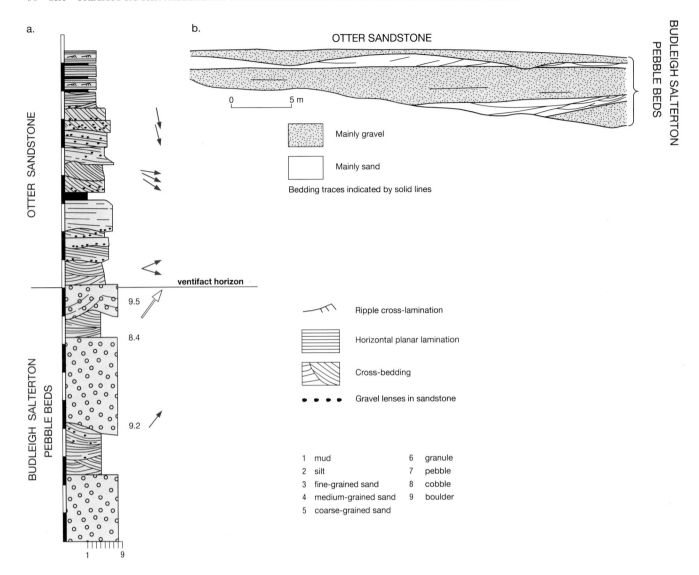

Figure 24
(a) Graphic sedimentary log of the Budleigh Salterton Pebble Beds and the Otter Sandstone at Beggars Roost Quarry [SY 062 941], with palaeocurrent azimuths and mean maximum clast sizes (in centimetres) indicated. Open arrows represent approximate orientations. The scale is in metres.
(b) Detail of an exposure of the contact between the Budleigh Salterton Pebble Beds and the Otter Sandstone at Beggars Roost Quarry, based on a photographic mosaic. Palaeoflow is obliquely into the face from right to left.

porphyries (some tourmalinised), or rhyolites. Dark grey to black tourmalinised hornfels forms 4 per cent or less of the total in the present district. Vein quartz forms up to 7 per cent at Venn Ottery (Table 12). There is little variation in the proportions of the pebble types within the district (Table 12). North of Talaton [SY 070 995], however, there is apparently an increase in the proportion of white vein quartz, for example at Clyst William [SY 0660 0277]; the lack of exposure does not enable this to be quantified.

The sources of the metaquartzite pebbles are formations in Britanny and Normandy or northward extensions of those areas into the English Channel. Laming (1982) considered that a possible source might have been a Palaeozoic ridge in the vicinity of Start Point. The contact metamorphic rocks (tourmalinised hornfels) suggest derivation from the aureole of the Dartmoor Granite (Henson, 1971). The porphyry clasts were compared by Henson (1971) with the quartz-porphyries of Cawsand Bay and Withnoe in south Devon, and also with the elvans of south-west England.

Thomas (1902, 1909) identified heavy minerals — for example, staurolite, kyanite, and sillimanite — that appeared to indicate derivation of the southern part of the Pebble Beds from a southern metamorphic terrain. However, Morton (1992) recorded heavy-mineral assemblages from a borehole [SY 0710 9558] near Straitgate Farm, which spanned the boundary between the Pebble Beds and the Otter Sandstone, that showed also the presence of minerals derived from the Cornubian massif. These assemblages were dominated by tourmaline (28.5 to 79.0 per cent), zircon (4.5 to 25.5 per cent), staurolite (9.0 to 30.0 per cent), and titanium minerals (5.5 to 21.0 per cent). Garnet, cassiterite, monazite, apatite, allanite, andalusite, calcic amphibole, topaz, anatase, rutile and brookite were also present. The abundant tourmaline, together with cassiterite, probable andraditic garnet, andalusite, topaz, allanite and monazite, indicate a Cornubian origin.

DEPOSITIONAL ENVIRONMENTS

The sedimentary features of the Budleigh Salterton Pebble Beds indicate deposition by low-sinuosity, highly braided rivers (Henson, 1971; Smith, 1990; Smith and Edwards, 1991). Evidence of a southerly and south-westerly provenance, indicated by the pebbles of fossiliferous quartzite and heavy minerals, is supported by palaeocurrent directions measured on the orientation of channel scours, rare cross-bedding, and imbrication (Smith and Edwards, 1991), and by the northward decrease in mean maximum clast size of the formation (Table 12). The southerly provenance is maintained as far north as Hillhead [ST 065 135], near Uffculme; north of there, the formation consists of the deposits of tributary river systems flowing from the west and carrying mainly locally derived pebbles of Devonian and Carboniferous sandstone and vein quartz (Thomas, 1902; Smith and Edwards, 1991).

Clyst William

A pit [ST 0660 0277] just north of Little Clyst William shows 9 to 10 m of reddish brown gravel, dipping at about 5° to 144°, composed of well-rounded pebbles generally less than 7 cm, some up to 10 cm, in a poorly sorted matrix of gritty silty sand. Locally, impersistent lenses of sand are present; one, up to 0.15 m thick and 2 to 3 m long, consists of moderately sorted, medium-grained sand. Lenses up to about 0.15 m thick of reddish brown silty clay are present about 3 to 4 m above the base of the section. The pebbles consist predominantly of white quartz, quartzite, black ?tourmalinised hornfels, weathered ?igneous rock (some completely rotted), and chert.

Larkbeare

At Larkbeare, a pit (Ussher, 1902, p.49 and fig. 12, p.48) shows the following section [SY 0720 9768]:

	Thickness m
Loam, dark brown, sandy, pebbly	c. 0.4

BUDLEIGH SALTERTON PEBBLE BEDS
Gravel, orange-brown, generally unbedded, comprising rounded clasts, mainly of metaquartzite < 7 cm (a few to 15 cm) in orange-brown to reddish brown gritty silty sand matrix	c. 1.5

Sand, colour-banded, yellowish brown to orange-brown, gritty, silty, medium- to coarse-grained, with scattered pebbles mostly < 3 cm. Lateral persistence uncertain owing to talus, but some small-scale faulting present	0.75
Gravel, as above	1.20
Concealed	c. 3.00

Rockbeare Hill Quarry

At the northern end of the quarry, a face [SY 0596 9492] shows up to 6 m of orange-brown and reddish brown gravel, containing predominantly rounded metaquartzite pebbles and cobbles, with small amounts (less than 10 per cent) of black schorl and white vein-quartz pebbles, in a gritty silty sand matrix. The gravels are locally frost-heaved in the top 3 m.

Beggars Roost Quarry

This quarry [SY 062 941] is now disused and is in process of being filled. Sections are present in the lower 8 m of the Otter Sandstone and in the upper 9 m of the Pebble Beds (Figure 24, Plate 7c). The latter consist predominantly of sheet-like, horizontally bedded gravel units with erosive bases, and lenticular beds of trough cross-bedded pebbly sand with eroded tops (Smith, 1989). The ventifact bed (p.88) is present at the top (Edwards and Smith, 1989) (Plate 7b). Details of mean maximum clast size and composition of clast types are given in Table 12.

Higher Metcombe

At Scott's Pollard, near Manor Farm, a pit [SY 0549 9218] shows 1.5 m of cross-bedded, colour-banded, reddish brown and orange-brown, fine-, medium- and coarse-grained, locally gritty sand with scattered fine gravel consisting of clasts of rather angular quartzite, quartz, and 'tourmalinite', generally less than 2 cm across. An impersistent bed up to 5 cm thick consists of sand with clay clasts. Several small faults affect the sequence.

A section [SY 0674 9204] at Southfield Farm is as follows:

	Thickness m
Loam, brown, sandy, with scattered quartzite pebbles	0.3
Loam, grey, sandy, pebbly	0.2
Gravel, orange-brown, with rounded pebbles mainly < 6 cm of quartzite, in gritty silty sand	1.0
Gravel, fine, with mainly well-rounded pebbles < 2 cm forming an open framework texture	0.3
Gravel, as bed above last	(seen) 0.2

Venn Ottery Common

Venn Ottery Quarry [SY 065 913 to 067 910] shows the basal 8 to 10 m of the Pebble Beds, although the junction with the Aylesbeare Mudstone is not exposed. The formation dips east at about 4°, and consists predominantly of horizontally bedded gravel with impersistent thin sand lenses. At the south-eastern end of the quarry, the gravels are periglacially disturbed to depths of up to 6 m below ground level, but at the north-western end only the upper 1 to 2 m are affected. Details of mean maximum clast size and composition of clasts are given in Table 12.

Colaton Raleigh Common–Woodbury Common–Bicton Common

Boreholes [SY 0392 8785; 0422 8798; 0494 8797; 0480 8723] on Colaton Raleigh Common penetrated Pebble Beds to depths of 13.9, 26.8, 20.6, and 31.6 m respectively.

A pit [SY 0318 8727] at Woodbury Castle (Ussher, 1902, p.52) shows up to 1.0 m of yellowish brown and pale grey laminated, poorly sorted, gritty, fine- to medium-grained sand, part of a sand bed within the Pebble Beds. The topmost 0.5 m contains slabs of ferruginous sandstone measuring 40 cm by 3 cm, orientated parallel to bedding; the lower 0.4 m is indistinctly cross-bedded.

West of Uphams Plantation, Woodbury Common, a fire-break [SY 0425 8675 to 0410 8644] yields wind-faceted and polished pebbles (ventifacts) (Leonard et al., 1982).

The Pebble Beds give rise to a feature overlying the Littleham Mudstone of the Bicton Common inlier (Plate 8a). A gully [SY 0388 8588] exposes the following section close to the base of the formation:

	Thickness m
Soil, grey, sandy, pebbly	0.4
Gravel, of pebble, cobble and boulder grades, to maximum diameter of about 30 cm, in a matrix of orange-brown silty sand, with a few impersistent sand beds; the basal 0.4 m iron-cemented into a hard conglomerate layer; sharp base	5.0
Gravel, dark reddish brown, pebbles generally less than c. 5 cm diameter, containing many clasts of reddish brown mudstone; matrix is reddish brown, very clayey, silty, fine- to medium-grained sand; gravel is locally iron-cemented	2.0

Kingston

A borehole [SY 0702 8773] near Pophams Farm, Kingston, penetrated the full thickness of the formation (22.4 m) between 12.19 and 34.59 m depth. The lithology was recorded by the driller as 'pebble beds' and 'sandstone and pebbles', with the basal 3.49 m as 'pebbles with bits of marl'.

Otter Sandstone

The Otter Sandstone sharply overlies the ventifact-bearing bed at the top of the Budleigh Salterton Pebble Beds (Edwards and Smith, 1989). The lower 100 m of the formation are probably present in the district, with a small outcrop area in the south-east of the district between Broad Oak and Bicton Gardens (Figure 22). The bulk of the formation consists of greyish orange, yellowish and reddish brown, fine- to medium-grained sandstone, which weathers to sand. The sandstones are planar-bedded and cross-bedded, variably cemented, and locally micaceous. Thin units of hard intraformational conglomerate are present at various levels, and lenticular beds of reddish brown clay or mudstone (commonly recorded in borehole logs as 'marl') occur locally, but are not of mappable extent. Mudstone beds are locally rich in muscovite where it has concentrated along bedding surfaces. Irregular layers of hard concretionary limonite or goethite ('iron pan') are developed in places. Intraformational conglomerate with a calcareous cement is present at a few localities, where it consists of metaquartzite and sandstone gravel clasts with mudstone clasts in a matrix of silty sand (Henson, 1971, p.241). Mappable pebble beds are present at Harpford Common [SY 068 901], and near Bicton College [SY 068 863]; they are lithologically similar to the Budleigh Salterton Pebble Beds, with well-rounded pebbles and cobbles (locally up to 0.2 m, mostly pebbles less than 5 cm), predominantly of metaquartzite, in a matrix of silty sand.

Exposures of the formation are few, and mainly occur along sunken lanes and in stream banks. The formation has been exploited for public water supply, mostly in the Otter valley, immediately east of the district. Within the district, the Otter Sandstone is mainly in faulted contact with the Littleham Mudstone or the Budleigh Salterton Pebble Beds. An unfaulted junction is visible at Beggars Roost Quarry [SY 062 941], Rockbeare Hill, where the Otter Sandstone rests on the ventifact-bearing bed at the top of the Budleigh Salterton Pebble Beds (p.88; Plate 7b).

No palaeontological evidence of the age of the Otter Sandstone has been found in the district. In the adjoining Sidmouth district, calcretes in coastal sections represent calcification around the tap roots of phreatophytic plants (Mader, 1990; Purvis and Wright, 1991). A varied macrofossil assemblage comprises plant debris, including an equisetalean (*Schizoneura*), arthropods including an insect and branchiopod crustaceans, fish (*Dipteronotus cyphus*), amphibians (*Eocyclotosaurus* sp., *Mastodonsaurus lavisi*, indeterminate capitosaurids and other remains) and reptiles (*Rhynchosaurus spenceri*, *Tanystropheus* sp., and indeterminate procolophonids, rauisuchians and a possible ctenosauriscid) (Whitaker, 1869; Huxley, 1869; Johnston-Lavis, 1876; Metcalfe 1884; Carter, 1888; Walker, 1969; Paton, 1974; Spencer and Isaac, 1983; Milner et al., 1990; Benton, 1990; Benton et al., 1994). Milner et al. (1990) considered an Anisian age most likely.

Little lithological variation is evident in the Otter Sandstone of the district. The basal 2 m of the formation at Beggars Roost Quarry [SY 062 941] consists of cross-bedded, medium-grained sandstone with mudstone drapes and clasts. It is overlain by about 4 m of pebble-grade gravel, gravelly sand, and very coarse-grained sand. The sands show low-angle (about 10°) planar sets or trough cross-beds with azimuths towards the east and south-east (Figure 24; Plate 8c). The sand contains angular feldspar grains, granitic rock fragments and aureole rocks (including tourmalinised hornfels). The clasts indicate derivation from the Dartmoor Granite and its aureole. The feldspathic sands and gravels are capped by a thin ventifact bed, which is sharply overlain by about 2 m of fine-grained sands with basal large-scale sets of low-angle cross-bedding, mudstone layers, and ripple cross-lamination (Figure 24; Plate 8c). The feldspathic sands also occur at Hillhead Quarry [ST 065 135], Uffculme (Smith and Edwards, 1991). They are absent at Budleigh Salterton, where the basal 10 m of the formation consist of fine- to medium-grained aeolian sandstone (Henson, 1971; Clarey, 1988).

The sandstones are largely composed of quartz, metaquartzite and feldspar grains, with flakes of muscovite, in a matrix of haematite-stained silt and clay. They were categorised by Henson (1971) mainly as subarkoses. In the south Devon outcrop, early calcite cements have locally been dissolved, and the sandstones are loosely cemented by calcite, ferric oxides and kaolinite (Milodowski et al., 1986; Strong and Milodowski, 1987). The heavy minerals

Plate 8a Feature of the Budleigh Salterton Pebble Beds overlying Littleham Mudstone on Bicton Common [SY 038 859]; base of the Pebble Beds indicated by a broken line (A 15342).

Plate 8b River cliffs [SY 0625 8865] near Stoneyford, Hawkerland, show up to 9 m of Otter Sandstone, consisting of reddish brown, weakly cemented, fine- to medium-grained sandstone, locally with gritty sandstone beds. The hammer is 0.3 m long (A 15286).

of the Otter Sandstone, studied by Thomas (1902; 1909) are not markedly different from those in the adjacent formations, but staurolite is less abundant than in the Budleigh Salterton Pebble Beds. The assemblage is typical of a metamorphic source area.

The Otter Sandstone in coastal and some inland exposures consists of sheet sandstones arranged in fining-upward cycles about 1 to 5 m thick. Each cycle commences with an intraformational lag conglomerate resting on an erosional surface, passing upwards through cross-bedded sandstone into laminated siltstone (Henson, 1971; Clarey, 1988; Mader, 1985). The intraformational conglomerates contain clasts of sandstone, mudstone and calcrete, and are

the principal source of vertebrate remains in the coastal exposures, but all the principal lithologies have yielded material (Spencer and Isaac, 1983).

DEPOSITIONAL ENVIRONMENTS

Henson (1971), Mader and Laming (1985), Mader (1985b), and Clarey (1988) concluded that the Otter Sandstone was deposited by shallow, low-sinuosity, highly braided rivers. Milner et al. (1990) noted that the vertebrate fauna included aquatic crocodile-like forms, and their presence suggests that there were some long-lived water bodies, containing fish, on the floodplain. The ter-

Plate 8c Basal Otter Sandstone at Beggars Roost Quarry [SY 061 941]. The hammer (0.3 m long) lies on a bed of pebbly gravel and gravelly sand containing angular feldspar and granitic rock fragments. The bed is overlain by a thin ventifact-bearing clay bed (marked by Xs), equivalent to that similarly indicated on Plate 7c, overlain by fine-grained sand (A 15282).

restrial animals included herbivores (rhynchosaurs and a procolophonid) and carnivores (rauisuchians).

Within the present district, the Budleigh Salterton Pebble Beds were overlain initially by gravelly feldspathic sands deposited by rivers flowing eastwards from the Dartmoor Granite. A period of reg soil formation is indicated by the presence above the feldspathic gravelly sands of a thin ventifact bed. The succeeding main portion of the Otter Sandstone was formed on an extensive sandy braidplain, with some aeolian dunes (Mader and Laming, 1985); calcrete palaeosols are widely developed. Palaeocurrents in the basal 6 m of the formation indicate derivation predominantly from the west (Smith, 1989). Outside the district, and in the higher parts of the formation, Henson (1971), Selwood et al. (1984), and Clarey (1988) indicate consistent transport directions from the south.

Beggars Roost Quarry

At Beggars Roost Quarry [SY 062 941], Rockbeare Hill, up to 8 m of Otter Sandstone rest on the Budleigh Salterton Pebble Beds. The sequence is described on p.92, shown graphically in Figure 24, and illustrated in Plates 7c and 8c.

Harpford Common

A section [SY 0641 9063] in a stream in Woolcombes Plantation shows exposures of up to 6 m of drift on Otter Sandstone, consisting of up to 0.6 m of dark reddish brown, very weakly cemented, moderately sorted, fine- to medium-grained sandstone. On Harpford Common, rounded metaquartzite pebbles and cobbles in a track [SY 0677 9015] suggest a pebble bed within the Otter Sandstone.

Goosemoor (Newton Poppleford)

A borehole [SY 0651 8954] near Hillside, Goosemoor, penetrated Otter Sandstone, logged as 'red sand, soft sandstone, and sandstone', to a depth of 41.5 m, resting on 'gravels', interpreted as Budleigh Salterton Pebble Beds, to a final depth of 50.6 m.

River cliffs [SY 0625 8865] near Stoneyford show sections in Otter Sandstone (Plate 8b), dipping at 6° to 082°, as follows:

	Thickness m
RIVER TERRACE DEPOSITS	0.8
OTTER SANDSTONE	
Sandstone, weakly cemented, and sand, orange-brown, yellowish brown and reddish brown, fine- to medium-grained; hard in basal 0.3 m	2.0
Clay, reddish brown, silty	0.5
Sandstone, yellowish brown	0.1
Clay, reddish brown	0.05
Sandstone, reddish brown, moderately to weakly cemented, very silty; mainly homogeneous, but locally with gritty sandstone beds; bedding picked out mainly by impersistent dark reddish brown units of hard silty clayey sand	(seen) 6.0

Kingston

An exposure [SY 0647 8815] near Kingston shows 1.5 m of reddish brown, weakly cemented, fine- to medium-grained sandstone, with a 0.1 m-thick lens of gravel. Nearby [SY 0651 8815] are exposures of 0.5 m of weakly cemented orange-brown fine- to medium-grained sandstone. Exposures in the lane [southwards to SY 0652 8802] show weakly cemented, orange-brown, fine- to medium-grained sandstone with gritty gravelly beds, locally iron-cemented. At Kingston, laneside exposures [SY 0650 8787] show about 1.5 m of yellowish brown to orange-brown gritty granular sand with iron-cemented layers.

A borehole [SY 0702 8773] near Pophams Farm, west of Colaton Raleigh, showed the following strata (driller's log):

	Thickness m	*Depth* m
Soil	0.45	0.45
OTTER SANDSTONE		
Sand and gravel	2.13	2.58
Sand and a few pebbles	3.19	5.79
Sandstone, soft, with pebbles	3.65	9.44
Sandstone, soft, with layers of ironstone	0.60	10.05
Marl with bands of sandstone	2.13	12.19
BUDLEIGH SALTERTON PEBBLE BEDS	(seen)	

Stowford

Near Stowford, up to 3 m of yellowish brown, and some orange-brown, fine- to medium-grained sand with some thin (2 to 3 cm) bands of reddish brown clayey sand and reddish brown clay are visible in stream-bank sections [SY 0624 8723].

About 200 m west of Stowford, a section [SY 0585 8720] in a field-bank shows about 1 m of orange-brown and yellowish brown, weakly-cemented, gritty medium-grained sandstone, locally with scattered subrounded clasts up to 4 cm across of metaquartzite, vein quartz and tourmalinite.

Bicton

Excavations [SY 0677 8642] at College Farm, Bicton College, show the following section, shown on the 1:50 000 scale map as pebble beds within the Otter Sandstone but possibly representing a partly faulted outcrop of Budleigh Salterton Pebble Beds:

	Thickness m
Gravel, rounded to well-rounded, with pebbles (mostly < 5 cm), predominantly of metaquartzite, in reddish brown gritty silty sand; possibly some polished clasts. Some impersistent beds of reddish brown sand pass laterally into pebble beds. Many quartzite clasts show vertical long axes owing to periglacial action	1.4
Sand, moderately persistent laterally in the north face of the excavation; contains isolated pebbles and cobbles. Not visible in the south face of the excavation	0.6
Gravel, rounded to well-rounded pebbles and cobbles, predominantly of metaquartzite, in reddish brown gritty silty to very silty sand; maximum mean clast size greater than uppermost gravel unit (up to 20 cm)	(seen) 0.7

SEVEN

Permian igneous rocks

Volcanic rocks associated with the Exeter Group sequence of the district include basalts and lamprophyres, with minor occurrences of rhyolite and agglomerate. They give radiometric dates (Table 13) indicating that they were erupted just before the intrusion of the Dartmoor Granite (dated at about 280 Ma). The oldest intrusion, a lamprophyre dyke in the Crackington Formation near Bridford, is dated at about 300 Ma, close to the age of the Carboniferous/Permian boundary.

Fragments of igneous rock are locally abundant in the Permian breccias; those in the Early Permian breccias consist of quartz-porphyry, tuffs and rhyolite, all untourmalinised, plus scattered pebbles of rotted lamprophyre. The Late Permian breccias also contain a variety of acid igneous debris, much of which is tourmalinised. Isotopic evidence suggests that this is derived from volcanic rocks formerly sited above the Dartmoor Granite pluton.

EXETER VOLCANIC ROCKS

Scattered outcrops of volcanic rock at or near the base of the Permian sequence around and south-west of Exeter, and interbedded within the Permian rocks of the Crediton Trough (Figure 25), have been referred to as 'Feldspathic Traps', 'Exeter Traps', or 'Exeter Volcanic Series'. The informal term Exeter Volcanic Rocks is used here for these geographically separate volcanic rocks, which occur within different sedimentary formations. Early descriptions were given by De la Beche (1835, 1839), Vicary (1865) and Hobson (1892). The first detailed account was in the Geological Survey Memoir for the district (Ussher, 1902). Further detailed work was reported by Tidmarsh (1932), and a major reappraisal of the rocks was carried out by Knill (1969). The geochemistry and petrogenesis have been examined by Cosgrove (1972), Exley et al. (1983), Thorpe et al. (1986), Grimmer and Floyd (1986), Thorpe (1987), and Leat et al. (1987). Radiometric age dates on the Exeter Volcanic Rocks of the district have been reported by Miller et al. (1962) and Miller and Mohr (1964). The palaeomagnetism of these rocks was studied by Cornwell (1967). Cornwell et al. (1990) reported on the magnetic evidence for their form and nature in the district, and identified a possible large concealed lava body near Alphington.

The Exeter Volcanic Rocks consist of two groups: alkaline olivine-basalts of within-plate type, and calc-alkali lamprophyres; there are also minor occurrences of rhyolite and agglomerate. In the Knowle area of the Crediton Trough, the earlier flows are lamprophyric but later eruptions produced basalts. Most of the volcanic rocks were erupted as minor local flows and vents. The lavas have undergone pervasive haematitic alteration during lat-

eritic weathering in a tropical climate, and reddish brown iddingsitic pseudomorphs after olivine phenocrysts are a characteristic feature of the rocks. X-ray diffraction analysis (Kemp, 1992) shows that the presumed original alkali volcanic mineralogy of feldspar, augite and quartz has been altered to include haematite, carbonates (calcite, dolomite and ankerite), clay minerals and zeolites. The clay mineralogy is dominated by smectite, which occurs not only interstitially but also as vesicle infills.

Field relations and radiometric age dates (Table 13) suggest that the Exeter Volcanic Rocks of the district are of Early Permian age.

Geochemistry and petrogenesis

The results of a detailed examination of the petrography and geochemistry of the Exeter Volcanic Rocks carried out during the resurvey of the district are presented by Fortey (1991, 1992). This study includes whole-rock and trace-element analyses of 17 new samples and earlier analyses by Cosgrove (1972), Thorpe et al. (1986) and Leat et al. (1987). The new analyses are given in Appendix 1.

The basalts comprise a cogenetic suite which can be divided into the alkaline to marginally subalkaline (contaminated) basalts of Dunchideock, Halscombe and Posbury, and the potassic trachybasalts of Pocombe, Columbjohn and other localities. The lamprophyres of Killerton and Knowle Hill show a calc-alkaline affinity, although examples from Holmead (outside the Exeter district) include more strongly trace-element-enriched analcimic varieties.

Alteration of the volcanic rocks has disturbed the major elements, notably calcium. However, trace element distributions confirm that the basaltic rocks vary from transitional alkaline-subalkaline basalt (Dunchideock and Halscombe) to more strongly alkaline and potassic trachybasalt (Pocombe and Budlake-Columbjohn). The lamprophyres are calc-alkaline minettes, olivine-minettes and analcime-bearing 'lamproites'.

The geochemistry suggests that the rocks are close to primary mantle melts, having undergone a degree of olivine and pyroxene fractionation, but lacking significant feldspar fractionation. The lamprophyres were probably derived from much the same mantle material as the basaltic rocks, but with the additional complication of incompatible element enrichment related to mantle metasomatism and/or low degrees of partial melting.

Thus far, the chemical data have shed little light on the origin of the numerous quartz and feldspar xenocrysts in the Dunchideock, Halscombe and other rocks. The generally coherent nature of the chemical variations makes it difficult to propose assimilation of a major amount of

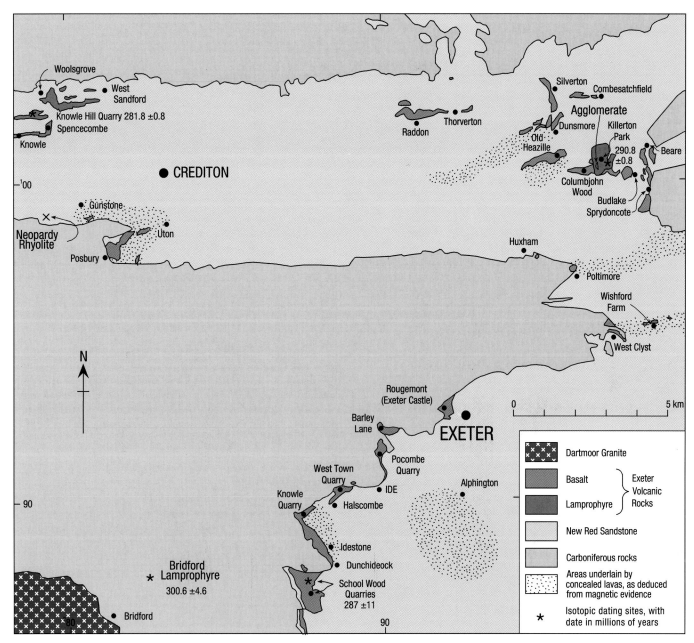

Figure 25 Distribution of the Exeter Volcanic Rocks and the Dartmoor Granite in the district, showing also some inferred extents of concealed Exeter Volcanic Rocks, based on magnetic survey evidence. The location of the Bridford Lamprophyre is also shown.

unrelated material such as Culm sedimentary rocks. One speculative possibility envisages assimilation of a small amount of strongly porphyritic rhyolitic magma formed by fractionation of cogeneric basalt. Mixing of the liquid part of this magma with the basalt could have taken place without greatly modifying the trace element and rare earth element profile. This interpretation is supported by the geochemical data from the rhyolite at Neopardy [SX 7938 9888] (p.111), which suggest a mantle origin and possible extreme fractionation of the parent magma. Leat et al. (1987) concluded that the Dartmoor Granite includes a

high proportion of granitic material formed by fractionation of the potassic magmas.

Magnetic anomalies

Lavas of the Exeter Volcanic Rocks have been shown to be characterised by very stable directions of magnetisation. Creer (1957) showed that they have mean directions of magnetisation of 190° (declination) and -10° (upwards inclination), consistent with their formation in a near-equatorial position. Studies by Cornwell (1967)

Table 13   Radiometric ages for igneous rocks of the district.

Rock type	Locality	Method	Authors	Age
Intrusive lamprophyre				
Minette	Bridford [SX 8263 8752]	K-Ar Biotite	3	300 ± 4.6
Exeter Volcanic Rocks				
Basalt	Dunchideock [SX 876 873]	K-Ar Whole rock	2	287 ± 11
Minette	Killerton [SS 975 005]	K-Ar Whole rock	1	285 ± 6
Minette	Killerton [SS 975 005]	K-Ar Biotite	4	283 ± 7
Minette	Killerton [SS 975 005]	Ar-Ar Biotite	6	290.8 ± 0.8
Lamprophyric biotitic-microsyenite	Knowle Hill [SS 789 022]	Ar-Ar Biotite	6	281.8 ± 0.8
Dartmoor Granite				
Biotite-granite	various localities	Rb-Sr Whole rock isochron	5	280 ± 1
Biotite-granite	Haytor Quarry [SX 759 774]	U-Pb Monazite	7	280.4 ± 1.2
Fe-skarn	Haytor Mine [SX 773 772]	Ar-Ar Hornblende	7	280.3 ± 1.0
Breccia clasts				
Granite clast in Heavitree Breccia	Kennford [SX 916 870]	K-Ar Biotite	3	281 ± 7
Quartz-porphyry clast in Newton St Cyres Breccia	Eastacott [SX 8444 9762]	Ar-Ar Biotite	6	279.7 ± 1.4

1. Miller et al. (1962), recalculated using constants of Steiger and Jäger (1977).
2. Miller and Mohr (1964), recalculated using constants of Steiger and Jäger (1977).
3. Rundle (1976).
4. Rundle (1981).
5. Darbyshire and Shepherd (1985).
6. Chesley, personal communication (1992).
7. Chesley et al. (1993).

suggested that the original magnetite had been largely destroyed by tropical weathering or alteration due to hydrothermal processes, and had been largely replaced by haematite. A chemical remanent magnetisation was thus acquired, but so soon after the extrusion of the lavas that the difference in direction of magnetisation was not significant. Cornwell (1967) deduced that the mean direction of magnetisation of the Exeter Volcanic Rocks was 188°–13°, and the corresponding palaeomagnetic pole lay at 46° N and 165° E. Owing to the widespread presence of haematite, the intensity of the total magnetisation is relatively low and is dominated by the remanent magnetisation component. With Q-values (the ratio of the remanent to the induced magnetisations) commonly greater than 10, the forms of the magnetic anomalies associated with the lavas are largely controlled by the remanence (cf. Cornwell et al., 1990, 1992 b).

The aeromagnetic anomaly map of the district and adjacent areas is shown in Figure 30 (Chapter 9). Residual anomaly maps of the Crediton Trough and Exeter areas (Figures 26 and 27) reveal anomalies associated with mapped lavas or their concealed extensions and, in a few cases, with probable concealed lavas (Cornwell et al., 1990, 1992 b). Areas likely to be underlain by concealed lavas are shown on Figure 25. A more extensive low-amplitude anomaly associated with the Crediton

Trough (Figure 26) suggests that the Permian rocks there may be underlain by additional lavas at greater depth.

The lavas at Killerton Park are unusual in that they retain a significant magnetite content, and the associated magnetic anomaly is correspondingly large (Figure 27). Ground magnetic surveys (Figure 28) indicate that the base of the lamprophyre is subhorizontal, suggesting that it is a lava sheet interbedded with the Thorverton Sandstone rather than an intrusive plug (Cornwell et al., 1992b). There is a marked contrast between the southern, more strongly magnetised, rocks and the more weakly magnetised rocks at the northern end of the park. Between these two areas is a col which is believed to be filled with Thorverton Sandstone. Despite the marked magnetic contrast between the north and south lamprophyres, there is no obvious lithological or petrological contrast between the two areas. Magnetic profiles across Columbjohn Wood west of Killerton Park indicate that the basalt has the form of a simple sheet thinning and dipping to the south. These near-surface anomalies are not apparent in Figure 28; the anomaly seen west of the fault forming the western margin of the Killerton lamprophyre could have a deep-seated source.

At Burrow Farm [SX 990 980], near Broadclyst, a group of aeromagnetic anomalies suggests the presence of one,

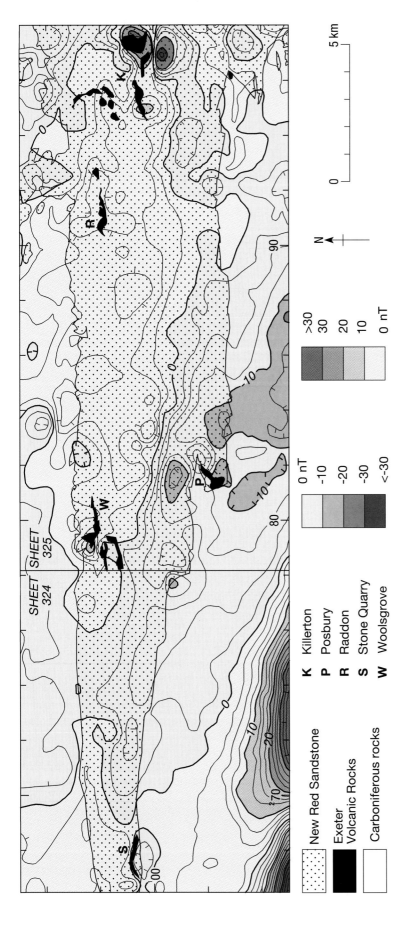

Figure 26 Residual aeromagnetic anomaly map of the Crediton Trough area, with contours at 2.5 nT intervals. The anomalies were derived from those shown in Figure 30 by removing a second order polynomial regional field. The map shows local anomalies associated with mapped lavas, and a broader anomaly indicating a more deep-seated source along the Crediton Trough.

New Red Sandstone

Exeter Volcanic Rocks

Carboniferous rocks

K Killerton   **P** Posbury   **R** Raddon   **S** Stone Quarry   **W** Woolsgrove

0 nT   -10   -20   -30   <-30

>30   30   20   10   0 nT

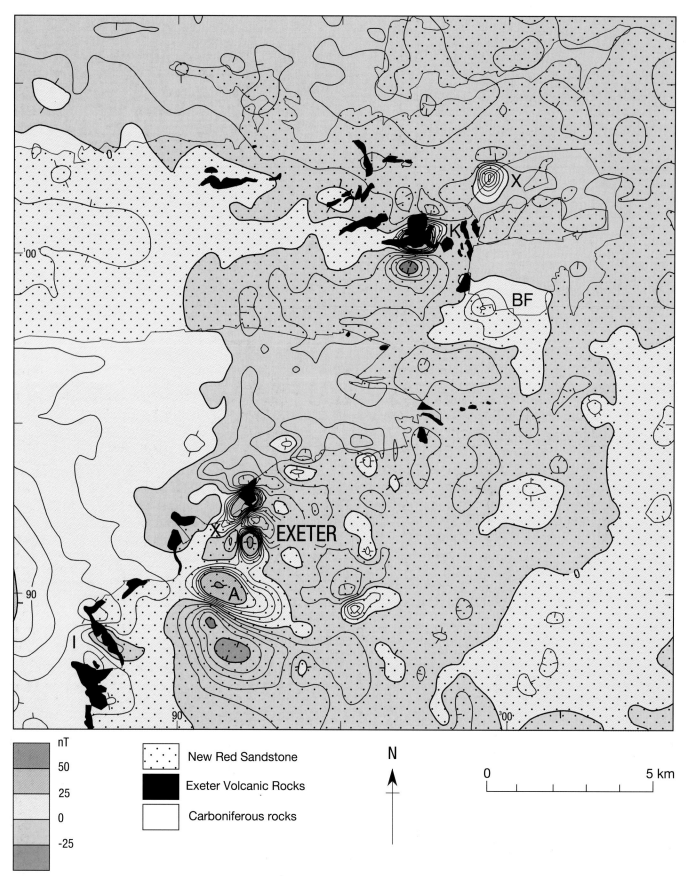

Figure 27   Residual aeromagnetic anomaly map of the area around Exeter and the eastern Crediton Trough, with contours at 5 nT intervals.

The anomalies were derived from those shown in Figure 30 by removing a third order polynomial regional field. Anomalies occur over mapped lavas at Killerton (K) and Idestone (I), and over possible concealed lavas at Alphington (A) and Burrow Farm (BF).

× indicates anomalies of probable man-made origin.   Key to geology as in Figure 26.

a.

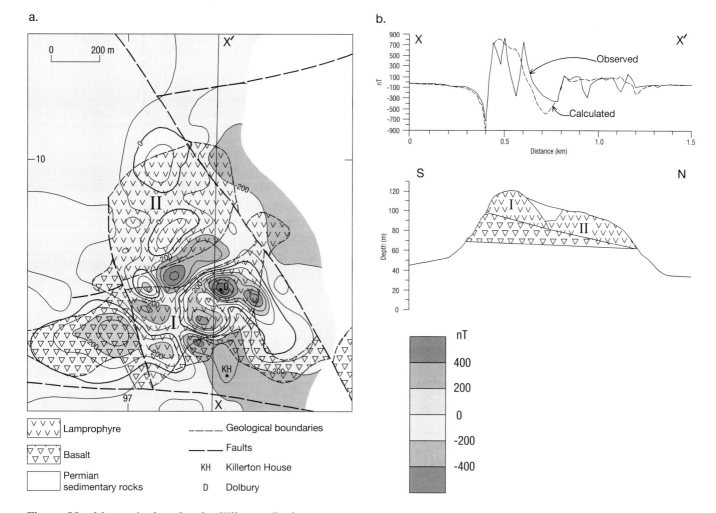

Lamprophyre

Basalt

Permian
sedimentary rocks

Geological boundaries

Faults

KH   Killerton House

D   Dolbury

b.

nT

400

200

0

-200

-400

Figure 28   Magnetic data for the Killerton Park area.

(a) Simplified magnetic anomaly map based on ground survey data with contours at 100 nT intervals, showing also surface geology. I and II refer to the magnetic and less magnetic components of the lamprophyre (see text).
(b) Magnetic profile XX' (location shown in (a)) and model producing the calculated curve. Properties of lavas (cf. Cornwell et al., 1990): lamprophyre I — vertical magnetisation with intensity of 3.8 A/m; lamprophyre II — magnetisation parallel with lava base and intensity of 1.9 A/m; basalt — nonmagnetic.

possibly two, concealed lavas extending westwards from the Ashclyst Forest inlier of Crackington Formation (Figure 29 a). The anomalies extend across the outcrop of Permian rocks to link up with the small outcrop of lava at Poltimore (Figure 25). One of the anomalies, which was confirmed by four ground magnetic profiles, is believed to be due to a faulted lava lying at a depth of about 60 m (Figure 29 b). The shape of the anomaly is consistent with the existence of a reversed, early Permian direction of magnetisation.

Just to the south, a less-well-defined aeromagnetic anomaly provides evidence for concealed lava between the isolated occurrence at Wishford Farm and the West Clyst lava (Figure 25) at the margin of the Carboniferous outcrop.

Near Alphington (Figure 27), a broad anomaly with reversed polarity (i.e. with negative anomaly to south) provides strong evidence for another, more extensive,

concealed lava covering an area of about 3 km². The form of the anomaly suggests that the lava lies at a depth of a few hundred metres, probably increasing in depth to the south-east. The east-south-east elongation of the anomaly could indicate either the original shape of the flow or the orientation of a feeder dyke, but the form of the lava could be partly defined by downward extensions of faults mapped at the surface. The relatively large amplitude of the magnetic anomaly suggests that it is caused by magnetite-bearing rock similar to that at Killerton Park.

In the Idestone area, two ground magnetic profiles and the aeromagnetic survey suggest a dipping sheet-like lava with an early Permian direction of magnetisation. The upper part of the lava appears to be particularly strongly magnetised (1.5 amperes per metre), as at Killerton and Alphington.

a.

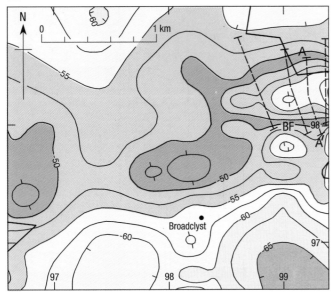

b.

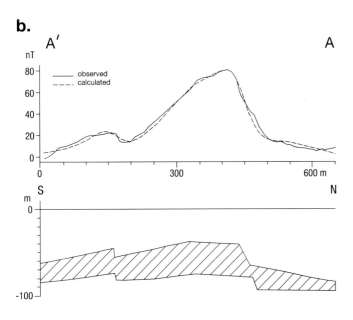

Figure 29   Magnetic data for the Burrow Farm area, Broadclyst.

a. Aeromagnetic anomaly map with contours at 2.5 nT intervals. Locations of ground profiles indicated by dashed lines.
b. Ground magnetic profile AA′ (location shown in a) and interpretation. Magnetic rocks indicated by diagonal shading are probably lava. Nonmagnetic sedimentary rocks unshaded. The magnetic rocks have a direction of magnetisation of 180°–10° and an intensity of magnetisation of 1 A/m.

BF — Burrow Farm

Lamprophyres

Lamprophyric rocks up to at least 20 m thick crop out west of Crediton between Brandirons Corner [SS 788 023] and West Sandford, at Killerton Park, and at Wishford near Broadclyst (Figure 25). The Brandirons Corner–West

Sandford lavas occur within the basal part of the Knowle Sandstone and in the Bow Breccia. They consist of lamprophyric biotitic microsyenite which, at Knowle Hill Quarry [SS 789 022], has yielded a $^{40}Ar/^{39}Ar$ plateau age of 281.8 ± 0.8 Ma (Chesley, personal communication, 1992). The strongly alkaline medium-grained rock contains abundant phenocrysts of unaltered biotite, accompanied by numerous iddingsite (haematite-rich) pseudomorphs after olivine and possible pyroxene (Plate 9a). Knill (1969) noted that the rock also contains phenocrysts of pale green-brown diopside, much altered to calcite. Apatite, magnetite and calcic plagioclase have been reported. The groundmass consists of interlocking allotriomorphic orthoclase grains accompanied by accessory quartz and possible analcime or altered nepheline. Minute plagioclase laths form an accessory constituent embedded within the potassium feldspar crystals.

A large mass of biotite-lamprophyre (minette) forms the prominent hill at Killerton Park [SS 973 004]. The minette has been dated by several authors (Table 13); the most recent determination, using the $^{40}Ar/^{39}Ar$ method, gives a plateau age of 290.8 ± 0.8 Ma (Chesley, personal communication, 1992). The field relations of the minette are not clear, but magnetic surveys indicate that its lower surface is subhorizontal (Cornwell et al., 1990), and it is probably a lava flow up to 35 m thick interbedded with the Thorverton Sandstone. On the west side of the Park, the minette is apparently faulted against the Columbjohn Wood basalts (p.107).

The minette is a fine-grained, somewhat massive, pale grey rock which has undergone pervasive alteration so that originally abundant clinopyroxene has mostly been replaced, although unaltered pyroxene was seen in some thin sections. In its slightly altered state, the minette consists of plates of biotite up to 3 mm across and altered augite granules dispersed abundantly through the groundmass of fine-grained interlocking orthoclase (Plate 9b). The groundmass also contains minute apatite prisms, a minor constituent in some specimens (E 947, 3229), but swarming abundantly through the orthoclase in others (E 3248, 28708), together with small amounts of sodic plagioclase and quartz (E 3228). Apatite also occurs as minute phenocrysts in several samples (E 28708, 28901, 28903, 28904, 65007). Minute zircon grains are inconspicuous, though their presence is indicated by the high Zr content (several hundred parts per million) of the rock.

Much of the feldspar and mica is darkened by minute granules of goethitic material, and it is commonly difficult to make out original opaque grains. However, the presence of magnetite (titanomagnetite with about 5 per cent MgO and Al-Mn-magnetite) was reported by Jones and Smith (1985) and confirmed in the southern part of the outcrop by Cornwell et al. (1990). Jones and Smith (1985) reported rare ilmenite grains with about 6 per cent MgO and 2.9 per cent Al_2O_3.

Small drusy cavities or vesicles are widespread, and are associated with ramifying patches of subsolidus, pneumatolytic character in which calcite is accompanied by late-formed, clear (unaltered) biotite, alkali feldspar and epidote (E 3248, 28708, 28901). Such patches may appear as

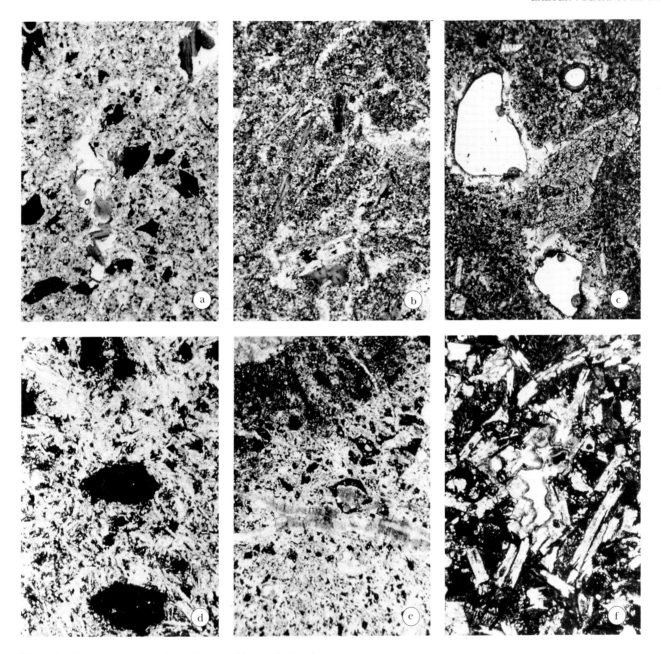

Plate 9   Photomicrographs of Exeter Volcanic Rocks.

(a) E 65019A, Knowle Hill Quarry, near Crediton [SS 789 022] (plane polarised light, 2.4 mm left–right): lamprophyre in which phenocrysts of biotite and haematitised olivine occur in a vuggy groundmass of turbid orthoclase.

(b) E 65008, Killerton Park Quarry [SS 975 005] (plane polarised light, 3.5 mm left–right): minette in which biotite phenocrysts occur in a turbid, fine-grained alkali-feldspar-rich groundmass containing irregular patches of clear, deuteric orthoclase.

(c) E 65001, Knowle Quarry, Ide [SX 874 895] (plane polarised light, 4 mm left–right): contaminated basalt in which irregular quartz and plagioclase xenocrysts occur in a turbid groundmass with sparse labradorite microphenocrysts.

(d) E 62298, School Wood Quarry, Dunchideock [SX 876 873] (plane polarised light, 2 mm left–right): olivine basalt in which iddingsite (haematite) pseudomorphs after olivine phenocrysts occur in a pervasively haematitised, holocrystalline groundmass.

(e) E 65003, Pocombe Quarry [SX 898 915] (plane polarised light, 4 mm left–right): haematitised olivine-basalt cut by dolomite veining, showing a skeletal haematite pseudomorph after olivine enclosing a central patch of carbonate.

(f) E 65014, Raddon Quarry [SS 908 020] (plane polarised light, 2 mm left–right): haematitised ophitic basalt showing a vuggy cavity rimmed by brown chalcedonic quartz and filled by clear quartz.

Photomicrographs by Dr N J Fortey.

pale spheroidal spots on hand specimens. The alkali feldspar is present typically as minute euhedral tabulae of unaltered orthoclase ('adularia') grown around the margins of the patches (E 28709, 28900, 28901), though some examples also show the presence of microcline in the centres of the patches (E 28902). One specimen (E 28905) is crossed by a cataclastic microcline veinlet. Calcite also occurs finely disseminated in many specimens. Flow foliation is uncommon, but where present can be seen locally to pass through calcitic pockets (E 28709), indicating that in some cases the subsolidus component was deposited by replacement of the earlier foliated mineral assemblage. A minor degree of late alteration is indicated by rare stylolitic films of chlorite and iron-oxide (E 3248) with which traces of muscovite and possible prehnite are associated. Xenoliths are uncommon, but a patch about 2 mm wide of microgranitic character (E 944) and polygranular quartz patches with sutured outlines (E 947) may be of exotic origin.

Knill (1969, p.123) considered that variation in the concentrations of apatite and augite was evidence that two bodies of minette were present, respectively characterised by augite-biotite and biotite-apatite assemblages, with the augitic variety occupying the central part of the hill and the augite-poor rock occurring on its eastern flank. No evidence has been found to substantiate this division.

Minette similar to that at Killerton is present in three small outcrops [SX 9860 9555; 990 956; 9945 9570] at Wishford Farm, near Dog Village, Broadclyst. On the assumption that the Wishford minettes are of similar age to the Killerton minette (about 290 Ma), it is likely that there is a substantial unconformity between the minette and the overlying Dawlish Sandstone, which is probably of Late Permian (about 260 Ma) age (p.32). Possibly the Wishford minettes lie at fairly shallow depth on Culm basement, and were buried beneath the aeolian dunes of the Dawlish Sandstone in the Late Permian.

Knowle Hill–Meadowend–West Sandford

Knowle Hill Quarry [SS 789 022] formerly exposed at least 30 m of lava (Ussher, 1902). The upper part was described as highly vesicular, the lower part more massive. Ussher (1902) noted that the upper vesicular part of the mass was trachytic and microlitic. The rock is oxidised, scoriaceous, lamprophyric biotitic microsyenite.

A quarry [SS 797 024] at Meadowend shows up to 6 m of weathered lamprophyric lava, very rubbly and friable at the top, but less weathered below. The lava is mostly massive and purple-grey, with scattered patches of vesicles filled by calcite or a zeolite mineral. Traces of banding on weathered surfaces suggest a dip of 35° to 130°; irregular jointing is present throughout. Tidmarsh (1932) described joints in the lava at Meadowend filled with red sandstone and lined with specular haematite.

At West Sandford Barton, a section [SS 8090 0283] shows up to 2.3 m of lamprophyric lava resting on breccia with thin interbeds of sandstone. The lava is purplish grey, vesicular, mostly very weathered and friable, fractured on red clay-filled joints and with common traces of manganese oxide minerals. Towards the base of the flow, the lava is extremely vesicular and scoriaceous; lobes and tongues of lava penetrate the underlying breccia for a distance of up to 30 cm. The base of the lava is very irregu-

lar, showing a general dip of 15° to 150°; beneath it, the sandstone and breccia are disturbed and include ash and lava fragments.

Killerton Park

A quarry [SS 9714 0018] near the Ice House in Killerton Gardens contains exposures of medium grey minette with biotite phenocrysts, and small irregular vesicles partly filled with ?haematite. Small disused quarries [SS 9701 0055; 9699 0059; 9698 0078; 9699 0088; 9700 0093; 9715 0100; 9706 0088] along the western side of the minette outcrop contain exposures or debris of medium grey minette with numerous biotite phenocrysts. A quarry [SS 9743 0100] on the north-east side of the park shows about 2 m of medium grey minette. Another quarry [SS 9765 0075] shows, in the west face, large irregular blocks of grey minette with irregular cavities partly filled with ?haematite. In the north face, irregular blocks of minette project downwards into rubbly reddish brown sand with rotted minette fragments.

The best exposures at present visible in the Killerton minette are in a quarry [SS 975 005] on the east side of the park where up to 15 m of medium grey minette with scattered black biotite phenocrysts are present. In the south-west of the quarry, a vein [SS 9751 0046], up to 0.4 m wide and dipping at 76° to 006°, is filled with black ?pyrolusite and reddish brown clay. The direction of the quarry face is locally controlled by near-vertical joints striking at 012°. On the north side of the quarry, a possible fault, striking 280°, separates brown-weathered minette on the north from less-weathered minette on the south.

Basalts

Basalt, possibly up to 70 m thick, is present at or near the local base of the New Red Sandstone between Ide and Dunchideock, south-west of Exeter (Figure 25). The field relationships of these rocks indicate that they form two distinct lava flows separated by the fault-controlled valley of the River Kenn. The northern flow extends from West Town Quarry [SX 886 903] (Ide) through Halscombe and Idestone to near Dunchideock Bridge [SX 884 878], and rests in most places on the Crackington Formation, except where a thin unit of breccia and sandstone (Knowle Sandstone) occurs beneath the lava between Idestone and Dunchideock Bridge. The southern lava flow extends from Dunchideock to beyond the southern edge of the district; it rests on Crackington Formation in its more northerly part, but Knowle Sandstone up to 20 m thick intervenes farther south. Both lava flows are similar, being characterised by xenocrysts of quartz and plagioclase, but iddingsite pseudomorphs after olivine phenocrysts are much more abundant in the southern flow, while the northern flow contains more abundant quartz and plagioclase phenocrysts. The lava in the southern flow at School Wood Quarry [SX 876 873], Dunchideock, has been radiometrically dated at 287 ± 11 Ma using the whole rock K/Ar method (recalculated from Miller and Mohr (1964), using the decay constants of Steiger and Jäger (1977)).

Basalts of the northern flow contain numerous xenocrysts of quartz and plagioclase (Plate 9c), together with a few of potassium-feldspar (E 946, 949, 65000); a small number of pseudomorphed phenocrysts of olivine are present in some specimens (E 29169). The quartz xenocrysts are rounded, ovoid single crystals, some more than 2 mm wide, surrounded by slender rims of dense, microscopically granular, augite-quartz and haematite formed by alteration of the pyroxene. Plagioclase xenocrysts are similarly rounded, some with re-entrant forms. Several retain relict cores of relatively calcium-poor plagioclase. However, the xenocrysts are dominated by broad margins of more calcic composition through which run networks of fine threadlike dissolution channels, the 'frit channels' described by Tidmarsh (1932). In some cases, there is an additional thin outer rim free of dissolution channels, which was evidently deposited as a shell surrounding what remained of the original crystal. Examples examined during the present study show that the cores consist of andesine. The surrounding margins show complex reverse zoning with labradorite developed in association with the dissolution channels. However, electron microprobe analysis (Fortey, 1991) has indicated that the channels are surrounded by rims of potassium-rich feldspar (about 60 per cent orthoclase (Or)) a few microns thick. The channels are variously open or filled with haematitic material probably deposited later than the initial formation of the channels. The outer rims of the xenocrysts consist of plagioclase slightly more calcic (anorthite (An) 60–73 per cent) than that formed in the enclosed margins and that found in the groundmass of the rock. Additionally, a xenocrystic patch of interlocking orthoclase was observed in one specimen (E 65001).

The holocrystalline groundmass is dominated (E 65000) by an interlocking mesh of slender labradorite laths (An 53–68 per cent) and interstitial patches of potassium-feldspar and quartz seen as a brown mesostasis. Minute ilmenite plates are common, as are secondary iron-oxide granules, calcite and chlorite. Minute augite grains are only rarely preserved (E 65000), but both quartz and orthoclase occur as minute euhedral grains locally. Fine-scale dissolution porosity is extensive. The rock differs from that of the southern flow in containing few iddingsitic pseudomorphs after olivine phenocrysts, consistent with the greater degree of hybridisation implied by the numbers of xenocrysts.

The lava of the southern flow consists of vesicular olivine-basalt which has undergone pervasive lateritic oxidative weathering. The upper part of the flow is highly vesicular and penetrated by fractures and fissures hosting neptunean sandstone veins. The central part of the flow is massive and hard, despite the alteration (Plate 10a). The marginal parts of the flow show weak flow foliation. Screens (up to 3 cm thick) of red sandstone and networks of mudstone veins occur within the lava. Locally, an 'inosculated' facies is exposed in which blocks of vesicular or rubbly lava are intimately associated with complex intermixtures of sandstone and lava; at School Wood Quarry, Dunchideock, about 6 m of relatively homogeneous basalt sharply overlie about 5 m of the 'inosculated' rock. The base of the flow is exposed in the southern quarry (Plate 10c).

In thin section, sandstone veins in vesicular basalt consist of fine-grained, quartz-rich sandstone in which calcite forms a patchy cement, and hairline calcite veinlets occur at the boundaries of sandstone veins and within the adja-

Plate 10a Exeter Volcanic Rocks, School Wood Quarry [SX 8755 8726], Dunchideock. The quarry face shows irregularly jointed greyish red-purple basalt; igneous layering dips at approximately 13° north-east. The hammer is 0.3 m long (A 15310).

cent basalt. Ovoid vesicles seen in many thin sections are generally filled by calcite, and where close to sandstone veins may also contain detrital sediment (E 28719). Fine-scale brecciation in which millimetre-scale basalt clasts sit in a matrix of comminuted basalt is seen in at least one specimen (E 29157).

The abundant olivine phenocrysts are preserved as reddish brown, millimetre-scale, iddingsite pseudomorphs (Plate 9d) which commonly consist of shells of haematite or goethite enclosing minute pockets of serpentinitic material. Ophitic augite and, more commonly, fine granular augite is seen in the greyish brown holocrystalline groundmass of some specimens. The composition of augite granules analysed from one sample (E 62999) is wol_{36}-ens_{46}-fer_{18}. However, haematitic granules remain abundant, and the augite is extensively replaced by iron-oxide and clay, while the dominant mesh of labradorite laths shows a degree of turbid argillic alteration.

In a little-altered sample (E 62999), fine-scale secondary dissolution porosity is present throughout the ground-

Plate 10b   Exeter Volcanic Rocks, Pocombe Quarry [SX 8985 9149], Exeter. The quarry face is in purple, locally vesicular, olivine alkali basalt, extensively veined by dolomite. The hammer is 0.3 m long (A 15314).

Plate 10c   Base of lava flow, School Wood Quarry [SX 8750 8708], Dunchideock, showing complex zone of mixing of blocks of purple basalt with white amygdales in reddish brown, micaceous, fine-grained sandstone. The hammer is 0.3 m long (A 15311).

mass, which otherwise consists largely of an interlocking mesh of plagioclase laths (56 to 65 per cent An) about 100 microns long, forming about 60 per cent of the rock. The accompanying subequant augite grains are 20 to 25 microns in width and form about 10 to 15 per cent of the groundmass. In some cases they are zoned, with irregular cores of relatively magnesium-rich augite mantled by more iron-rich augite. Ilmenite blades 5 to 10 microns long are also common. The interstices contain pockets of inter-grown potassic feldspar (40 to 60 per cent Or) and quartz. Optically, this interstitial material appears as a colourless to brown mesostasis which Knill (1969) considered to include zeolitic material or analcime. Fine calcite grains

are disseminated in some specimens and accessory grains of possible tremolite are also present locally (E 17173).

Rounded, corroded xenocrysts of quartz, plagioclase and, rarely, potassium-feldspar occur in small numbers. The feldspar xenocrysts are distinct from the euhedral tabular labradorite microphenocrysts also present, both in their rounded, commonly re-entrant forms and the characteristic patchy distribution of minute 'fritted' haematite-filled dissolution channels (E 17173, 28718, 62997, 62998). One example (E 62999) consists of 'fritted' labradorite (An 55 to 65 per cent) surrounded by a more calcic rim (An 60 to 71 per cent). The dissolution 'frit' channels are surrounded by rims of potassium-rich calcium-poor feldspar, as is also the case in the northern lava flow.

Basalt of a more alkaline, trachytoid character than that described above crops out at West Town Farm [SX 888 904] near Ide, around Pocombe on the western side of Exeter, and at Exeter Castle (Rougemont) (Figure 25). Similar basalt also forms isolated outcrops farther north around West Clyst [SX 975 955; 976 953], near Poltimore [SX 962 972], and near Huxham [SX 9515 9755] (Figure 25). In the Ide and Exeter city area, the lavas rest on weathered Crackington Formation. The present disconnected outcrops probably represent the remnants of one flow which were separated by erosion before being overlain by the Whipton Formation in the Late Permian.

At Pocombe, the lava is estimated to be up to 20 m thick and consists of moderately massive, somewhat vesicular, purple basalt with subhorizontal white dolomite veins (Plates 10b and 14b). At Crossmead Quarry [SX 8992 9145], the lower 8 m are of purple basalt with white dolomite veins; the top 4 m are in a rubbly facies of lava blocks with irregular blocks and interlacing 'veins' of pale reddish brown sandstone. The Pocombe lava is rich in olivine phenocrysts completely made over to skeletal red iddingsitic pseudomorphs which enclose patches of carbonate and chlorite (Plate 9e; E 65003). Secondary dissolution porosity is extensive. Haematitic granules are numerous, and disseminated dolomite granules are common to abundant. The flow-foliated groundmass is dominated by laths and tabular microphenocrysts of labradorite (E 3236, 65006) with, in some cases, slender selvedges of oligoclase (compositions of about 30 per cent An and 10 per cent Or are recorded), which is abundantly mantled by potassic feldspar (about 50 per cent Or). Quartz occurs only sparsely in interstices within the feldspar mesh, but it also forms localised discrete minute pockets.

At the eastern end of the Crediton Trough, outcrops of strongly altered, vesicular, alkaline basalt form the ridge of Columbjohn Wood west of the minette outcrop at Killerton Park, and also outcrop around Budlake, Beare and Sprydoncote east of Killerton (Figure 25). Thicknesses probably range up to 30 m. The lavas are of similar trachytic character, and may belong to the same flow; they are interbedded with the Thorverton Sandstone. Complex admixtures of lava with sandstone, and the presence of eroded igneous clasts in sandstones immediately overlying the lavas, all point to extrusion into unconsolidated sediments. Trace elements (Fortey, 1992) indicate that the lavas are akin to alkali basalts and are similar chemically to the Pocombe lava. The Columbjohn rock is a trachytoid lava rich in microphenocrysts of olivine, which has suffered strong, pervasive haematitic alteration. The groundmass is dominated by orthoclase crystals, and contains opaque platelets, probably ilmenite blades and/or altered biotite flakes (E 28710). Two generations of vesicle fillings, the earlier of cryptocrystalline silica mineral and the later of mosaics of interlocking calcite grains, can be distinguished in thin section (E 28907). The siliceous material also occurs in isolated orthoclase-rich patches in rock in which interstitial patches of a green clay mineral occur within the groundmass (E 28908). Interstitial analcime is also present (E 28909). In its upper part, the lava is invaded on fractures by tongues of fine-grained carbonate mosaic carrying a few grains of quartz and orthoclase (E 28910). At Budlake Quarry [SS 980 001] the rock varies from moderately vesicular to rich in white-green amygdales, and admixture with reddish brown sandstone is also seen. In thin section, iddingsite pseudomorphs indicate that olivine was plentiful as minute phenocrysts in a groundmass dominated by minute crystals of orthoclase accompanied by oligoclase, oxide granules and an interstitial mesostasis. Opaque platelets in the groundmass include ilmenite blades, but oxidised biotite flakes may also be present (E 958, 65009). The apparently chalcedonic or zeolitic interstitial mesostasis displays a fibrous radiating habit and locally contains a vermicular clay mineral (E 29191).

Around Raddon, Thorverton and Silverton, olivine-basalt and olivine-dolerite lavas are present within the Thorverton Sandstone (Figure 25) and are up to 20 m thick. In the Dunsmore area [SS 957 016], basalts occur at the top of the formation, whereas at Silverton another flow, apparently separate from that at Dunsmore and at a lower stratigraphical level, spans the boundary with the underlying Cadbury Breccia. Similar lava is also present at Spencecombe [SS 794 017] and near Knowle village west of Crediton, where it occurs near the top of the Knowle Sandstone and is up to 21 m thick (Edmonds et al., 1968, p.143). The lavas are typified by exposures at Raddon Quarry [SS 908 020], thin sections of which show ophitic olivine-dolerite (E 951, 65012, 65013) in which pseudomorphs after olivine phenocrysts about 0.2 mm wide are made up of haematite shells enclosing minute pockets of carbonate and chlorite. The abundant ophitic pyroxene is made over to pale brown clay with iron-oxide granules and platelets. Laths of andesine-oligoclase have rims of alkali feldspar (e.g. E 29183), and are in part made over to carbonate and brown to pink amorphous material (E 17306). The altered interstitial/ophitic groundmass contains abundant granules of iron-oxide and carbonate, together with minute cryptogranular patches of possible secondary quartz or zeolite. Platelets of opaque material in many cases mark pyroxene cleavage planes, though some probably represent primary ilmenite blades. Intense alteration is seen in highly vesicular rock (E 29184, 29187, 65014) in which amygdales are made up of concentrically banded rims of brown-stained quasi-amorphous siliceous material surrounding pockets of quartz and carbonate (Plate 9f). Deep green clay is present in some vesicles (E 17306, 28714) and occurs interstitially in the groundmass. Finely disseminated carbonate is also present, and the rock is crossed by carbonate veinlets. In the 'pegmatitic facies' of Knill (1969), the labradorite crystals are up to 0.6 mm wide, and are commonly microscopically veined by colourless material, 'probably analcime'.

On the southern side of the Crediton Trough between Posbury and Uton, basalt lavas up to 40 m thick occur at the junction between the Knowle Sandstone and the Crediton Breccia (Figure 25); the stratigraphical position of the lavas is thus similar to those in the southern part of the district between Ide and Dunchideock. At Posbury Clump Quarry [SX 815 978], veining and brecciation of the basalt by dolomite is so extensive in places that the rock has an almost agglomeratic appearance (Knill, 1969). The rock is a strongly altered vesicular olivine-basalt in which olivine phenocrysts are altered to iddingsite pseudomorphs. A minor content of interstitial potassium-feldspar within the weakly flow-foliated groundmass (E 3222) suggests an alkaline variety. Carbonate occurs in veinlets, disseminated through the groundmass and, with quartz, infilling vesicles. One sample (E 28717) contains a patch of interlocking allotriomorphic orthoclase which may be a granitic or syenitic xenolith. Another (E 65016) contains what may be a similar xenolith: this rock, which displays abundant secondary dissolution and associated clay and iron-oxide alteration of olivine phenocrysts and groundmass feldspar, contains an ovoid 'xenolith' of altered feldspar. An outer rim encloses a microscopically porous core of clay, areas of possible relict potassic feldspar, and swarms of minute granules of an Sr-Al-P-rich mineral. This sample also displays what is more probably an altered plagioclase xenocryst.

Red sandstone which underlies the base of the lava is exposed at Uton [SX 823 985]. The rock contains numerous basaltic pumice fragments, many of which are collapsed, while others are undeformed; some vesicles contain quartz amygdales, while the host sediment has infilled others (E 28716, 65015).

Knowle–West Sandford

Quarries [SS 786 016] near Hartland Cross, now infilled, showed, in 1987, a small face with up to 0.5 m of grey-purple, weathered and rubbly basalt resting on reddish brown, indurated, silty fine-grained sandstone with scattered pale greenish grey streaks and lenticles. The basalt is amygdaloidal, with some vesicle infilling by quartz with chalcedony or opal cores. A section [SS 7865 0158] recorded during the survey of the Okehampton district showed 3 m of purple and dark grey vesicular lava with bright green weathering 'zeolite'. The flow is estimated to be up to 21 m thick (Edmonds et al., 1968, p.143).

Exposures in a quarry [SS 794 017] at Spence Combe show strongly altered ophitic olivine-dolerite (E 29194, 65018) petrographically similar to the lavas at Raddon Quarry [SS 908 020] and Uton [SX 823 985].

Posbury–Uton

Strongly altered olivine-basalt occurs in the quarry at Posbury Clump [SX 815 978], and at Folly Farm, Uton [SX 815 985]. The Posbury Clump Quarry shows faces up to 18 m high in purplish brown basalt. In the north-western face, up to 3.5 m of highly weathered, rubbly basalt pass down into scoriaceous lava with prominent near-vertical joints. In the vertical face of the north-eastern part of the quarry, up to 4 m of rubbly lava overlie up to 14 m of pale purplish brown basalt with many large tabular and flat-lying vesicles. Flow laminae dip at 25° to 130°. In the eastern faces, abundant angular to subrounded fragments of red sandstone are present in the basalt. A section in Posbury Quarry described by Ussher (1902), showing sandstone beneath the basalt, is no longer visible. Throughout Posbury Quarry, the abundance of vesicles is variable and patchy; in places they are filled with opaline silica, white clay or calcite.

The Uton Quarry [SX 8235 9854] exposes the base of a basalt lava flow resting on an irregular surface of crudely interbedded sandstone and breccia. The lava is purplish brown, with irregular layers of vesicles filled with white clay, black manganese oxides, or haematitic material. Many inclusions of sandstone and breccia are present within the basalt.

Raddon–Thorverton

The best exposures of the Raddon lava are in Raddon Quarry [SS 9096 0202 to 9102 0201], about 2.5 km west of Thorverton. The contact with the overlying Thorverton Sandstone is exposed [SS 9078 0200] in the south-west corner of the quarry; the Thorverton Sandstone contains eroded pieces of the lava (see p.66). The top of the lava is a very amygdaloidal, greyish red basalt with amygdales containing white ?montmorillonoid minerals. In the west face [SS 9077 0204], sheets of reddish brown to grey-green sandstone occur within the lava (Knill, 1969, p.119). The base of a 12 to 15 m face [SS 9079 0206] is greyish red-purple dolerite with few amygdales. Farther east, a 15 m face [SS 9085 0206] shows, at the base, greyish red-purple dolerite with amygdaloidal bands: one up to 0.2 m thick grades down into non-amygdaloidal dolerite, but is very sharply overlain by non-amygdaloidal dolerite. In the main eastern bay of the quarry, an 8 m face [SS 9095 0202] consists of greyish red-purple, sporadically amygdaloidal dolerite.

Up to 2 m of greyish red-purple-weathered amygdaloidal basalt are exposed [SS 9115 0198] in the lane between Raddon and Chapel Farm, but the contact of the basalt with underlying sandstone, described by Ussher (1902, p.70) is no longer exposed. A quarry [SS 9172 0211] 200 m north-north-east of Canns Farm, Raddon, shows about 3 m of greyish purple dolerite. An exposure [SS 9237 0215] in a quarry in Thorverton shows about 2.0 m of weathered, rather friable, extremely amygdaloidal pale red-purple basalt.

Silverton–Stumpy Cross–Dunsmore–Red Cross–Stockwell

Quarries [SS 9467 0145 to 9497 0155] at Greenlands Plantation yield scattered blocks of amygdaloidal basalt.

Near Great Pitt Farm, Silverton, a section [SS 9516 0313] shows 0.2 m of pale red-purple vesicular basalt resting on a 0.1 m-thick indurated bed, partly vesicular, with basalt, breccia and Culm sandstone fragments, which rests in turn on indurated breccia. Nearby exposures [SS 9518 0313] are of up to 1.2 m of pale red-purple vesicular basalt with some vesicles lined with a soft white mineral. Rubbly vesicular lava is visible by the roadside [SS 9548 0300] in Silverton. On the south side of the village, a section [SS 9558 0254] near Channons was figured by Ussher (1902, fig. 16), who described 'grey Trap rock, much decomposed and split up by numerous irregular joints, [which] rests upon red sand-rock, in places stained blackish and brecciated towards the base, upon rubbly breccia'. The contact between lava and sedimentary rock is no longer visible. The top 0.2 m of the currently visible exposure is sandstone, possibly with basaltic material incorporated, upon 1 m of hard breccia.

A quarry [SS 9568 0160] near Dunsmore exposes about 10 m of pale red-purple and greyish red-purple moderately vesicular to slightly vesicular basalt or fine dolerite. Some vesicles have a brown earthy fill; others are lined with a white mineral. Veins of coarse ophitic dolerite have been reported by Knill (1969, p.119).

Sections [SS 9562 0195] at Babylon Farm show about 0.9 m of rubbly weathered, pale purple-red, slightly vesicular basalt or fine dolerite, probably in fault-contact with sandstone. The rock contains irregular vesicles, filled or lined with a pale greenish grey mineral (locally dark green). At the south-west end of the exposure [SS 9558 0192], up to 2 m of basalt and dolerite are completely weathered to soft clayey 'sand' in the top 1.5 m; the bottom 0.5 m is very rubbly loose rock.

A quarry [SS 9704 0247] at Combesatchfield, Stockwell, shows about 4 to 5 m of vesicular basalt, overlain by Thorverton Sandstone (p.66). The west face [SS 9702 0248] shows pale red-purple basalt or very fine dolerite with scattered amygdales filled with a green mineral within an initial lining of white mineral. At the southern end of the quarry [SS 9705 0247], just beneath sandstone, pale red-purple basalt contains numerous vesicles lined with a white to very pale green mineral.

Old Heazille Farm–Columbjohn Wood–Budlake–Beare–Hollis Head

Basalt and dolerite form a ridge extending eastwards through Old Heazille Farm [SS 9575 0070]. Exposures of basalt in a quarry [SS 948 005] contain vesicular banding that dips at 27° to 152°. Surface debris [SS 9562 0080] is of pale greyish red-purple slightly vesicular basalt, with vesicles filled with a pale green mineral. About 2 m of grey dolerite are exposed by the roadside [SS 9590 0078] between Old Heazille Farm and Riverside; Ussher (1902, p.66) noted an exposure, at the same locality, showing the igneous rock '... resting on red loamy sand, succeeded by earthy gravels containing no apparent trace of igneous rock and dipping persistently southward'. Exposures [SS 9609 0086] of weathered dolerite are visible at the edge of the alluvial floodplain east of Riverside.

At the western end of Columbjohn Wood [SS 9619 0026], reddish brown, silty, very fine-grained sandstone with disturbed bedding contains irregular pieces of reddish brown vesicular lava. Old quarries [SS 9635 0027 to 9644 0028] show exposures of pale red to pale red-purple, vesicular to very vesicular lava, with vesicles commonly lined with a white to very pale green mineral. Complex admixtures of sandstone with lava are common in these quarries; a sandstone bed [SS 9636 0027] within lava dips at 14° to 208°. Exposures [SS 9683 0053] on the north side of Deodara Glen, Killerton Park, are of pale red-purple basalt with very vesicular layers.

Budlake Quarry showed in the north-east corner [SS 9807 0012] up to 10 m of pale red-purple, moderately vesicular basalt in which banding dips at 22° to 172°. In the south-east face [SS 9808 0009], greyish red basalt contains numerous amygdales filled with white and green minerals; near the top of the face are complex mixtures of vesicular basalt with reddish brown sandstone.

Outcrops of lava occur near Beare Farm [SS 985 007] and about 250 m west of Lower Comberoy Farm [SS 989 008]. Exposures in the entrance to a quarry [SS 9861 0090] at Beare Farm are of pale red-purple, moderately vesicular basalt. Exposures in the quarry showed 4.6 m of sand and sandstone, with 'Trap' fragments in the lower part, resting on 'finely vesicular Trap' (Ussher, 1902, p.66).

Small exposures are present between Budlake and Hollis Head. The easternmost patch [SS 9875 0030] forms a small feature with abundant dolerite debris in the soil. Two further exposures [SS 9864 0020; 9858 0019] of rubbly dolerite, probably fault-bounded, are visible in the lane side. Stream exposures [SS 9886 0005] of pale greyish red-purple slightly vesicular basalt are part of an outcrop of lava that extends southwards towards Sprydoncote House.

Huxham–Poltimore–Broadclyst

Volcanic rocks occur north of Sprydoncote [SX 9880 9995], Hazel Wood [SX 987 997], at Bampfylde House [SX 962 972], near Huxham [SX 9515 9755], and at West Clyst [SX 975 955 and 976 953] (Figure 25). There is no good exposure, but surface material indicates pale reddish brown vesicular basalt with small mafic phenocrysts. The Pinhoe rocks were described by

Ussher (1902) as hard grey basalt, in part vesicular, cut by sandstone dykes and pipes. A sample (E 28913) from West Clyst Farm is a non-vesicular olivine-basalt which has undergone strong pervasive haematitic alteration. The Bampfylde House rock is a strongly altered, vesicular olivine-basalt in which the groundmass is rich in orthoclase and plagioclase altered to turbid material and clear secondary albite (E 29193).

Exeter

Purplish brown basalt, partly massive and partly vesicular, with calcite and celadonite infillings and local calcite veins, forms the hill on which Rougemont Castle (Exeter Castle) [SX 921 930] is sited. A borehole [SX 9230 9293] in Exeter city centre, at the southern end of Longbrook Street, showed 10.2 m of purple-red and purple basalt beneath 15 m of Whipton Formation (Figure 10).

An exposure [SX 9212 9318] adjacent to the railway line below Blackall Road shows highly weathered, rubbly, spheroidally weathered, purplish brown lava with few vesicles and sparse calcite veining.

Barley Lane–Pocombe–Crossmead

A road cutting [SX 8999 9231] east of Barley Lane reveals 2 m of greyish purple basalt. A quarry [SX 8989 9177] north of The Quarries shows 3 to 4 m of purple, rather rotted, fractured lava, with amygdales containing a white or green mineral, and purple rubbly, non-vesicular lava with some hard sandstone admixtures. Purple Culm shale was reported beneath the lava in temporary excavations in the quarry floor. An exposure [SX 8986 9127] on the north side of Pocombe Hill shows 4 to 5 m of greyish purple, roughly banded basalt dipping at 8° to 052°; it rests on about 0.4 m of hard rubbly lava and sandstone, which in turn rests on purple-stained Crackington Formation shales and sandstones.

Exposures are still visible in the Pocombe Quarry (the Crossmead locality of Knill, 1969). The contact with the overlying Whipton Formation is visible in the entrance [SX 8994 9133] (p.45). A face [SX 8993 9139] on the east side shows 4 m of moderately massive, somewhat vesicular, purple basalt with subhorizontal white dolomite veins and near-vertical joints striking at 319°. The upper 4 to 6 m of the face are inaccessible, but apparently consist of lava. Another face [SX 8984 9154] is about 8 m high; the lowest 6 m are of purple dolomite-veined basalt; the inaccessible top 2 m are of rather rubbly lava, possibly with admixtures of basalt and sandstone. A face [SX 8977 9145] on the west side of the quarry shows purple, locally amygdaloidal basalt, extensively veined by dolomite and with local admixtures of hard, reddish brown, fine-grained sandstone and lava at the base. Near-vertical joints strike at 305°. Nearby [SX 8979 9143], clasts of purple lava are present in a matrix of reddish brown clayey sand. Near the western entrance [SX 8989 9129] are 5 to 6 m of purple finely vesicular basalt. Joints dip at 85° to 221°.

Opposite the quarry entrance, a cutting [SX 8990 9126] shows about 4 m of purple basalt with some thin dolomite veins. The lowest 1.2 m are hard and somewhat vesicular, and are overlain by 1.2 m of purple vesicular basalt; above are at least 1.6 m of purple basalt. Near-vertical joints spaced at 0.1 to 0.4 m intervals strike at 323°.

A quarry [SX 8992 9145] bordering on the grounds of Crossmead Hall has a lower face showing 8 m of purple basalt with white dolomite veins. The top 4 m or so of the face are inaccessible, but apparently consist of a rubbly facies of lava blocks with irregular blocks and interlacing veins of pale reddish brown sandstone.

West Town (Ide)

In the West Town Quarry, about 1.5 m of Halscombe-type lava consisting of irregularly jointed, pale greyish purple basalt with rounded xenocrysts of glassy quartz and plagioclase, and a small number of pseudomorphed phenocrysts of olivine (E 29169), are visible on the north side [SX 8855 9030]. The rock is pervasively altered, and relict pyroxene has not been recorded. On the south side of the quarry [SX 8860 9025], layering in rubbly purple-grey basalt dips at 24° to 147°.

Cuttings [SX 8874 9039] on the old railway west of West Town Farm show about 6 m of greyish purple, strongly altered, ophitic olivine-dolerite (E 65004), locally vesicular, with brown iddingsite phenocrysts (Pocombe-type lava). The groundmass of the lava is rich in orthoclase as well as plagioclase, indicating an alkaline character (E 65005). Alteration products include pseudomorphs of haematite and calcite after olivine pseudomorphs, iron-oxide and calcite disseminated through the groundmass, and calcite fracture veinlets. Igneous layering dips very approximately at 12° to 117°. Near-vertical joints strike at 207° and 255°. The nature of the contact with the Halscombe- type lava exposed in the nearby West Town Quarry (see above) is uncertain. Near West Town Farm, crags [SX 8887 9037] show 2 to 3 m of purple to greyish-purple-weathered, locally rather rubbly, irregularly jointed, locally veined basalt. A sheet of pale green, mottled pale red, hard sandstone 2.5 to 6 cm thick dips very approximately at 10° to 148°.

Halscombe–Knowle Quarry–Idestone

A quarry [SX 8827 8989] at Halscombe shows 2 to 3 m of massive greyish red-purple basalt with scattered 1 to 2 mm glassy quartz xenocrysts, and with local areas of hard reddish brown brecciated rock, possibly admixtures of sand and lava.

The lava in the base of the face of Knowle Quarry [SX 8740 8955] is hard, massive, greyish red-purple basalt with scattered quartz xenocrysts. The upper face shows more vesicular basalt. Tidmarsh (1932, p.723) recorded an upward gradation from the normal (i.e. basaltic) type through a flow-banded facies to a porphyritic or andesitic rock; xenocrysts of quartz and feldspar were abundant in both types. He noted that vesicles were developed in the upper part of the flow, and were up to 1.2 m long; they were filled with carbonates, agate, quartz, zeolites, and iron and manganese oxides. Tidmarsh calculated the direction of flow from the elongation of vesicles to be towards 035°. He also recorded tear faulting, indicated by horizontal slickensides, bringing the porphyritic facies into contact with the normal facies at the western end of the quarry.

Temporary exposures near the quarry entrance showed the highly amygdaloidal and vesicular base of the lava flow, resting with an irregular junction on about 1 m of thinly bedded breccia, sandstone and mudstone; the dip of bedding and the base of the flow is to the east at 10–20° (written communication, Dr R T Taylor, July, 1994).

Exposures of basalt [SX 8761 8879; 8768 8868] near Great Knoll, about 400 m north-north-west of Idestone, are of greyish to very dusky red-purple, very hard, brittle basalt with fairly common 1 to 2 mm glassy quartz xenocrysts. Basalt with some intermingled sandstone is visible intermittently [SX 8802 8845 to 8792 8853] in the banks of the lane leading to Idestone. Intercalations of sedimentary material are indicated by hard breccia [SX 8802 8845] and sandstone with lava [SX 8800 8843]; abundant reddish brown sandstone debris associated with vesicular lava is present in soils [SX 8812 8840] about 300 m east of Idestone. A quarry [SX 8826 8808], about 500 m south-east of Idestone, shows pale red to greyish red amygdaloidal basalt overlying breccia, although the contact is not exposed.

Dunchideock–School Wood Quarries–Lakeham

Just west of Dunchideock Bridge, lava apparently overlies cemented breccia [around SX 8824 8775]. Farther west, however, the breccia beneath the lava is apparently absent and does not reappear until a place [SX 8701 8795] about 270 m west of Ashlake Cross.

Crags [SX 8780 8765] of basalt are present in the hillside about 200 m east-north-east of Dunchideock Barton. Some 2 to 3 m of hard greyish purple basalt are overlain by about 2 to 3 m of vesicular basalt. An exposure [SX 8767 8752] just east of Biddypark Lane, near Dunchideock Barton, shows heavily fractured and jointed, pale greyish purple basalt, locally vesicular, and with some impersistent veins of reddish brown sandstone.

Exposures [SX 8740 8771] near a quarry in Orchard Copse, Dunchideock, show greyish red-purple, hard, compact basalt with reddish brown iddingsite phenocrysts. Exposures [SX 8742 8792] in a copse about 150 m east-south-east of Ashlake Cross show pale red-purple to greyish red-purple basalt with iddingsite phenocrysts.

A quarry [SX 8697 8787] at Rockclose Plantation, 700 m west-north-west of Dunchideock Church, shows 2 to 3 m faces in pale brownish grey to brownish grey basalt with 1 to 2 mm reddish brown iddingsite phenocrysts.

Exposures [SX 8743 8752] in a field about 200 m west-south-west of Dunchideock Barton are of pale purple basalt with iddingsite phenocrysts; joints dip at 47° to 267°.

The best exposures of the Exeter Volcanic Rocks around Dunchideock are in two disused quarries in School Wood (the Webberton Cross quarries of Knill, 1969).

School Wood: northern quarry   Near the entrance [SX 8761 8721], a face shows 3 to 4 m of hard purple-grey basalt, irregularly jointed, with some veins up to 3 cm thick of reddish brown fine-grained sandstone. At one point [SX 8758 8723], about 4 m of basalt show a crude layering which dips at 13° to 027°. The most prominent set of joints is near-vertical and strikes at 085°. Just to the north-west, layering dips at 24° to 035°. Nearby [SX 8755 8726], about 4 m of massive, irregularly jointed, greyish red-purple basalt with iddingsite (Plate 10a) pass up fairly sharply into 2 to 3 m of very vesicular purple lava, locally with an intricate network of mudstone 'veins'. Near the northern end of the northern quarry, faces [SX 8749 8730] show 8 to 10 m of massive, hard, purple-grey basalt. Layering dips at 26° to 042°. Two main sets of joints are present, dipping at 74° to 255° and at 83° to 155°. Other joints are subparallel to igneous layering. At the northern end of the quarry, on the south side, faces [SX 8748 8748] and large tumbled blocks are of purple vesicular to very vesicular lava (with some elongated amygdales), and complex admixtures of reddish brown sandstone and lava. Large blocks of vesicular lava are locally enclosed in sandstone. The appearance of this mixed facies is much more rubbly, and more reddish brown in colour, in contrast to the massive purple-grey basalt present all along the east face. Just south of the above, a face shows the fine-grained basalt facies (typical of the east face) overlying the mixed facies, and dipping into the east face (explaining the lack of mixed facies in the east face). About 6 m of basalt facies rest on about 5 m of mixed facies. The contact between the two facies is fairly sharp, and dips at about 20° to 045°.

School Wood: southern quarry   In the western corner [SX 8750 8708] of the quarry, sandstone and breccia are visible beneath lava; the sequence is given on p.40. The base of the flow consists of 2.8 m of purple basalt with white amygdales, complexly intermingled with reddish brown, somewhat flaggy, micaceous fine-grained sandstone, dipping at 20° to 065°; sandstone is dominant in the lowermost 0.6 m (Plate 10c). Along the face

[SX 8754 8701 to 8758 8704] that parallels the road are up to 10 m of massive, irregularly jointed, purple-grey basalt with iddingsite.

A quarry [SX 8765 8711] on the opposite side of the road to School Wood shows about 2 to 3 m of grey and purple basalt with a few near-vertical to steeply dipping sheets of hard reddish brown sandstone 1 to 3 cm thick. The basalt is grey, with a few reddish brown iddingsite phenocrysts, and a few glassy quartz xenocrysts. A 1.5 m face [SX 8762 8711], just to the west, shows complex admixtures of lava and reddish brown sandstone.

West of Penn Hill are exposures [SX 8779 8684] of about 1 m of rubbly weathered, purple basalt with moderately well developed joints which dip at 70° to 275°.

Lavas are exposed in the lane to Higher Ashton, near Lakeham, 200 m west of Lawrence Castle. On the east, the first exposures [SX 8737 8693], on the south side of the lane, are of reddish brown hard sandstone. The exposures from there westwards show soft, rotted, purple, vesicular basalt, locally with intercalations of hard splintery sandstone. At the bend in the lane [SX 8728 8612], about 2 m of purple hard basalt are exposed.

Agglomerate

Agglomerate has been mapped on the evidence of surface debris in a small area [SS 9713 0047 to 9715 0030] on the west side of the Killerton Park minette outcrop. The rock is pale red to greyish red and consists of angular pieces of feldspathic Culm sandstone and finely vesicular lava (some with bronze-coloured mica) in a somewhat vesicular matrix with scattered mica.

Earlier authors have described occurrences of agglomerate not located during the resurvey: Ussher (1902) noted at Killerton a quarry exposing breccia or volcanic agglomerate consisting of light red indurated sandstone with buff spots associated with numerous fragments of vesicular 'trap', and also described an old quarry at Posbury Chapel in which two horizons of 'trap' are separated by sediments overlying a body of agglomerate. Knill (1969) reported a small vent of coarse agglomerate at Crossmead, and also referred to possible agglomerate in a temporary exposure [SX 985 990] south of Sprydoncote.

Rhyolite

Small exposures [SX 7938 9888] of pale grey flow-banded rhyolite are present in the south side of a cutting at Summer's Lane, Neopardy. The outcrop is too small to show on the map. A trial pit [SX 7933 9887] showed 0.15 m of head on 0.55 m of rhyolite, with subhorizontal flow-banding, underlain by 2.57 m of purplish red clayey breccia. Rhyolite was absent in another pit dug 20 m west of the first, suggesting that the outcrop is of very limited lateral extent. The rhyolite is probably a fragment of a formerly much more extensive outcrop that has been preserved from erosion in the fault zone at the southern margin of the Crediton Trough. Its stratigraphical position is uncertain owing to poor exposure; it may be underlain by Bow Breccia, and faulted against Newton St Cyres Breccia. A thin section (E 66092) shows a very fine-grained matrix of quartz and argillised feldspar with irregular flakes of partly sericitised biotite, enclosing scattered shards of devitrified cryptocrystalline banded

material. This material is comparable in texture to rhyolite fragments from the Late Permian breccias of the Crediton Trough (e.g. E 65564) and is also geochemically similar (Table 15). The rhyolite outcrop at Neopardy is the only in-situ representative of the major acid igneous effusion associated with the emplacement of the Dartmoor Granite.

IGNEOUS CLASTS IN THE PERMIAN BRECCIAS

The breccias of the Exeter Group contain a variety of clasts, including much igneous debris (Table 11). Surprisingly, in view of the proximity of the Exeter Volcanic Rocks, basic lava clasts are rare, being almost entirely confined to the strata overlying the Exeter Volcanic Rocks. In contrast, acid igneous material is very common, particularly in the younger breccia formations; such debris provides a record of the nature, and history of uplift and erosion, of the Cornubian batholith from which it is derived. The presence of quantities of rhyolite and acid tuff provide evidence for episodes of suprabatholithic volcanism that are only sparsely represented by in situ rocks at the present day.

The oldest Permian rocks containing igneous clasts are the uppermost parts of the Cadbury Breccia, which include rare small pebbles of quartz-porphyry. Such material is commoner in the overlying Bow Breccia and Knowle Sandstone, where it occurs together with pebbles of rhyolite and acid tuff. None of these rocks is tourmalinised, but they are commonly affected by argillic alteration. Rare pebbles of rotted and bleached lamprophyre are locally present; they appear to have been minette originally, but their state of alteration is extreme, and there is some uncertainty about their original composition.

In the Late Permian part of the Exeter Group, the Crediton and Alphington breccias contain a variety of igneous clasts including flow-banded rhyolite, ignimbrite, acid tuff and quartz-porphyry. In the lower part of the Crediton Breccia, the clast assemblage includes untourmalinised pebbles and cobbles of acid lava, tuff and quartz-porphyry (e.g. E 65564, 65565, 65567), mostly pale grey or cream in colour, but in some cases with pervasive purple-red or brown iron oxide staining. Fragments of brecciated and tourmalinised sandstone (e.g. E 65566) and shale are also present, probably having originated as vent breccias. In the upper parts of the Crediton and Alphington breccias, a variable proportion of the acid igneous clasts is tourmalinised, most commonly with irregular anhedral fragments of blue tourmaline with fibro-radiate blue or blue-green tourmaline overgrowths (e.g. E 65506, 65568). Textural evidence suggests that the tourmaline is replacive, and that the bodies of lava or quartz-porphyry were invaded by boron-rich fluids prior to erosion and transport as pebbles.

In the Newton St Cyres and Heavitree breccias, the proportion of tourmalinised acid igneous clasts increases (e.g. E 65569, 65571) and sparse pebbles and cobbles of granite (E 54629) and aplogranite (E 1438) are present in the higher parts of both formations. The potassium feldspar 'murchisonite' fragments that are a distinctive feature of these breccias were considered by Dangerfield and Hawkes (1969) to have been derived from potassium-feldspar megacrysts in the roof zone of the Dartmoor Granite rather than from volcanogenic sanidine. In the present survey, this has been confirmed by specimens collected from 'Cromwell's Cutting', Pitt Hill [SX 8165 0020], near Crediton, where rock material adhering to murchisonite clasts (e.g. E 66093) is brecciated and tourmalinised biotite-granite.

The nature, source and age of the clasts in the breccias of the Exeter Group have been investigated by geochemical techniques including major- and minor-element analyses (Table 14) and isotopic investigations (Table 15). The major-element geochemistry confirms the very high levels of silica in all of the clasts analysed and their general granitic nature. In comparison with the analysis quoted for a typical biotite-granite from Dartmoor, the generally lower levels of Na and Ca and the high loss-on-ignition can be considered to be due to argillic alteration. Much of the variation in minor elements can be ascribed to alteration due to tourmalinisation, sericitisation and argillising processes.

A study of the neodymium and samarium isotope geochemistry has shown a contrast between clasts from the older and younger parts of the Exeter Group. The data presented in Table 15 demonstrate three distinct groups by epsilon neodymium parameter (ϵ_{Nd}), a factor (Faure, 1986) relating the $^{143}Nd/^{144}Nd$ ratio for a rock to the same ratio for a chondritic uniform reservoir. Positive ϵ_{Nd} values for igneous rocks indicate a contribution from differentiated mantle material, the proportion of such material increasing with the value of the epsilon parameter; in contrast, negative ϵ_{Nd} values show a contribution from crustal rocks, increasing with more negative values. Of the rocks analysed, the most negative are the acid clasts from the Bow Breccia, which have values showing an essentially crustal origin. Their ϵ_{Nd} values are close to those for the Bodmin Granite and for volcanic and acid intrusive rocks of the east Cornwall coast. Clasts from the Upper Permian have, in contrast, ϵ_{Nd} values showing some mantle contribution to the melt, lying close to the value quoted (Mrs D P F Darbyshire, personal communication, 1992) for the Dartmoor Granite.

The least negative ϵ_{Nd} value was obtained from the Exeter Volcanic Rocks, from material from the rhyolite outcrop at Neopardy, and is problematic, suggesting that this rock crystallised from acid magma substantially of mantle origin. ϵ_{Nd} values for the Exeter Volcanic Rocks of basic composition range from + 1.1 to - 1.5, and it is possible that extreme differentiation of a mantle melt may have given rise to a bimodal volcanic suite of which the Neopardy rhyolite is the only in-situ acid survivor. Bimodal Permian volcanic rocks of this type outcrop elsewhere in Europe, for example at the margin of the Sudetic Basin in northern Bohemia (Tasler et al., 1979).

Radiometric age determinations (Table 13) on acid igneous clasts from the Heavitree and Newton St Cyres breccias have given K-Ar and Ar-Ar ages, respectively, that are congruent with Rb-Sr, U-Pb and Ar-Ar ages for the Dartmoor Granite and its envelope.

Table 14 Major- and minor-element analyses of acid igneous rocks from the district and the adjacent Dartmoor Granite.

%	E 65567	E 65568	E 65569	E 66092	E 38430	E 50396
Major elements (weight per cent oxide)						
SiO_2	73.95	74.01	74.86	78.34	71.16	72.74
Al_2O_3	12.63	12.67	12.14	13.87	14.31	16.54
TiO_2	0.16	0.28	0.35	0.14	0.41	0.01
Fe_2O_3	2.29	3.29	3.83	1.82	3.05	0.95
MgO	0.18	0.47	0.34	0.18	0.49	0.11
CaO	0.18	0.22	0.10	0.08	1.33	0.27
Na_2O	1.46	0.70	0.57	0.25	3.24	0.46
K_2O	8.08	5.65	5.28	0.35	5.03	6.42
MnO	0.04	0.10	0.02	0.04	0.07	0.03
P_2O_5	0.10	0.17	0.15	0.01	0.19	0.24
Loss on ignition	0.68	2.01	2.01	4.87	1.08	2.22
Trace elements (parts per million)						
As	14	15	185	—	18	4
Ba	135	167	451	—	229	79
Cs	40	93	37	45	40	31
Nb	15	17	17	15	19	55
Rb	489	492	364	121	367	488
Sr	25	77	953	6	96	26
Sn	13	21	28	34	15	< 2
Th	6	14	25	7	21	3
V	7	25	31	13	30	< 2
Y	18	33	39	8	33	7
Zr	65	106	134	49	138	33

Analyses E 65567–66092 and E 50396 by T Brewer, University of Nottingham; analysis E 38430 from Darbyshire and Shepherd (1985). Specimens and localities as follows: E 65567 rhyolite clast, Crediton Breccia [SS 8910 0035]; E 65568 quartz-porphyry clast, Crediton Breccia [SS 8323 0080]; E 65569 quartz-porphyry clast, Newton St Cyres Breccia [SX 8444 9762]; E 66092 rhyolite outcrop [SX 7933 9887], Neopardy; E 38430 granite, Blackingstone Quarry [SX 784 857], east Dartmoor; E 50396 quartz-porphyry 'elvan', Sousson's Wood [SX 678 798], Dartmoor.

Table 15 Samarium/neodymium (Sm/Nd) isotope data for acid igneous rocks, including clasts from the Exeter Group of the Crediton Trough.

	Sm/ppm	Nd/ppm	$^{147}Sm/^{144}Nd$	$^{143}Nd/^{144}Nd$	Assumed age (Ma)	ϵ_{Nd}
RCSB	6.47	31.3	0.1249	0.512160	295	-6.7
RCSC	6.27	33.0	0.1150	0.512124	295	-7.0
SD 244	7.70	42.6	0.1093	0.512143	295	-6.4
SD 248	7.83	42.2	0.1123	0.512148	295	-6.4
SD 251	7.77	41.0	0.1144	0.512117	295	-7.1
E 65564	0.569	2.95	0.1168	0.512229	280	-5.2
E 65567	1.07	4.03	0.1614	0.512330	280	-4.8
E 65568	3.96	17.8	0.1345	0.512299	280	-4.4
E 66092	0.253	1.32	0.1158	0.512464	280	-0.5
Bodmin Granite (Median)	—	—	—	—	295	-7.1
Dartmoor Granite (Median)	—	—	—	—	280	-4.7

Samples RCSB and RCSC, felsite (?acid tuff) pebbles from the Bow Breccia [SS 722 018 and 7764 0144, respectively]; SD 244 rhyolite, Withnoe, east Cornwall [SX 4035 5178]; SD 248 quartz-porphyry dyke 'elvan', Withnoe, east Cornwall [SX 4037 5177]; SD 251 rhyolite, Kingsand, east Cornwall [SX 4415 5201]; E 65564 rhyolite clast, Crediton Breccia [SS 8873 0179], near Raddon Cross; E 65567 rhyolite clast, Crediton Breccia, Shute [SS 8910 0035]; E 65568 quartz-porphyry clast, Crediton Breccia, Deep Lane [SS 8323 0080], Crediton; E 66092 rhyolite outcrop, Summer's Lane [SX 7933 9887], Neopardy. All analyses by Mrs D P F Darbyshire, Natural Environment Research Council Isotope Geology Laboratory. Values for the Bodmin Moor and Dartmoor granites, personal communication, Mrs D P F Darbyshire, 1992.

The available evidence suggests that the Bow Breccia and Knowle Sandstone accumulated before and during the eruption of the Exeter Volcanic Rocks, between about 291 and 282 Ma. Acid igneous debris included in the Bow Breccia as pebbles was derived from a distant source having an essentially crustal origin probably from around the Bodmin Granite, which has given a Rb-Sr mineral isochron age of 287 ± 2 Ma (Darbyshire and Shepherd, 1985), or related to the Cawsand-Withnoe acid igneous rocks of east Cornwall. During an interval to 280 Ma, in which the Dartmoor Granite with its associated acid volcanic rocks was emplaced and uplifted, substantial erosion of the Exeter Volcanic Rocks and their enclosing sedimentary rocks took place.

DARTMOOR GRANITE

The north-eastern edge of the Dartmoor Granite occupies approximately 5 km² of high moorland in the south-west of the district. The ground forms a very small part of the total area of the intrusion, which occupies about 625 km² in central and west Devon and is the largest granite pluton in Britain. A whole-rock Rb/Sr isochron age of 280 ± 1 Ma (Lower Permian), initial $^{87}Sr/^{86}Sr$ ratio 0.7101, was determined for the Dartmoor pluton by Darbyshire and Shepherd (1985) (Table 13). This age for the cooling of the granite has been substantiated from U/Pb and Ar/Ar data (Table 13) by Chesley et al. (1993).

The margin of the Dartmoor Granite in the district is characterised by short intrusive segments and long faulted intervals with displacement along faults oriented northwest to south-east. Along the northern margin, northwest of Heltor Rock [SX 7997 8703], the intrusive contact trends east–west and is near-vertical, passing undeflected across the incised valley north of Leign Farm [SX 788 878]. The contact is displaced by a 1 km-long fault seg-

ment north–east of Burnicombe Down [SX 805 873], and trends approximately north-south along the eastern intrusive margin near Bridford where the contact is near-vertical or inclined steeply toward the east.

Lithology and petrology

The Dartmoor Granite is a granite sensu stricto in the terminology of Streckeisen (1976); it contains more than 20 per cent quartz and a modal proportion of potassium feldspar to total feldspar greater than 35 per cent. In the present district, it is a mottled pale grey to buff-weathering, pinkish white, coarsely porphyritic biotite-granite that forms part of the megacrystic facies or 'giant' granite of Brammall and Harwood (1923). This lithology represents the main intrusive phase of the pluton; there is no in-situ occurrence of the poorly megacrystic or 'blue' granite, formerly described as a separate intrusive phase (Brammall and Harwood, 1923), but now regarded as a gradational facies of the megacrystic granite (Hawkes in Edmonds et al., 1968; Hawkes, 1982; Hawkes in Selwood et al., 1984).

In the Bridford area, the granite contains feldspar megacrysts that are typically 40 mm in length and range up to 150 mm; those above 15 mm in length constitute between 9.5 and 21 per cent of the rock by volume (Hawkes in Selwood et al., 1984 and written communication, 1992; Table 16). Contoured megacryst distribution maps of the Dartmoor Granite, which show the present district to fall within the most richly megacrystic facies, are given by Dangerfield and Hawkes (1981) and Hawkes (1982, p.94). The feldspar megacrysts are predominantly greyish white or pinkish white orthoclase-perthites showing a patchy development of microcline-perthite. There is a lesser abundance of very pale greenish white plagioclase crystals, probably albite-oligoclase, that range up to 20 mm in size. Quartz occurs as composite megacrysts, up

Table 16   Summary of the feldspar megacryst data from the Dartmoor Granite.

Locality	SX grid reference	Volume per cent[1]	Megacryst preferred orientation
Pixie Rock, Middle Heltor Farm	7920 8760	18.5	56° to 016° 40–60° to 345°
Lower Heltor	7945 8763		subvertical strike 068° or absent
Westcott	7881 8716	20.0	15° to 060° or absent
Westcott Quarries	792 874	17.0	
Little Westcott	7922 8714		50–70° to 277° or absent
Heltor Rock	7997 8703	21.0	34° to 005° 79° to 278°
Burnicombe Down	8025 8700	17.0	5–20° to 070° 45° to 345°
South of Westacott	7873 8619	13.5	60–76° to 115°
Rowdon Rock	8121 8603	12.0	10–20° to 120°
Blackingstone Quarry[2]	7839 8578	9.5	
Blackingstone Rock[2]	7865 8560	16.0	

1 Data obtained from feldspars more than 15 mm in length provided by J R Hawkes (written communication, 1992).
2 In Newton Abbot district (Sheet 339); data from Selwood et al. (1984).

to approximately 25 mm in size, and also as a principal component of the matrix together with potassium feldspar, plagioclase and biotite.

Detailed petrographic descriptions of the Dartmoor Granite on the adjacent Okehampton (324) and Newton Abbot (339) sheets, given by Edmonds et al. (1968) and Selwood et al. (1984), indicate that the matrix grain size is mostly in the range 0.5 to 4 mm, and that muscovite, tourmaline, apatite, zircon, magnetite, and monazite occur in accessory abundances.

Modal analyses of megacrystic granite, poorly megacrystic granite and aplogranite from areas immediately adjacent to the present district are given by Edmonds et al. (1968) and Morton (1958), and are summarised in Table 17.

Table 17   Modal analyses of the Dartmoor Granite (volume per cent).

	1	2	3	4
Potassium feldspar	38	36	17	29
Sodium feldspar	29	23	36	24
Quartz	26	34	25	42
Biotite	7	7	21	4
Colourless mica, Zircon and tourmaline	—	—	1	1

1. Megacrystic granite, Blackingstone Quarry [SX 7839 8578] (Edmonds et al., 1968); 2. Poorly megacrystic granite, Merrivale Quarry [SX 5458 7530] (Edmonds et al., 1968); 3. Granodiorite, Clampitt Down [SX 819 848] (Morton, 1958); 4. Aplogranite, Clampitt Down [SX 819 848] (Morton, 1958).

Small quartz-tourmaline pods, a few centimetres in size, are relatively abundant at outcrop and as a component of the regolith that forms a thin cover to much of the pluton. There is some indication that large pods of quartz-tourmaline-haematite breccia are associated with the marginal faults, and this lithology is abundant near Burnicombe [SX 801 876]. Quartz-tourmaline rock with specular haematite occurs as brash south-west of Bridford and south of Hedgemoor [SX 805 855], just south of the district.

Greyish-cream-weathering fine-grained aphyric biotite-granite (aplogranite) with patchy pegmatitic enclaves forms two sill-like bodies 15 to 25 cm thick at Heltor Rock [SX 7997 8703] (Plate 11a). The upper body is 15 to 25 cm thick, and the lower is 45 to 90 cm thick; sill margins are parallel or subparallel with the megacrystic layering and inclined north.

Igneous enclaves

Sparse isolated spheroidal mafic segregations, 5 to 30 cm in size, occur along the northern granite margin at Westcott Quarries [SX 792 874] and at Pixie Rock, Heltor [SX 7920 8760]. They comprise plagioclase- and biotite-rich granitoid rocks with compositions in the approximate range granodiorite to quartz diorite. Inclusion margins are diffuse and most are ovoid in shape. The segregations appear to be typical of mafic granitoid rock enclaves that represent an early intrusive phase of many granite plutons. Mafic inclusions of the same type were described within the Dartmoor Granite by Brammall and Harwood (1923; 1932) and interpreted by them as xenoliths. Xenoliths of country rock are absent from the present district, but have been described from the Okehampton (324)

Plate 11a Heltor Rock, Dartmoor [SX 7997 8703], a classical Dartmoor tor with subhorizontal sheet joints in megacrystic granite. A conspicuous sill of pegmatitic aplogranite cuts the granite and is inclined north. The hammer is 0.3 m long (A 15280).

Plate 11b Megacrystic layering in Dartmoor Granite, Burnicombe Down [SX 8025 8700]. The feldspar megacrysts average 4 cm in length and define a subhorizontal layering dipping between 8° to 070° and 20° to 074°. The megacrysts account for about 17 volume per cent of the total granite. The coin is 2 cm in diameter (A 15281).

and Dartmoor Forest (338) districts (Edmonds et al., 1968; Reid et al., 1912).

Megacrystic layering

The feldspar megacrysts are tabular in shape and exhibit preferred planar orientations at the scale of an outcrop (Plate 11b; Table 16). Planar fabrics typical of the north-eastern pluton margin are inclined toward the east or north-east at low angles. By contrast, steep north-inclined fabrics or near-vertical megacryst fabrics that strike north–south occur within 1 km of the northern pluton margin. While most outcrops contain a single megacryst layering,

at Burnicombe Down [SX 8025 8700] a steeply inclined fabric in which individual feldspars are closely stacked or imbricated is patchily superimposed on the dominant low-angle foliation. Elsewhere, particularly around West-cott [SX 789 869], low-angle fabrics are absent and steeply inclined fabrics are not laterally persistent. Although there is no exposure adjacent to intrusive or faulted margins, brash suggests that the megacrystic layering persists to the margin of the pluton. Aphyric medium-grained granite occurs as brash in the extreme north-east of the pluton near Burnicombe [SX 803 876], and the lithology may be representative of the intrusive margin in that area. The lack of continuity between outcrops precludes a regional analysis of megacryst fabrics, and their structural significance is unknown.

Geochemistry

Modern geochemical data from the Dartmoor Granite are scarce. The geochronological study of south-west England granites by Darbyshire and Shepherd (1985) includes major- and trace-element determinations of seven Dartmoor Granite samples, one of which is from Blackingstone Quarry, only 200 m south of the district (Table 5). All samples are peraluminous high-silica granites enriched in Rb and with relatively low abundances of heavy rare earth elements like Y and Zr. According to Darbyshire and Shepherd (1985), the high initial $^{87}Sr/^{86}Sr$ ratio of 0.0710 and the granite mineralogy are consistent with an 'S-type' or crustal origin. When plotted on a Rb versus Y + Nb diagram of the type described by Pearce et al. (1984), the Dartmoor Granite samples form a tight cluster within the field defined for collision granites. The high Rb/Zr ratios (2.3 to 6.7) place the samples within the syn-collision field described by Harris et al. (1986); this signature discriminates between them and the post-collision and volcanic arc magmas.

Extensive exposures of pale grey-weathering, coarsely porphyritic biotite-granite occur at Pixie Rock [SX 7920 8760] near Middle Heltor Farm, and show a well-defined megacrystic layering of feldspars oriented at 56° to 016° and 32° to 025° on a scale of a few metres. A representative preferred planar orientation dips at 40–60° to between 340° and 350°. Vertical joints strike at 150°, and subhorizontal sheet joints with a 20 cm vertical spacing are present throughout. Exposures nearby on Lower Heltor [SX 7945 8763] are very highly megacrystic, but generally contain no preferred igneous layering; a vertical planar fabric striking 068° is present locally. Feldspar megacrysts average 40 to 50 mm in length (maximum 100 mm), and locally form up to 50 per cent of the granite volume.

Numerous small outcrops of porphyritic biotite-granite occur between Westcott Wood and Heltor Rock, but these rarely contain a conspicuous megacrystic layering. Near Westcott [SX 7881 8716], porphyritic granite with approximately 20 per cent of large megacrysts has, in large part, no preferred orientation or has a weakly defined subhorizontal layering inclined north-east. Exposures 200 m north-east from Little Westcott [SX 7922 8714] contain approximately 30 per cent megacrysts with relatively large cross-sectional widths; megacrysts 30 mm wide and 60 mm long are typical. The feldspars do not define a continuous layering, although a high-angle fabric inclined west is present locally. A single aplitic granite vein, 2 m long and 50 mm wide, cuts the granite and dips at 85° to 160°.

Heltor Rock [SX 7997 8703] is a classical Dartmoor tor with vertical joint sets striking at 010° and 095°, and subhorizontal sheet joints with 10 to 20 cm vertical separation (Plate 11a). The granite is grey or buff weathering, and contains approximately 21 per cent megacrysts that average 40 mm (maximum 100 mm) in length. The predominant feldspar layering is oriented at 34° to 005° and a sub-fabric at 70° to 278° is patchily developed. Two sill-like bodies of greyish cream to pale buff-coloured pegmatitic aplogranite, 15 to 25 cm and 45 to 90 cm thick, occur parallel or subparallel with the principal megacrystic layering. Granite at the eastern foot of the tor is highly megacrystic, but a preferred orientation is less well defined; where present, the layering is oriented at 40–65° to 360°.

Highly megacrystic granite crops out on Burnicombe Down [SX 8025 8700] and closely resembles that of Heltor Rock. Megacrysts average 40 mm in length and define a subhorizontal layering that is usually inclined in the range between 8° to 070° and 20° to 074°. Imbricate zones of feldspar laths inclined at 45° to 345° occur within poorly defined subhorizontal layers. Megacryst-rich granite that contains more than 30 per cent feldspar megacrysts is patchily developed.

Sparsely megacrystic granite is exposed 800 m south-south-west from Westcott, where the volume of feldspar megacrysts is less than 15 per cent. The maximum feldspar length is 8 cm and averages approximately 5 cm. The predominant megacryst layering is a high-angle fabric oriented at 76° to 115°; an exposure [SX 7888 8638] 200 m farther north-east exhibits a similar layering oriented 60° to 084°.

Rowdon Rock [SX 8121 8603] is a small granite tor located 300 m from the intrusive margin of the pluton. The principal lithology is a pale grey to buff-weathering sparsely megacrystic biotite-granite in which the megacrysts are typically small (maximum 5 cm with a single isolated 10 cm-long crystal), and commonly account for less than 10 per cent of the volume. A well-defined subhorizontal layering is inclined at 10–20° to 120°, toward the pluton margin. A sill-like tourmalinised pegmatite, 10 cm wide, is inclined subparallel with the megacrystic layering at 18° to 130°. Sheet joints, 20 cm apart, are present throughout.

METAMORPHIC AUREOLE OF THE DARTMOOR GRANITE

The increase in temperature and fluid activity associated with emplacement of the Dartmoor Granite resulted in extensive recrystallisation and growth of new mineral phases in adjacent country rocks. The degree of metamorphism is determined by the distance from the pluton margin, the composition and disposition of the country rocks and the distribution of introduced fluids. The presence of fluids rich in boron, carbon dioxide and chlorine can be inferred from a patchy distribution of metasomatic mineral phases that include tourmaline, axinite and scapolite. The source of the pronounced magnetic anomalies associated with the aureole of the granite is discussed in Chapter 9 (p.120).

The mapped width of the aureole along the northern margin of the granite, where a structurally ordered sequence has been exhumed, is in the range 1.5 to 2 km. Allowing for the effects of tectonic rotation, the stratigraphical thickness of metamorphosed strata in the pluton roof is in the approximate range 0.8 to 1.2 km in this part of the intrusion. However, the permeability of rocks in the roof-zone of the granite is thought to have influenced the width of the aureole, which expands to 2 km just west of the district in ground where Teign Chert is absent. A narrower aureole, 1 km wide, occurs along the eastern granite margin. There, the outer margin of the aureole is defined by major faults, possibly formed in a strike-slip regime, which separate a structurally complex Early Carboniferous sequence within the aureole from unmetamorphosed Late Carboniferous strata.

The compositional diversity of sequences along the eastern granite margin allows a subdivision into an inner, higher temperature, and an outer, lower temperature aureole (Morton, 1958). The boundary, defined principally by the degree of amphibolitisation in dolerites, is parallel with, and approximately 500 m east of, the granite margin. In this account, the effects of metamorphism on argillaceous and siliceous compositions, as well as on dolerites, are described in turn.

Argillaceous hornfels

The lowest grade of metamorphism of argillaceous rocks occurs within the outer 500 to 850 m of the aureole, with the development of massive to splintery hornfelses. The hornfelses occur interbedded with sandstones in Crackington Formation north of the River Teign in Cod Wood [SX 785 892] and around Steps Bridge [SX 803 884], in the upper part of the Ashton Shale in Bridford Wood [SX 7985 8819], and in some intervals of the Combe Shale in Stone Copse Quarry [SX 8237 8602], Bridford. A thin section (E 65526) shows that original sedimentary laminations are preserved and, although there is variable metamorphic mineral growth, particularly of sericite, metamorphic spotting is absent.

Closer to the granite, dark grey to black, spotted, chiastolite hornfelses are developed, in which the original claystone-siltstone layering is sometimes difficult to recognise, or is obscured by finely divided carbonaceous material. Spotted hornfelses of this type expand in thickness east of the 79 easting along the northern margin of the granite, where they are present in the Crackington Formation at least 1 km from the pluton margin. The Teign Chert, which may have acted as an impermeable cap-rock farther west, is absent from the granite roof in this ground. Typical thin sections (E 30833, 30835 and 30836) of hornfels reveal very dark grey claystone or siltstone with parallel or cross-laminations defined by sericite and carbonaceous clay, and distinct spots of chiastolite or chiastolite-sericite intergrowths.

Biotite-cordierite hornfels is present in Combe Shale brash near Bridford [SX 818 865] and on Christow Common [SX 822 852] (in the adjacent Newton Abbot district), approximately 100 to 400 m east of the pluton margin. Thin sections described by Morton (1958) show subidioblastic crystals of spongy biotite and porphyroblasts of cordierite, in part replaced by sericite, in a fine-grained, quartz-sericite matrix.

The highest grade argillaceous hornfelses crop out at only one locality in the district. Pale brown to buff-weathering quartzose hornfelses with ferruginous laminations occur 100 m south-east of Higher Lowton [SX 8080 8741]. A thin section (E 65534) consists of a very fine-grained and complex mixture of quartz, sericite and chlorite, with a layering enriched in opaque mineral grains.

Meta-cherts

The Teign Chert has a varied lithological composition that includes near-monomineralic silica rocks, siliceous shales, limestones, and tuff bands. The metamorphosed equivalents of this sequence were grouped together as 'calc-flintas' by Morton (1958); the description following is for the most part based on the petrographic data for different compositional layers given by that author.

Cryptocrystalline silica within high-purity cherts outside the aureole is recrystallised to a granoblastic quartz mosaic, 0.02 to 0.04 mm in grain size, together with chlorite and sericite. Microbiotic debris, including radiolaria casts that retain some ornament, are preserved at Stone Copse Quarry [SX 8242 8600], approximately 750 m from the granite margin. Chlorite and sericite in the groundmass are replaced by small amounts of biotite, epidote, tourmaline and allanite within about 400 m of the granite margin.

Low purity cherts, which contain larger original abundances of sericite, chlorite, calcite, dolomite and iron ores, produce colour-banded cherts on metamorphism. Fine examples of cherts with green, pink and grey layers occur near Higher Lowton [SX 8086 8760] and in Stone Copse Quarry. Morton (1958) reports assemblages that include: quartz-biotite; quartz-diopside; quartz-tremolite/actinolite; quartz-diopside-tremolite/actinolite; quartz-diopside-tremolite/actinolite-ferrohastingsite; and quartz-diopside-epidote-tremolite/actinolite-ferrohastingsite. Quartz is present as a fine-grained granoblastic mosaic, and the mafic phases in part form monomineralic layers.

Very low-purity (quartz-poor) cherts were also reported from Stone Copse Quarry by Morton (1958), who described amphibole-rich layers with the assemblages: ferrohastingsite-plagioclase-magnetite; ferrohastingsite-diopside-plagioclase; ferrohastingsite-epidote-plagioclase-(chlorite-calcite); tremolite/actinolite-zoisite-biotite-(chlorite-calcite); and pyroxene-rich layers with the assemblages: diopside-plagioclase-(tremolite) and diopside-plagioclase-ferrohastingsite-(allanite-sphene).

Cherts dominated by metasomatic phases occur along the northern crop near Birch [SX 8225 8725], where coarse-grained banded hornfelses contain axinite-amesite-spinel and lesser abundances of biotite-chlorite-tremolite/actinolite-calcite-pyrite. The predominant phase is axinite, which occurs as 1 mm-size subidioblastic blades. A second metasomatic chert variety occurs on the southern edge of Bridford Wood [near SX 801 877], approximately 30 m from the granite margin, where the assemblage petalite-allanite-pistacite-ferrohastingsite-axinite-sphene-apatite was recorded by Morton (1958). Petalite plates, up to 25 mm in size, poikilitically enclose diopside.

Metadolerites

Compared with the unmetamorphosed albite-dolerites east of the granite aureole, dolerites close to the pluton margin have different mineralogical compositions and record a progressive destruction of original igneous textures. Morton (1958) proposed a two-fold division of the aureole along the eastern granite margin as follows:

an *outer aureole* (about 0.5 to 1 km from the granite margin), in which igneous textures in dolerites are perfectly preserved or little altered, but which contain cloudy, inclusion-rich plagioclase of composition Ab_{95}–Ab_{46} (albite to sodic labradorite);

an *inner aureole* (less than 0.5 km from the granite margin), in which increasing destruction of igneous textures in dolerites is recorded, most or all of the original pyroxene is recrystallised, and plagioclase contains few inclusions and is Ab_{36} to Ab_{50} (labradorite) in composition.

Typical metadolerite in the outer aureole occurs in the Bridford quarries [SX 822 866] and in Stone Copse Quarry. Aphyric or ophitic textures are preserved in which plagioclase contains abundant minute inclusions of sericite, pale green amphibole, biotite or rare diopside, augite is at least partly altered to either actinolite or pale green hornblende and biotite, and opaque grains are replaced by leucoxene. Veins filled with chlorite and tremolite are common; patchy alteration to fine-grained, blue-green amphibole, turbid clinozoisite, dumortierite and schorlite occur locally. According to Morton (1958), apatite is less abundant and sphene more abundant that in unmetamorphosed albite-dolerites.

Extensively amphibolitised metadolerite within the inner aureole occurs in Scatter Rock Quarry [SX 8205 8550] (just south of the district boundary), and can also be seen north-west of Bridford near Seven Acre Lane [SX 811 868]. Within this metamorphic zone, recrystallisation has at least partially destroyed igneous textures; amphibolitic hornfelses occur within about 300 m of the granite margin. Plagioclase is recrystallised to a granoblastic aggregate of labradorite that is either clear or contains well-crystallised inclusions of actinolite, biotite and hornblende. Pyroxene is scarce and almost all replaced by colourless and well-formed amphibole. Amphibole either is zoned hornblende, which averages 0.3 mm and is up to 1 mm in size with a tremolite/actinolite core, or occurs as small grains of ?hastingsitic composition.

BRIDFORD LAMPROPHYRE

A temporary exposure [SX 8263 8752] near Weeke Barton, 1.6 km north-east of Bridford, exposed the margin and interior of a lamprophyre dyke. Biotite K/Ar determinations on a suite of samples from the intrusion gave a mean age of 300.6 ± 4.6 Ma (Rundle, 1976).

The Bridford Lamprophyre is a medium-grained minette; a modal analysis (volume per cent) is given by Dangerfield (written communication, 1991):

K-feldspar	36.6	Opaque oxides	3.1
Plagioclase	8.6	Apatite	2.1
Phlogopite	30.5	Chlorite	7.8
Pyroxene	4.0	Carbonate	5.5
Olivine	0.2	Quartz	1.6

According to Hawkes (petrographic description, 1972), the plagioclase is oligoclase in composition, and is in part replaced by granular and irregular developments of potassium feldspar. Phlogopite occurs in large ragged crystals which are both internally and marginally altered to green and colourless chlorites; pale green chlorite also occurs in fibrous interstitial patches and is present within some oligoclase crystals. Accessory apatite is abundant.

EIGHT

Cretaceous and Palaeogene rocks

CRETACEOUS: UPPER GREENSAND

Extensive outliers of subhorizontal Upper Greensand form the flat-topped Haldon Hills between the valleys of the River Teign and the River Exe in the adjacent Newton Abbot district. The northernmost tip of the hills extends into the south of the present district and forms a small outlier at Lawrence Castle [SX 875 862].

The Upper Greensand here is capped by Palaeogene flint gravels, and gravelly wash and head commonly obscures its outcrop. Exposures are rare; they show unconsolidated, locally glauconitic, fine-grained sands, which range up to an estimated 30 m in thickness, and rest unconformably on the Permian Alphington Breccia. The only exposure is on the east side of the hill, where a cutting [SX 8756 8617] shows yellowish green, silty, fine-grained sand. Faunas from the Upper Greensand of the Newton Abbot district indicate a range from Upper Albian to Cenomanian (Selwood et al., 1984). Hamblin and Wood (1976) divided the Upper Greensand succession of the Haldon Hills into four lithological units, but none of these can be recognised in the present district because of the lack of exposures.

The Upper Greensand was probably deposited in a shallow sea periodically subject to strong current and wave action. The fauna includes thick-shelled bivalves and corals, and is a nearshore marine assemblage (Selwood et al., 1984).

PALAEOGENE: TOWER WOOD GRAVEL

In the Newton Abbot district, the Haldon Hills are capped by three units of flint gravel, collectively termed the Haldon Gravels. Two of these, the Tower Wood Gravel and the Buller's Hill Gravel, are of Palaeocene to Eocene age (Selwood et al., 1984); the third is Head Gravel of Pleistocene age. The Tower Wood Gravel, named from Tower Wood Quarry [SX 8768 8567], just south of the district, is the oldest of the gravels and rests on the Upper Greensand; it is the only unit to crop out in the district, where it caps the outlier of Lawrence Castle.

The formation consists of up to 8 m of flint gravel with large unabraded flints set in a matrix of clay with a little sand. Pickard (1949) recorded a basal bed containing quartz and schorl pebbles at Tower Wood Quarry, but this is no longer visible. Some white rounded quartz pebbles are, however, locally present amongst the surface debris of mainly unworn flints. Selwood et al.(1984, p.124) noted that the flints of the Tower Wood Gravel are pale grey inside and white outside, and unabraded by transport. Their margins have been shattered in situ by frost action. The matrix clay is generally white with local brown staining. A section [SX 8755 8611] 45 m south of the castle shows about 0.6 m of unabraded angular flints in a matrix of yellowish brown to brown silty sandy clay.

The Tower Wood Gravel contains flints showing no evidence of transport, and probably formed by the in-situ solution of chalk. The basal bed consists of sand with quartz pebbles, and has been interpreted (Hamblin, 1973 a) as a residual deposit, formed by the solution of flint-free Lower Chalk and deposited as a basal gravel during the westward transgression of the Chalk sea. The clay matrix of the Tower Wood Gravel was probably derived from a hydrothermal kaolinite deposit in the Dartmoor Granite area, rather than by solution of chalk (Hamblin, 1973 a, b). The presence of a granite-derived suite of heavy minerals supports derivation of the Tower Wood Gravel from the Dartmoor area.

The Chalk fossils from the flints in the Haldon Gravels indicate that the earliest possible age for their formation is the Senonian stage of the Upper Chalk. By mineralogical comparison with deposits farther east, the Tower Wood Gravel was dated by Hamblin (1973 a) as pre-Bagshot Beds and possibly pre-Reading Beds, and thus might in part be of Palaeocene age.

NINE

Structure

The Devonian and Carboniferous rocks of the district were folded and faulted during the Variscan Orogeny. The Carboniferous rocks are unconformably overlain by Permo-Triassic rocks that are little deformed and have a relatively simple structure. In the north of the district, Permian rocks occupy the fault-bounded Crediton Trough; elsewhere in the district they form an eastward-dipping sequence. Information about the deeper structure of the district has been obtained from magnetic and gravity surveys.

DEEP BASEMENT

Evidence for a deep magnetic basement in Devon is provided by an east–west-trending magnetic high north of the Dartmoor Granite, which is terminated to the north by a pronounced gradient zone extending from the Tiverton Trough to Bideford Bay. The source of the long-wavelength anomaly is unknown, but must be deep seated and is probably of pre-Devonian age. Many alternative interpretations are possible (Cornwell, 1991); one model indicates a steep northern margin at depths of 3 to 4 km beneath the area of Culm rocks and, to the south, a deep-seated continuation of the magnetic body beneath the Dartmoor Granite. The upper part of the body appears to abut the northern edge of the non-magnetic granite. The form of this magnetic basement may be determined partly by Variscan thrusting as well as by the form of the granite. The basement rocks indicated by the magnetic evidence extend westwards under much of Devon and into the offshore region. Their eastern margin extends in a north–south direction through the Exeter district, very broadly paralleling the edge of the main outcrop of the Permian rocks, and suggesting that this line might reflect a major crustal structure.

The Dartmoor Granite is associated with a pronounced gravity low, part of which is shown on Figure 31. The form of the low suggests that the granite surface dips steeply to depths of 10 to 12 km adjacent to the Exeter district.

MAGNETIC ANOMALIES ASSOCIATED WITH CARBONIFEROUS ROCKS

The aeromagnetic data show elongated large-amplitude anomalies within about 3 km of the margin of the Dartmoor Granite (Figure 30) and closely associated with outcrops of Lower Carboniferous rocks (Cornwell, 1991). The main part of the anomalies is negative, suggesting a strong remanent magnetisation, and the form of the anomalies commonly indicates source rocks at depths of more than 50 m. It is most likely that the anomalies are due to a zone of pyrrhotite formed from pyrite-bearing shales during the intrusion of the Dartmoor Granite (Thomson et al., 1991). The most northerly anomaly (A1 in Figure 30) coincides in part with the Drewsteignton Anticline and indicates an eastward extension of 4 to 5 km into the Exeter district. The anomaly A2 (Figure 30) probably indicates a concealed structure involving Lower Carboniferous rocks.

FORM OF THE CREDITON TROUGH

The density contrast between the Carboniferous and the Permian rocks has enabled the shape of the Crediton Trough to be modelled using Bouguer gravity anomaly data. Data from Edmonds et al. (1968), and Davey (1981 a and b) and BGS regional gravity surveys have been supplemented by additional surveys during the present work (Cornwell et al., 1992 a).

The Bouguer gravity anomaly map (Figure 31) shows that the Crediton Trough is associated with an elongated gravity low with an amplitude of up to 7 milligal (mGal). North of the trough, the background levels over the Culm are approximately constant, but south of it they are more variable. The gravity low associated with the Dartmoor Granite appears towards the western end of the trough. To the east, the change in background values across the Crediton Trough is probably related to a density contrast within the Culm, the Crackington Formation having a higher density than the more arenaceous Bude Formation. The effects of the Exeter Volcanic Rocks are uncertain; although they appear to be localised at the surface, magnetic data suggest that they are more extensive at depth, and may be continuous at the bottom of the trough (p.98; Cornwell et al., 1992 b).

The results of the three-dimensional modelling of the Crediton Trough (Cornwell et al., 1992 a) are shown in Figure 32. The regional field data have been adjusted to include both the effects of the Dartmoor Granite and the regional density change in the Culm, and the residual anomalies have been modelled using three density contrasts (- 0.15, - 0.20 and - 0.25 Mg/m^3), the value of - 0.20 Mg/m^3 being the most realistic. Although the maximum depth to the base of the Permian varies in the three models, between 0.6 and 1.0 km, the shape of the trough is similar in all three cases, with the maximum thickness of Permian rocks in the widest part of the basin.

More detailed models for the trough are obtained from the interpretation of individual profiles, and two examples are shown in Figure 31 (b and c). Profile AA' illustrates the steep, faulted southern margin to the trough and a possible concealed fault in the Permian to the north. The second profile (BB') passes close to Ashclyst Forest and illustrates the contrast in the thickness of Permian rocks to the north and south of the Culm inlier.

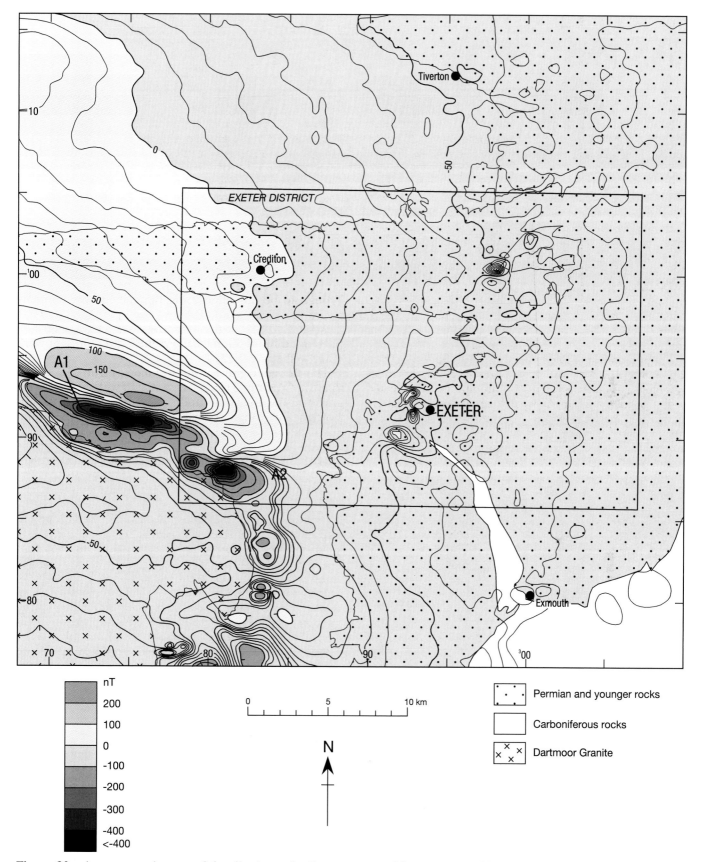

Figure 30   Aeromagnetic map of the district and adjacent areas, with contours at 10 nT intervals.

The map is based on data acquired by a survey flown at a mean terrain clearance of 150 m along north–south flight lines 400 m apart. Aeromagnetic survey data east of the 00 easting were acquired at a flight height of 305 m and are not included.

Key to geology: New Red Sandstone stippled; Dartmoor Granite shown by crosses; other rocks (mainly Carboniferous) unshaded.

a.

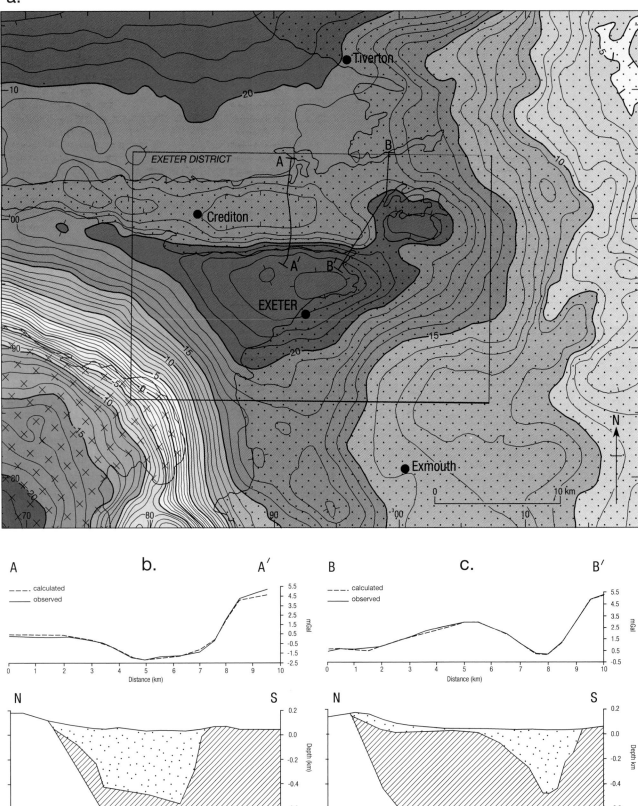

Figure 31 (a) Bouguer gravity anomaly map of the district and adjacent areas, with contours at 1 mGal intervals. Density for Bouguer reduction 2.55 Mg/m³. Key to geology as in Figure 30.
(b) and (c) Residual gravity anomaly profiles AA′ and BB′ (see (a) for location) and models for the Crediton Trough. Background field 19 mGal. Crackington Formation (density 2.65 Mg/m³, diagonal lines), Bude Formation (density 2.60 Mg/m³, unshaded), Permian rocks (density 2.45 Mg/m³, stippled) (cf. Cornwell et al., 1992a).

a.

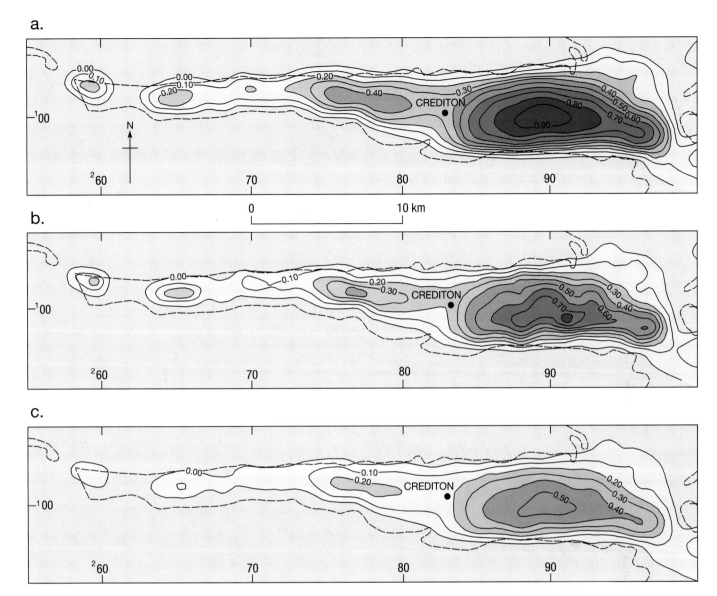

b.

c.

Figure 32   Alternative models for the form of the Crediton Trough, based on three-dimensional interpretation of residual Bouguer gravity anomalies. Smoothed contours at intervals of 0.1 km below OD (0) for density contrasts of: (a) - 0.15 Mg/m^3, (b) - 0.20 Mg/m^3, (c) - 0.25 Mg/m^3.

Broken lines indicates boundary between New Red Sandstone and Carboniferous rocks.

The aeromagnetic map (Figure 30) for the Crediton Trough area shows that short-wavelength anomalies due to near-surface lavas are superimposed on more regional changes due to deep-seated magnetic bodies. There is also evidence of an intermediate-wavelength anomaly associated with the trough, indicated in Figure 30 as slight bends in the 20 to - 40 nanotesla (nT) contours at about the 00 northing. This magnetic feature has been isolated by removing a simple second-order polynomial field, and the residual anomaly is plotted in Figure 26. Possible explanations for the anomaly are: the presence of lavas near the base of the Permian sequence; a susceptibility contrast between the Permian and the Culm; or the existence of a thrust in the underlying magnetic basement.

Durrance (1985) noted the close resemblance of the shape of the Crediton Trough to the model described by Gibbs (1984), in which the development of extensional basins is related to the reactivation of listric faults curving down to the mid-crust; down-stepping faults may be present on both margins of the basin. Faults of this type appear to be present along the north side of the Crediton Trough (Davey, 1981 b) and along the south side between eastings ^{2}83 and ^{2}96. Durrance (1985) proposed that the Crediton Trough had formed by extensional reactivation of a south-dipping Variscan thrust, on which the Crackington For-

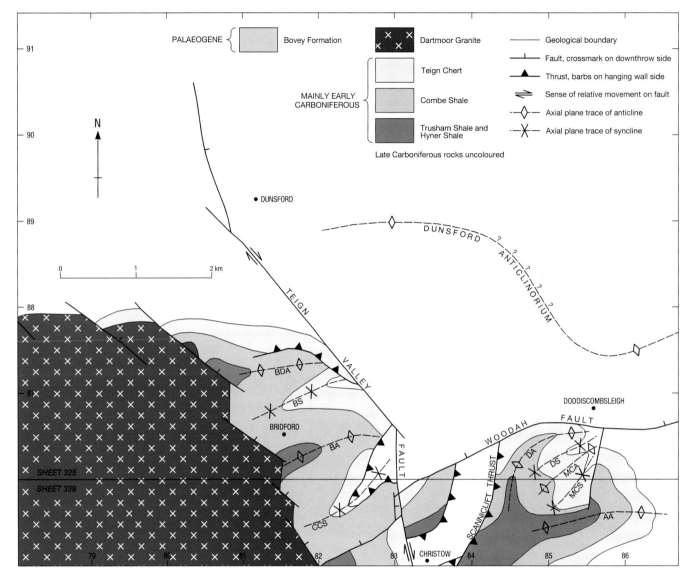

Figure 33 Simplified geological map of the Dunsford–Bridford–Doddiscombsleigh area showing locations and names of the major folds and faults. Igneous rocks (except for Dartmoor Granite) omitted for clarity. Key to abbreviations for folds:

BDA	Birch Down Anticline	DA	Doddiscombsleigh Anticline	BS	Bridford Syncline
DS	Doddiscombsleigh Syncline	BA	Bridford Anticline	MCS	Mistleigh Copse Syncline
CCS	Christow Common Syncline	MCA	Mistleigh Copse Anticline	AA	Ashton Anticline

mation had been displaced northwards over the Bude Formation; this model would explain why the trough coincides along much of its length with the boundary between the Crackington and Bude formations.

FORM OF THE PERMO-TRIASSIC ROCKS OUTSIDE THE CREDITON TROUGH

The main outcrop of the Permo-Triassic rocks dips east at 5° or less. Dips in the breccia formations south of Exeter range up to 20°, but this figure includes original depositional dips. The easterly dip affects the Jurassic

rocks of districts farther east, but not the Upper Greensand; it was therefore probably formed in the Late Jurassic or Early Cretaceous. In the eastern part of the district, the eastward decrease of Bouguer anomaly values reflects the increasing thickness of low-density Permo-Triassic rocks into the Wessex Basin. South and southeast of Exeter the gravity data indicate the existence of an additional low-density body. The most likely explanation of the low is that it represents an unexpected thickening of Permo-Triassic rocks of density 2.4 Mg/m^3 in the western end of a concealed trough (the Otterton Trough of Gregory and Durrance, 1987) analogous to that at Crediton. Alternatively, the anomaly could be due to a

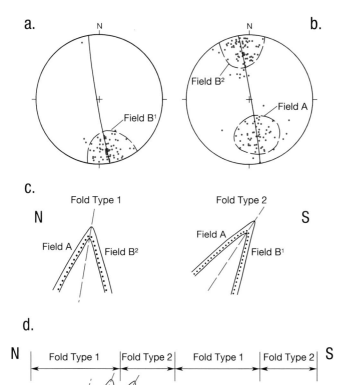

Figure 34   (a) and (b): stereograms of poles to bedding planes in the Crackington Formation of the Whitestone area (1:10 000 sheet SX 89 SE), (a) inverted bedding, (b) right-way-up bedding, (c) diagram of fold types in the Crackington Formation, (d) diagram showing variation in fold type and dip of axial planes due to fanning around large-scale folds. See text for explanation.

southward thickening of pre-Permo-Triassic rocks with a density of 2.55 Mg/m^3, possibly associated with a fault or thrusts.

STRUCTURE OF THE DEVONIAN AND CARBONIFEROUS ROCKS

The Devonian and Carboniferous rocks of the district were folded on mainly east–west axes during the Variscan Orogeny. The youngest rocks affected by this folding in south-west England are of Late Carboniferous (Westphalian C) age, dated at about 310 Ma. Folds in the Crackington Formation in the Exeter district range between upright and overturned to the south. In the northern part of the district, the folds in the Bude Formation are mostly upright. In the Middle Teign Valley, there is an anticlinorial structure in which the fold axial planes are upright in the central part, but fan over progressively to the north and south; all the fold axes plunge steeply to the east off the Dartmoor Granite. In that part of the structure lying within the Exeter district, the folds are overturned to the north.

Folds in Lower Carboniferous strata

The Devonian and Early Carboniferous rocks in the south-west of the district have been folded into a series of tight, easterly plunging, kilometre-scale anticlines and synclines (Figure 33). The folds are upright to inclined, with axial surfaces inclined to south or south-east; for the most part, the folds plunge at 27° to 50° towards the east or east-north-east. The regional fold plunge is attributed to uplift that accompanied emplacement of the Dartmoor Granite (Morton, 1958). The locations and names of folds in the Bridford and Doddiscombsleigh areas are shown on Figure 33. Measured orientation data for each fold are given by Barton (1992) and Selwood et al. (1984).

Folds in Upper Carboniferous strata

The shales and sandstones of the Crackington Formation are folded on approximately east–west axes. The folds visible in the field vary in wavelength from 0.5 m, or less, to about 10 m.

The stereograms in Figure 34 distinguish poles to inverted bedding from poles to right-way-up bedding in the Crackington Formation of the Whitestone area (1:10 000 sheet SX 89 SE). Plotting of poles to bedding produces similar stereograms for the Tedburn St Mary area (1:10 000 sheet SX 89 SW) and the Dunchideock–Bridford area (1:10 000 sheet SX 88 NE and part of SX 88 NW). Two fold types (Types 1 and 2), which grade continuously into each other, can be distinguished and are shown diagrammatically on Figure 34. Type 1 folds are upright and both limbs are right-way-up. The northern limbs of anticlines are represented by the more steeply dipping part of field A on the stereogram (Figure 34), and the southern limbs by field B^2. Type 2 folds are overturned to the south. The northern limbs are right-way-up, less steeply dipping than those of Type 1 folds, and are represented by the less steeply dipping part of field A; the southern limbs are inverted, and are represented by field B^1 (Figure 34). The variation in axial plane dip from very steeply north-dipping in Type 1 folds to less steeply north-dipping in Type 2 folds can be explained by fanning around larger folds, as shown in Figure 34.

The Carboniferous rocks locally show a single cleavage. Grainger (1983) noted that the cleavage tends to be parallel to the northward-dipping limbs of the folds. Hinge zones of the folds are generally more intensively cleaved and jointed than the adjacent limbs. Sandstones may show a radial fracture cleavage. In the aureole of the Dartmoor Granite, the cleavage is masked by the effects of thermal metamorphism.

The Ashton Shale crops out in the core of an anticline that extends from Tedburn St Mary to Nadderwater (Figure 3). The outcrop is truncated by a north-west-trending fault

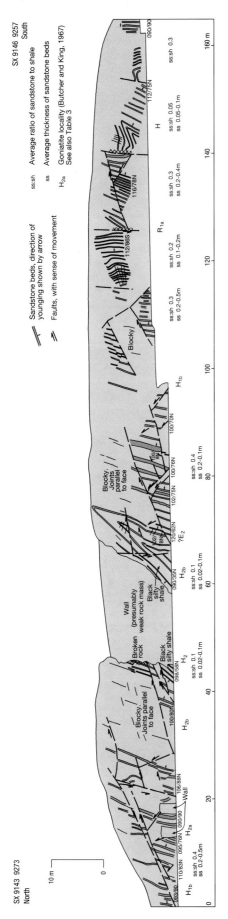

Figure 35   Vertical north–south section in the river-cliff at Bonhay Road, Exeter, showing structure of the Crackington Formation (after Grainger, 1983, fig. 47).

along the valley of the Nadder Brook. Grainger (1983) has shown that the anticline can be traced into Exeter where the Bonhay Road section [SX 9143 9273 to 9146 9257] (Figure 35) shows the southern limb of the anticline. A syncline containing Ashton Shale is present east of Longdown. Minor outcrops of Ashton Shale south of Newton St Cyres were interpreted by Scrivener (1988) as a possible thrust succession resting on Crackington Formation turbidites; alternatively, they may lie in the core of an anticline. The Bonhay Road section (Figure 35) shows mainly overturned (Type 2) folds and numerous fractures, both steep and flat-lying.

In the Dunsford area, the Dunsford Anticlinorium (Barton, 1992) trends east or east-south-east across the Crackington Formation (Figure 33). It links with a similar structure north of Doddiscombsleigh (Edwards, 1991). North of the axis, the strata are near-vertical, inclined north, and are upward-facing; very rarely the beds are vertical or inclined south at high angles and inverted. South of the anticlinorial axis, the strata are inclined south and are upward-facing. A regional asymmetry, defined by vertical or near-vertical strata in the north limb and rather shallower dips in the south limb, suggests that the axial surface of the Dunsford Anticlinorium is inclined south, consistent with deformation during translation towards the north; the fold may reflect an important structure at depth.

Selwood and McCourt (1973) described a low-angle (less than 10° dip) thrust, which they termed the 'Bridford Thrust', located 1 to 1.5 km south of the Dunsford Anticlinorium, and trending subparallel to its axis. According to these authors, east of the Teign Valley Fault the thrust passes north-east from Birch [SX 8220 8731], through the railway cutting near Sheldon [SX 835 875], to Idestone. Although minor structures are present at the two first-named localities, similar structures are not uncommon in the Crackington Formation elsewhere in the Bridford area, and there is no evidence for low-angle deformation anywhere along this line. Grainger (1983) examined clay-mineral assemblages from areas north and south of the postulated Bridford Thrust and found no significant differences between the samples. In particular, samples from immediately either side of the supposed fault showed a close match, suggesting that there had been virtually no dislocation of strata.

No axial plane can be defined for the Bude Formation owing to the lack of way-up data, but the average fold axis, plunging gently east-south-east, is defined by the spread of bedding poles. Sparse field evidence indicates that upright folds are commoner in the Bude Formation than in the Crackington Formation. Durrance (1985) proposed that the two formations have been juxtaposed along a thrust on which the Crackington Formation has been transported northwards over the Bude Formation, and which now lies concealed beneath the Crediton Trough.

Faults in Lower Carboniferous strata

In the Bridford area, north- or north-west-directed movement of several kilometres has occurred along faults associated with folding of the Lower Carboniferous and Devonian sequences. The Scanniclift Thrust (Figure 33),

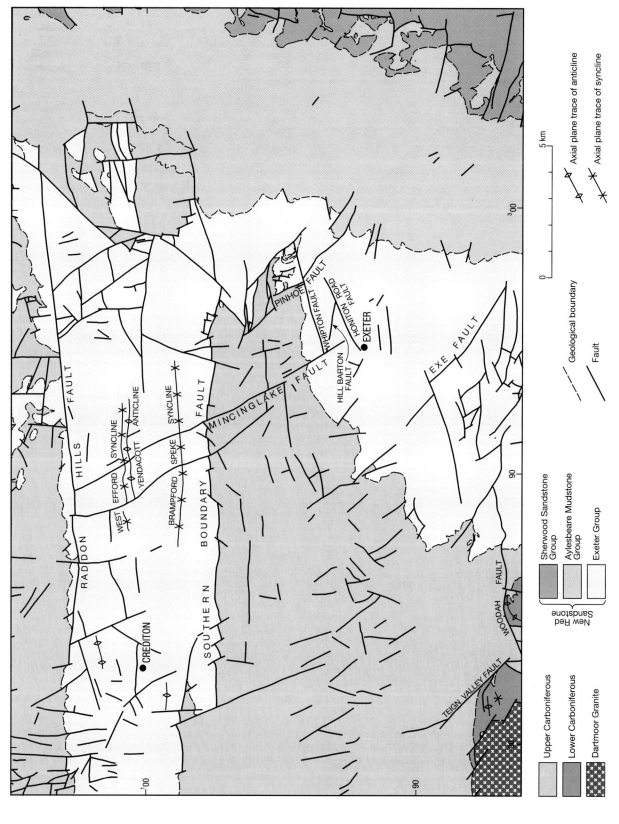

Figure 36   Distribution of faults in the district, showing also minor folds in the Permian rocks of the Crediton Trough.

near Scanniclift Copse [SX 843 860], Doddiscombsleigh, has carried inverted Teign Chert and Combe Shale across Crackington Formation; facing evidence inferred from sedimentary structures shows that the Crackington Formation below the thrust is the right way up. Around Woodah Farm [SX 848 866], the tectonic junction between the Teign Chert and Crackington Formation is a major east-north-east-trending normal fault, the Woodah Fault (Figure 33).

Faults in Upper Carboniferous strata

The distribution of faults in the Upper Carboniferous rocks of the district is shown on Figure 36. Faulting is thought to be more common in the Carboniferous rocks than is apparent from the number of faults marked on the map, because it is only where lithological contrasts are present that displacements can be recognised. During folding, there was slip along bedding planes, which commonly culminated in fracturing parallel to bedding. Such strike faults are likely to be much more common than is indicated on the map, but few can be identified with certainty, except where brecciated fault rock is present.

The main faults in the Crackington Formation are part of a regional set of fractures formed in response to north–south-directed compression during the Variscan earth movements (Dearman, 1963). Most of the faults are oriented north-west to south-east, and to a lesser extent north-east to south-west. Many show dextral wrench movement as well as normal movement. Fault movement on the north-westerly trend persisted into the Palaeogene in the adjacent Newton Abbot district. The age of the strike-slip movement on the Teign Valley Fault (Figure 33) is uncertain; more than one phase has been documented elsewhere from faults with this orientation. At least one earlier phase, synchronous with emplacement of the granite, is inferred from the present data.

Faults in the Dartmoor Granite

The spatial association of north-west-trending faults with the margins of the Dartmoor Granite, and with its aureole, suggests that intrusion was accompanied, or immediately followed, by large vertical and lateral displacements. There is a marked contrast between the northern margin of the granite, where stratigraphical units are broadly parallel with the pluton roof and the aureole is relatively wide, and the eastern margin, where tightly folded wall-rocks are intersected by the granite at large angles, and the aureole is relatively narrow and bounded by faults. The field relations suggest that, during intrusion, a tectonic stress regime of northward compression operated and that this was accompanied by translational faulting along the eastern margin of the pluton.

STRUCTURE OF THE PERMO-TRIASSIC ROCKS

The Variscan Orogeny was succeeded by lithospheric extension which produced fault-bounded basins throughout southern England (Chadwick, 1985). The Crediton Trough is one of these extensional basins, and gravity data

suggest it is a half-graben that was probably formed by tensile reactivation of a southward-dipping Variscan thrust. Basin formation was followed by a period in which later sedimentary sequences overlapped the earlier faulted basin margins; this more widespread depositional phase is represented in the Exeter district by the Aylesbeare Mudstone Group, which overlaps on to the Variscan basement at the eastern end of the Crediton Trough.

Faults

The Crediton Trough is bounded on its southern side by an east–west-trending normal fault with a throw of up to 800 m. At the eastern end of the trough, faults partly define the Ashclyst Forest inlier of Crackington Formation. Adjacent to the inlier, the outcrops of the Cadbury Breccia and Thorverton Sandstone are repeated around Killerton Park, indicating the presence of another uplifted block, bounded by east-west and east-north-easterly faults and underlain at relatively shallow depth by Culm. This conclusion is supported by gravity surveys (Cornwell et al., 1992 a).

Along the north side of the Crediton Trough, the Raddon Hills Fault can be traced from near Sandford to Bradninch (Figure 36). The field evidence is poor, but a gravity survey (profile 6 of Davey, 1981) across the Crediton Trough about 5 km east of Crediton shows it to coincide with one of several faults that form step-like features along the northern margin of the trough.

North-west-trending faults that extend into the Crediton Trough (Figure 36) probably represent reactivated Variscan faults. Gravity surveys (Cornwell et al., 1992 a) suggest that thickness variations in the Permian sequence take place across these faults. These variations are particularly evident across the Sticklepath Fault at the western end of the Crediton Trough, in the Okehampton district.

The distribution of faults elsewhere in the Permo-Triassic outcrop is shown on Figure 36. Faults with east–west to east-south-east trends are common, and are probably related to the tensional event that initiated the Crediton Trough. Two main fault trends are discernible in the Permo-Triassic rocks south of Broadclyst: in the Pinhoe area (Bristow and Williams, 1984) and in the Exminster area (Bristow, 1984 a), south-east or east-south–east-trending faults are displaced by later east-south-easterly, east-north-easterly, or east-trending faults.

Few faults have been detected in the Aylesbeare Mudstone outcrop because of the lack of lithological contrast. Faults cutting sandstones at the top of the Exmouth Mudstone and Sandstone near Woodbury trend between north-west and north-north-west. Faults cutting the Budleigh Salterton Pebble Beds trend between east-west and east-south-east, downthrowing to the south; they appear to postdate a set of north–south-trending faults. The boundary between the Budleigh Salterton Pebble Beds and the Otter Sandstone is faulted along most of its length by north–south faults locally offset by east–west faults downthrowing to the south.

In the Whipton-Pinhoe area, Permian sedimentation was probably influenced by contemporaneous fault movements (Bristow and Scrivener, 1984). The Monkerton Formation is present only to the end of the Mincinglake

Fault, while the Whipton Formation, Alphington Breccia, Monkerton Formation and Heavitree Breccia are confined to the west side of the Pinhoe Fault (Figure 36). The Hill Barton Fault may have influenced the deposition of the Heavitree Breccia and the Monkerton Formation; the Heavitree Breccia is not found north of the fault and the Monkerton Formation is much thinner on its south side. The east-north-east-trending Honiton Road Fault is presumed to have limited the deposition of the Monkerton Formation, which does not occur south of the fault.

Folds

Permian rocks occupy the broad asymmetrical Brampford Speke Syncline in the south of the Crediton Trough. The axis of the syncline trends east-west from Rewe to near Newton St Cyres (Figure 36). Gentle east–west-trending folds, the West Efford Syncline, Yendacott Anticline and other small folds around Crediton, affect the Permian rocks (Figure 36). Few folds occur in the Permo-Triassic outcrop outside the Crediton Trough, but at Exminster the Heavitree Breccia forms a broad eastward-plunging faulted syncline.

Resistivity surveys (Henson, 1971, fig. 6.6; Sherrell, 1970, figure 7) have indicated east–west and north-west to south-east ridges and troughs in the base of the Budleigh Salterton Pebble Beds at Woodbury Common. These are thought to be depositional features formed by channeling of the upper surface of the Littleham Mudstone, not incompetent folds of the Alpine Orogeny (as was suggested by Henson, 1971, p.214).

130

TEN

Quaternary

INTRODUCTION

During the Quaternary (Pleistocene and Recent) period, which represents the last 2.3 million years, there were marked climatic variations in the British Isles. Glacial and periglacial phases alternated with temperate intervals (interglacials and interstadials). The stages of the Quaternary recognised in Great Britain (Mitchell et al., 1973) are shown in Table 18.

Ice sheets covered extensive parts of Britain during the Anglian, Wolstonian, and Devensian stages. During its maximum advance in western Britain, an ice sheet of probable Wolstonian or Anglian age reached the coast of north Devon and Cornwall and also impinged on the Scilly Islands (e.g. Mitchell and Orme, 1967), but did not override the mainland. Consequently, the only glacial deposits known on the mainland of south-west England are patches of ice-transported material (till) on the north coast, for example at Fremington near Barnstaple and Trebetherick Point in north Cornwall, and in the Isles of Scilly.

Large variations in sea level occurred, with sea levels up to 150 m below the present level at times of glacial maxima, when huge volumes of water were locked up in the ice sheets. The rivers draining the district were thus graded to varying base levels; the deposits of the flood-plains that correspond to these differing base levels are now preserved as river terrace deposits.

No interglacial deposit is known from the Exeter district, but deposits attributed to the Ipswichian Stage are present at a site [ST 162 006] near Honiton, about 10 km east of the district boundary. There, mammalian remains included hippopotamus, elephant, giant ox and red deer (Turner, 1975).

The most recent (Devensian) ice sheet, reaching a maximum extent about 18 000 years ago, probably did not extend any farther south than South Wales. Nevertheless, because the peninsula of south-west England lay in the periglacial zone, features formed during earlier glacial stages were mostly obliterated (Cullingford, 1982). There was probably widespread formation of solifluction deposits (head) in the district during this period. Buried-channel deposits, probably graded to low Devensian sea levels, are present in the Exe estuary.

As a result of the melting of the Devensian ice sheets in the latest Pleistocene, sea level rose and drowned many estuaries in south-west England, including the Exe estuary. Sea level continued to rise in the Flandrian, but more slowly, and the estuary became filled with sediment.

Three types of Quaternary deposits are present in the Exeter district: head deposits, which formed largely under periglacial conditions; fluvial deposits; and marine deposits formed in the Exe estuary. The distribution of the fluvial and marine deposits is shown in Figure 37. In addition, Harrod et al. (1973) have recorded small areas of wind-blown silt (loess) on the outcrop of the Budleigh

Table 18 Stages of the British Quaternary.

Series	Stage	Climate	Deposits in Devon
RECENT	Flandrian	I	*Peat, Alluvium *Marine deposits of Exe estuary
PLEISTOCENE	Devensian	G	*Head *Loess *Buried channels of Exe estuary
	Ipswichian	I	Honiton *Hippopotamus* site *River terrace deposits
	Wolstonian	G	Fremington Till (north Devon) *River terrace deposits
	Hoxnian	I	Not known
	Anglian	G	Not known
	Cromerian and older	I	Not known

Abbreviations for climate: I = interglacial;
G = glacial

* Deposits are present within the Exeter district.

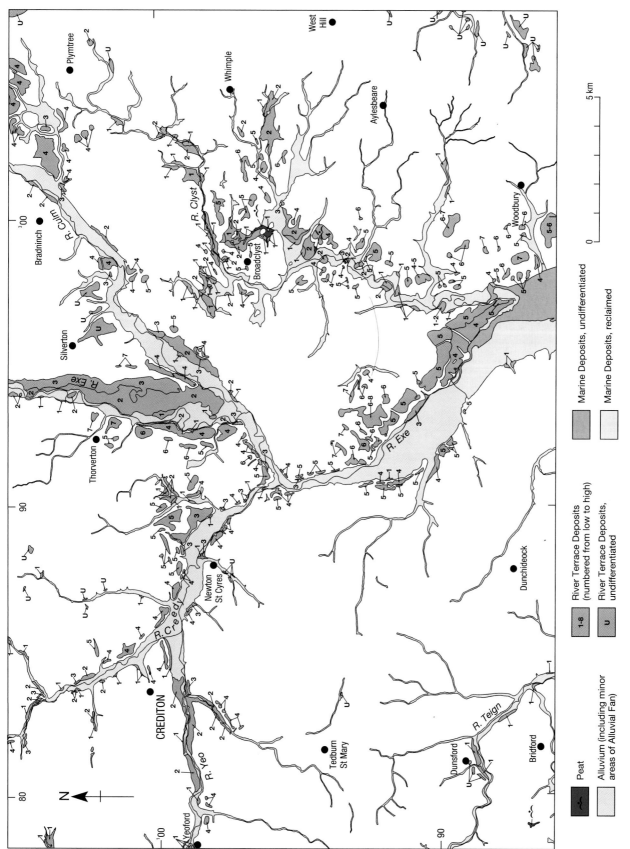

Figure 37   Distribution of peat, alluvium, river terraces and marine deposits in the district.

Salterton Pebble Beds at Lympstone Common [SY 040 854], just south of the district. This material was considered to have been blown in from the east during the coldest phase of the Devensian.

Dating of the Quaternary deposits of the district is hindered by their fragmentary nature, and the lack (except in the Flandrian deposits) of fossils or other material that can be dated by radiometric methods. Consequently, it has not been possible to correlate accurately the older Quaternary deposits with the sequence of stages recognised elsewhere in Britain.

HEAD

The term 'head' was introduced by De la Beche (1839) to describe rubbly slope deposits in coastal sections of south-west England. It has since been widely used to describe structureless or poorly bedded mixtures of clay, silt, sand and stones which are believed to have moved downslope mainly by solifluction (soil-flow) in a periglacial climate. In these conditions, alternate freezing and thawing of the surface layers results in a slow downslope flow of waterlogged soil and other unsorted material. Grainger (1983, pp.271–272) noted that: 'Periglacial processes contributing to soil development consist of congelifraction (the fracturing and opening up of fissures in parent rock by freeze–thaw stresses), and congeliturbation (the mixing or differentiation of soil and weathered rock profiles by differential heaving due to freezing and thawing). On slopes, the combination of these processes with gravitational stress results in accelerated mass movement of weathered material by gelifluction, thaw slumping and frost creep...' Soil creep, hill wash, and other mass movement processes are still active in the movement of surface materials.

Three types of head have been recognised in the district: blanket head and regolith, older head, and valley head (the latter termed 'Head' on the published 1:50 000 scale map).

Blanket head and regolith

Much of the outcrops of the Carboniferous and, to a lesser extent, the Permo-Triassic formations of the district are locally mantled by a deposit, mostly less than 2 m thick, which is composed partly of in-situ weathering products of the parent rock ('regolith'), and partly of material that has been subject to mass-movement processes (including solifluction). It is not possible to determine to what extent, if at all, regolith material has been transported by solifluction or other mass-movement processes, and the term 'blanket head and regolith' is used for this material. The deposit is so widespread that it has been omitted from the map for the sake of clarity. The thickness of the blanket head and regolith varies with topographical position. It is generally absent on sandstone ridges on the Bude Formation outcrop and on hilltops and ridges on the Crackington Formation, where weathered shale and sandstone are at or close to surface. Thicker head deposits are present on shallower slopes adjacent to ridges and hilltops.

Blanket head and regolith is very poorly sorted, and shows little or no evidence of bedding or grading; it is particularly well developed on the Crackington Formation and Bude Formation. The clay matrix and sandstone fragments may be deeply red-stained close to the boundary with the Permian. On the Crackington Formation, it comprises brown, reddish brown or grey silty clay, with angular sandstone, silty shale and siltstone fragments, resting on weathered in-situ siltstone, shale and sandstone.

An exposure [SX 8659 9213] (Plate 12) at Springdale landfill site, near Whitestone, showed a section typical of many in the blanket head and regolith on the Crackington Formation, consisting (beneath 0.3 m of soil) of up to 1.2 m of yellowish brown silty clay with fragments of shale, predominantly angular, and some sandstone fragments. The basal 0.3 m is gradational into weathered in-situ shale and sandstone.

Blanket head and regolith on the Bude Formation is generally less than 2 m thick, but several metres are locally present on the north-facing slopes of the Raddon Hills. A cutting [SS 9163 0468]

Plate 12   Temporary exposure [SX 8659 9313] at the Springdale landfill site, Whitestone; up to 1.2 m of blanket head and regolith, consisting of yellowish brown silty clay with angular shale and some sandstone fragments, overlies weathered in-situ shale and sandstone of the Crackington Formation (A 15319).

near Terley, Cadbury, shows a section typical of blanket head and regolith developed on the Bude Formation, consisting of 1.0 m of reddish brown clayey sandy silt with scattered angular sandstone fragments up to 0.1 m across, on 1.0 m of reddish brown silty sandy clay with abundant purple-stained shale fragments generally less than 40 mm across.

Blanket head and regolith on the Permo-Triassic outcrop generally closely resembles the solid formation from which it derives. The breccia-derived blanket head and regolith consists of a brown or brownish red, variably sandy, silty or clay-rich matrix with gravel-size clasts from the parent breccia formation. A typical section showing blanket head and regolith developed on Crediton Breccia was seen in an excavation [SX 8823 9984] at Higher Rew, Newton St Cyres, which showed 1.4 m of reddish brown, extremely silty and sandy clay, with abundant gravel of angular and subrounded sandstone, hornfels, quartz-porphyry and potassium feldspar. In Exeter, 3.66 m of blanket head and regolith resting on Alphington Breccia were recorded in a borehole [SX 9396 9227] at Heavitree Brewery.

Blanket head and regolith derived from the Permian sandstones comprises brown or brownish red, clayey or silty, fine-grained sand, with an admixture of very coarse-grained sand to fine gravel clasts derived from the thin beds of fine breccia, where these are present in the parent formations. A section [SS 8095 0212] at Aller Barton, West Sandford, showed 1 m of reddish brown sandy and silty loam with some gravel, resting on red clayey and silty sands of the Creedy Park Sandstone.

In the Dunchideock area, flints and cherts derived from the Upper Greensand and Tertiary deposits of the Haldon Hills are present in soils for about 2 km away from the hills, but are present only rarely north-east of the valley of the River Kenn. A section in a cutting [SX 8817 8620] near Lyalls shows 2 to 2.5 m of reddish brown, gritty, gravelly, unbedded loam, with large (0.1 m) angular flints and some Greensand chert, and smaller angular fragments of flint, chert, shale, sandstone and porphyry in a clayey sandy silt matrix, resting on Alphington Breccia.

Details of the blanket head and regolith are given in the Technical Reports which cover the district (p.x).

Older head

The outcrop of the Aylesbeare Mudstone is locally mantled by a dissected thin layer (generally less than 1.0 m) of gravelly clay or clayey gravel containing rounded quartzite pebbles and cobbles derived from the Budleigh Salterton Pebble Beds. Where this deposit is thicker and more extensive, it has been mapped as 'Older head'. The deposits occupy upland areas on the Aylesbeare Mudstone outcrop south of Whimple. The surfaces of the deposits have a low relief and dip gently westward away from the ridge formed by the Pebble Beds. The older head consists of pebbles and cobbles, mainly of rounded quartzite set in a yellowish brown, greyish orange, or reddish brown matrix that varies from silty clay to clayey silty fine-grained sand. The concentration of clasts varies, and gives rise to stony clay where clasts are few, and to clayey gravel where they are abundant. The thickness of the older head is usually less than 3 m but exceptionally may reach 4 m.

The abundant pebbles prove that the deposit has been transported, but its structureless, poorly sorted character rules out a fluvial origin. The older head is, therefore, believed to have originated as a solifluction sheet which moved westwards from the Pebble Beds ridge. The present outcrops are probably the dissected remnants of a formerly more widespread sheet.

Older head in the Plymtree area consists of reddish brown to yellowish brown clayey gravel, with rounded quartzite pebbles; 1 m of the deposit was seen in an excavation [ST 0553 0351] at Norman's Green.

In the Whimple area, the older head descends and passes westwards into river terrace deposits. The passage is gradational, and an arbitrary boundary has been taken between them in the Rockbeare area. Bristow (1984 b) considered the older head to be younger than the Second Terrace of the River Clyst. In the Hand and Pen area, a borehole [SY 0456 9581] showed up to 2.5 m of sandy clay with gravel (probably older head) overlying more than 2.5 m of dense cobble gravel (probably Second Terrace). Boreholes near Strete Farm [SY 038 953], Strete Ralegh, showed the older head to consist of up to 2.8 m of sandy clay with gravel.

Near Rockbeare, boreholes [e.g. at SY 0086 9467] through the older head that caps the ridge north of Treasbeare Farm showed the deposit to consist of 3.0 to 3.6 m of brown to reddish brown and yellow silty and sandy clay with some gravel and traces of organic matter. Older head forms a well-defined cap to the ridge east of Rockbeare House; typical sections of up to 1.1 m of reddish brown stony clay are present in ditches [SY 0371 9391 to 0369 9380]. A similar ridge capped by older head is present about 1 km north-west of Aylesbeare, and ditch sections [SY 0313 9302 to 0357 9289] show stony orange-brown and reddish brown clay. East of Aylesbeare, very pebbly soils [SY 044 922] and 1 m or more of clayey gravel in ditches [SY 0456 9209 to 0488 9173] indicate a high proportion of clasts in the older head.

Older head in the Woodbury area is typified by an excavation [SY 0177 8743] which showed 0.15 m of brown clayey loam soil with pebbles, overlying 1.4 m of reddish brown to orange-brown silty clay, clayey silt, and clayey silty fine-grained sand with rounded pebbles scattered fairly abundantly throughout.

Details of the older head deposits are given in the Technical Reports which cover the district (p.x).

Valley head

Valley head consists of locally derived rock debris and may comprise every variation between sandy and silty clay, and clayey and silty sand, with a variable pebble content. The valley head occupies valley sides and bottoms, and probably formed by a combination of solifluction, soil creep and slopewash; the latter two processes continue at the present time. Thicknesses of valley head deposits vary greatly: up to about 6 m have been measured in the district, but 1 to 2 m is a more usual range. On the published 1:50 000 scale map the deposit is referred to simply as 'Head'.

On the outcrop of the Crackington and Bude formations, the valley head consists of brown, greyish brown or grey clay with angular fragments of sandstone. Close to the boundary with the Permian rocks, the clay is reddish or purplish brown in colour. Numerous small streams on the Carboniferous outcrop have cut small sections, mostly overgrown, in valley head. In the Newton St Cyres area, a typical section [SX 8643 9644] near Sherwood shows 3.0 m of brown clay with a very poorly sorted mixture of angular and subangular sandstone and siltstone fragments, and flakes of grey and brown weathered shale, resting on Crackington Formation. A thicker section [SX 8890 9652] north of Bailey shows 5.0 m of brown clay, with subangular and angular sandstone fragments, resting on shale.

Farther south on the Crackington Formation outcrop, a stream bank section [SX 8846 9119] near Webby's Farm, Long-down, shows valley head, consisting of 0.6 m of grey silty clay and yellowish brown loam, on 0.7 m of grey and yellowish brown clayey gravel and gravelly clay with a shale-rich matrix and scattered clasts of angular Culm sandstone, resting on Crackington Formation.

On the south side of Ashclyst Forest, the valley head occupies fan-shaped areas [SY 001 984 and 011 985] and consists of angular fragments of Crackington Formation sandstone and some unweathered shale clasts in a sparse matrix of sandy clay.

Valley head typical of that developed on the Bude Formation of the district is exposed in a stream-bank [SS 8806 0393] near Lake Farm, Stockleigh Pomeroy, and consists of 2.2 m of reddish brown, mottled orange-brown, silty and sandy clay with abundant fragments of red- and purple-stained Culm sandstone and siltstone ranging from fine gravel to cobbles; some vein quartz pebbles are present. Crude horizontal bedding is suggested by the orientation of flattened pebbles.

Cuttings [SS 9053 0472] at Higher Coombe Farm, Cadbury, show about 2.0 m of an unbedded rubbly deposit consisting of hard purple to grey angular sandstone blocks, up to 0.15 m across, in a matrix of pale reddish brown silty sandy clay, probably in part weathered shale and siltstone, resting on purplish grey Bude Formation sandstone.

The composition of valley head developed on the Permo-Triassic formations reflects the local lithologies. In many areas it includes rounded and flattened sandstone and vein quartz pebbles derived from river terrace deposits.

Stream exposures [SS 8921 0189] near West Raddon are typical of valley head on Crediton Breccia; they consist of 1.0 to 1.5 m of brown clayey gravel that thins rapidly away from the stream course; the gravel consists of angular to subrounded Culm sandstone clasts up to 0.12 m across in a matrix of silty sandy clay and clayey sand, resting with an irregular base on the breccia.

A typical valley head section [SX 8808 8709] on the Alphington Breccia of the Dunchideock area near the Lord Haldon Hotel shows 1.8 m of unbedded clayey gravel and gravelly clay with subangular clasts mainly less than 7 cm across of sandstone, hornfels and porphyry, with a few flints in the top 0.4 m, in a matrix of silty sandy clay. Farther south, valley head sections [SX 8876 8654; 8891 8659] near Lower Byes Plantation show up to 0.7 m of loam, overlying up to 1 m of gravel containing large angular to subangular pebbles and cobbles of flint and Greensand chert, and angular to subangular clasts of sandstone, shale, chert and hornfels, overlying Alphington Breccia.

A thickness of 6.1 m of valley head, comprising red silty and sandy clay, has been proved above Heavitree Breccia in a borehole [SX 9185 8809] near Peamore House, and 3.7 m of reddish brown sand and gravel were present in another borehole [SX 9185 8744] in a shallow valley to the south.

In the Pinhoe area, valley head up to 3.2 m, and possibly up to 5 m, thick has been proved. A representative borehole [SX 9703 9431], which typifies the valley head on Dawlish Sandstone, showed the following sequence:

	Thickness m
VALLEY HEAD	
Clay, sandy, soft to firm, friable, reddish brown	0.50
Sand, brown, mottled	1.00
Sand and gravel, loose, reddish brown	0.50
Clay, very sandy, yellowish brown	0.50
Sand, fine-grained, reddish brown, with a little fine and medium gravel	1.20
DAWLISH SANDSTONE	
Clay, reddish brown, hard	0.50

On the Aylesbeare Mudstone outcrop, valley head varies between gravelly clay, clayey gravel, and gravel. It is typically a poorly bedded and very poorly sorted deposit of rounded quartzite pebbles and cobbles in a clayey matrix. Typical sections are seen in incised stream courses ('goyles'), as in the Rockbeare area, where a goyle section [SY 0546 9364] about 600 m north of Little Houndbeare Farm shows about 1.4 m of brown, unbedded, clayey sandy gravel with rounded pebbles and cobbles from the Budleigh Salterton Pebble Beds, resting on 0.4 m of reddish brown silty clay (Aylesbeare Mudstone).

Valley head on the Otter Sandstone outcrop is mainly variably stony sand, and sandy gravel; it contains quartzite and other clasts from the Budleigh Salterton Pebble Beds, together with flints and Greensand cherts derived from river terrace deposits. Soils on the Otter Sandstone outcrop are locally prone to gullying in heavy storms, and considerable volumes of sand can be washed down into the dry linear valleys that are floored by valley head. A section [SY 0642 9063] in a deeply cut stream-cliff in Woolcombes Plantation, Harpford Common, shows exposures of up to 6 m of valley head on Otter Sandstone, consisting of orange-brown clayey gravel and gravelly silty sandy clay, with rounded quartzite pebbles.

Details of valley head sections are given in the Technical Reports which cover the district (p.x).

PERIGLACIAL STRUCTURES

Fossil periglacial structures recorded from the district include oriented stones, ice-wedge casts, and possible patterned ground.

Oriented stones, including many with vertical long axes, have been recorded at a few localities within the district, on the outcrop of the Budleigh Salterton Pebble Beds. Examples were noted in a pit [SY 0326 8780] just south of Woodbury Beacon, at Venn Ottery Quarry [SY 065 913], and in Blackhill Quarry [SX 031 852] just south of the district. This phenomenon is attributed to the rotation and tilting of pebbles by differential frost heaving, and is generally considered to be an indicator of a periglacial climate. There is also a tendency for stones to move upwards through a deposit during repeated episodes of frost heaving.

At Venn Ottery Quarry [SY 065 913 to 067 910], bedding in the Budleigh Salterton Pebble Beds is periglacially disturbed to depths of up to 6 m below ground level, but at the north-western end of the quarry, only the upper 1 to 2 m are affected.

Fossil ice-wedge polygons near Newton St Cyres, at an altitude of 76 m above OD, suggest that permafrost in Devon was not confined to the upland areas (Cullingford, 1982). Gregory (1969, p.36 and plate 8) has recorded ice-wedge casts filled with head at Blackhill [SY 027 854], just south of the district. Patterned ground, possibly of periglacial origin, is present on Colaton Raleigh Common [SY 0473 8714 to 0500 8710] (Edwards, 1984 c). It occurs on the outcrop of the Budleigh Salterton Pebble Beds as hummocky ground forming cells typically 50 by 40 m in size separated by linear areas of low relief.

RIVER TERRACE DEPOSITS

River terrace deposits are developed in the valleys of the Exe, Clyst, Creedy and Culm, and their tributaries (Figure 37). They represent the deposits laid down on the former floodplains of these rivers, and range in height from 0.5 m to about 60 m above the present floodplains. Eight levels have been mapped, and are numbered in order of increasing altitude; a few terraces are shown as undifferentiated where there is insufficient evidence to allow their correlation. The upper surfaces of the terraces are commonly eroded and degraded. It has proved possible to correlate the terraces of the various river-systems, so that the same numbering system is applied throughout the district. There is no good evidence for dating the terraces. Durrance (1969, 1974) considered that buried terraces in the Exe estuary were of Devensian age.

Most of the terrace deposits consist predominantly of gravel, with a thin loam capping locally present, particularly on the Second and Third Terraces of the Exe Valley. On the higher terraces, erosion has removed any loam that might have been present. The variations in lithology and clast types within the river terrace deposits are summarised below; the distribution of the numbered terraces referred to is shown in Figure 37. The thickness of most terrace deposits does not generally exceed about 2.5 m, but recorded thicknesses range between 0.6 and 7.6 m. Details of thicknesses of terrace deposits are given in the Technical Reports covering the district (p.x).

The **Eighth Terrace**, about 60 m above the floodplain, has been recognised only in the Polsloe area of Exeter; it forms part of an extensive gravel spread which extends from St Leonard's to near Polsloe Priory and is probably composite, including material belonging to the Sixth and Seventh terraces in its lower part. It is at its highest (68 m above OD, 58 m above the River Exe floodplain), in the north. A section [SX 9328 9294] at Polsloe Road showed 1.3 m of brownish red sandy clay and interbedded gravel with rounded and flattened pebbles of sandstone and vein quartz, resting on Alphington Breccia. A borehole [SX 9335 9220] at Exeter School proved 2.13 m of clay-rich gravel resting on Alphington Breccia.

In the Exe Valley, the **Seventh and Sixth terraces** are present between Thorverton and Brampford Speke and north of Rewe, and in the Exeter city area; their surfaces lie at about 44 m and 33 m respectively above the floodplain. A few patches are also present in the Clyst Valley around Exton. The Seventh and Sixth Terraces in the Exe Valley north of Stoke Canon consist of gravel with Culm sandstone pebbles and cobbles. Two patches of Seventh Terrace east and south-east of Thorverton [SS 931 022; 927 015] attain 77 m above OD, and give rise to stony soils with abundant flat Culm sandstone pebbles and cobbles. North of Rewe, two patches of Seventh Terrace form a ridge [SS 949 014 to 949 006] attaining 69 m above OD.

Between Yellowford and Brampford Speke, three patches of Sixth Terrace [SS 926 005; SX 924 995; 921 988] consist of Culm sandstone gravel with some vein quartz. The Sixth Terrace is present beneath part of Exeter city centre, where it consists of discontinuous thin beds and lenses of gravel within clayey silts. The Sixth Terrace in the Rockbeare area consists mainly of gravel containing rounded quartzite pebbles. Subrounded flints also occur in association with the quartzite pebbles in the Sixth Terrace of the Clyst Honiton area [e.g. SX 9995 9270].

The **Fifth Terrace**, with its surface at about 22 m above the floodplain, is present at only a few places in the Exe Valley in the Thorverton and Brampford Speke areas, but is widely developed between Exeter city centre and Topsham on the east side of the valley and, to a lesser extent, on the west side of the valley in the St Thomas area. In the city area, it consists of gravel containing Culm sandstone and vein quartz clasts, in a matrix varying from sand to clay. A typical section [SX 9114 9423] above Cowley Bridge Road showed 2.3 m of gravel resting on weathered Crackington Formation shales, and consisting of rounded and flattened pebbles and cobbles of sandstone and vein quartz with inter-

stitial brown clay. Another section [SX 9073 9370] at Exwick showed 2.3 m of coarse gravel with abundant sandstone cobbles. A borehole [SX 9065 9204] near Merrivale Road proved 4.03 m of gravel resting on Whipton Formation and consisting of flattened and rounded pebbles of Culm sandstone and vein quartz in reddish brown sand. In Exeter city centre, an excavation [SX 9220 9225] showed 4 m of gravel, consisting of rounded pebbles of Culm sandstone in a red clay matrix.

The Fifth Terrace is present in the Culm Valley at a locality [SX 962 993] near Columbjohn where it consists of poorly sorted pebbles and cobbles up to 10 cm in diameter. The commoner constituents are well-rounded quartz and quartzite, with subsidiary pebbles of angular and subangular sandstone, lava and flint. The maximum thickness of the deposit is 2 m.

The Fifth Terrace between Countess Wear and Topsham is composed mainly of coarse-grained pebble and cobble gravel with a sparse, reddish brown, sand matrix. A borehole [SX 9428 9074] near Southbrook School proved 3.2 m of gravel, consisting of cobbles and pebbles of sandstone in reddish brown clay and sand.

Scattered Fifth Terrace remnants are present along the valleys of the Clyst and Creedy. Near Rockbeare in the Clyst Valley they consist of clayey gravel; rounded quartzites are the commonest constituents, with flints forming a minor proportion of the pebbles. The maximum thickness is estimated to be 1.5 m. Small scattered patches of Fifth Terrace around Dog Village consist mainly of well-rounded quartzite pebbles. Small areas of Fifth Terrace in the Clyst St George area are predominantly gravel with a high sand content. The Creedy Valley terraces consist of pebbly loam and gravel containing Culm sandstone and vein quartz pebbles.

The **Fourth Terrace** is developed to a varying degree along all the main river valleys, and its surface lies about 12 m above the floodplain. In the Exe Valley, it is present on either side of the river in the Rewe and Brampford Speke areas, and is also widely developed between Countess Wear and Topsham.

Brampford Speke is sited mainly on Fourth Terrace; a typical section [SX 9286 9927] in a river cliff near Fortescue Farm shows 2.1 m of gravel consisting of rounded and flattened Culm sandstone pebbles and cobbles, with some vein quartz and interstitial reddish brown sand, resting on weathered red sandstone. The Fortescue Farm Borehole [SX 9287 9938] proved 4.5 m of Fourth Terrace deposits. A cutting [SX 9273 9850] for a footpath at Brampford Speke exposes 3.5 m of gravel similar to that near Fortescue Farm. Rewe is sited on the Fourth Terrace; house footings [SX 9450 9913] exposed 1.1 m of pebble gravel with a brown sandy matrix.

A chain of five Fourth Terrace remnants extends between Woodrow, near Brampford Speke, and Pynes Water Works, near Cowley. A typical section [SX 9221 9688] near Woodrow Barton shows 0.3 m of brown clayey soil resting on 1.6 m of tightly packed, flattened pebbles and cobbles of sandstone and vein quartz, with interstitial, pale brown and yellow, sandy clay showing some localised red staining. The gravel is partly iron-cemented in the basal 0.25 m.

A large spread of Fourth Terrace extends along the east bank of the Exe from Countess Wear to Topsham. It consists of pebbly sandy gravel in which rounded quartzite predominates; angular to subangular flints are also present.

Extensive areas of Fourth Terrace occur in the north-east of the district between the River Culm and River Weaver. The largest outcrop, between Weaver Bridge [ST 014 034] and near Middle Bolealler [ST 029 041], is mainly gravel containing abundant pebbles of Greensand chert (up to 0.2 m across), metaquartzite, black, fine-grained, tourmalinised hornfels, vein quartz, Culm sandstone, and flint. Only small (less than 0.1 m) sections were visible in a ditch [ST 0265 0359] north of Shuffshayes Farm, Langford. Between Shutelake and Higher Weaver, the Fourth Terrace remnants have mainly flat upper surfaces between 63 and 67 m above OD, and are mainly of gravel, with pebbles and cobbles of Greensand chert, rounded quartzite, vein quartz and Culm sandstone. The thickness of the deposit is estimated to be between 1 and 3 m. Shallow (less than 1 m) grassed-over pits [ST 0455 0472] are present near Higher Weaver Cross.

There are a few remnants of the Fourth Terrace in the Clyst Valley, where it consists of sand and gravel dominated by pebbles and cobbles of rounded quartzite, with a minor proportion of subrounded flints. Patches of sand and gravel of the Fourth Terrace cap several low hills in the Clyst Valley around Broadclyst [e.g. SX 975 970; 974 966; 9815 9785; 9885 9818; 9922 9817]. They consist of coarse cobble gravel with common pebbles of quartz and quartzite and a minor proportion of flint, sandstone and lava. Exeter Airport is sited partly on the Fourth Terrace; boreholes proved between 1.15 and 2.2 m of sand and gravel north-east [SX 9935 9435] of Hayes Farm. Four scattered patches are present north of Blackhorse [e.g. SX 981 940] and east of Clystlands [SX 9978 9473], near Blue Hayes. All are very gravelly, with a dominance of well-rounded quartzites and only a minor proportion of subrounded flints; thicknesses probably do not exceed 1.5 m.

A few remnants of Fourth Terrace are present in the valleys of the Creedy and Yeo, and consist of gravel with Culm sandstone clasts, locally overlain by up to 2 m of pebbly silt. The Fourth Terrace forms a series of strong features along the northern side of the valley of the River Creedy. The mansion of Downes [SX 8515 9977] stands on a spread of yellowish and reddish brown silt, with abundant pebbles and cobbles of Culm sandstone. Between the Shobrooke Lake and Norton Farm are three remnants of Fourth Terrace [SX 872 993, 876 991, and 8775 9905]; an exposure [SX 8707 9924] in the westernmost remnant shows 0.9 m of reddish brown silt with abundant sandstone pebbles, overlying 1.4 m of gravel composed of subrounded and flattened pebbles and cobbles of sandstone and some vein quartz, resting on 0.2 m of Shute Sandstone.

The **Third Terrace**, about 6 m above the floodplain, is extensively developed in the Exe Valley upstream of its confluence with the River Culm, but hardly at all south of the confluence. It is separated from the Second Terrace by slight features, generally about 2 m high. Three boreholes penetrated the Third Terrace near Nether Exe. One [at about SS 944 008] penetrated 6.1 m of gravel; a second [SS 9395 0057] showed 0.30 m of soil on 5.19 m of sand and gravel; and a third [at about SS 941 001] showed 3.05 m of medium sandy gravel. The Burrow Farm Borehole [SX 9408 9958] proved 4.3 m of pebble gravel resting on Bussell's Mudstone.

Only a few Third Terrace remnants are preserved in the Culm Valley. One patch [SX 952 984], about 5 m above the Culm floodplain near Bussell's Farm, consists of sand and coarse cobbly gravel of quartz, quartzite, sandstone and flint.

In the Clyst Valley, the Third Terrace is also not much preserved. Third Terrace deposits are present east [SX 9845 9330] of Blackhorse, Clyst Honiton, where they consist of gravel composed of rounded quartzites and subrounded flint pebbles.

More extensive deposits are present along the valley of the River Creedy, where they consist of pebbly silt overlying gravel with Culm sandstone pebbles. The largest area of Third Terrace extends southwards from Hayne Barton [SX 8905 9944] to Norton Farm [SX 8883 9907], from there south-east to a place [SX 8998 9802] in the lane running southwards from Oldoven Cross, near Winscott Barton, and then northwards to the east of Jackmoor Cross [SX 8988 9888]. Much of this area is covered by reddish brown silty and clayey loam with a variable content of flattened, rounded sandstone pebbles.

The most extensive remnants of **Second Terrace** in the district are in the Exe Valley between Chitterley and Stoke Canon, with their surfaces at about 3 m above the floodplain. The sequence of deposits is generally in two parts; the lowest is of gravel and sandy gravel consisting of pebbles and cobbles, mostly of Culm sandstone and vein quartz, with some sand beds, while the upper is of silty clay and clayey silt averaging 0.8 m, but up to 1.3 m, in thickness. A ferruginous cement is locally present. River bank exposures [SS 9406 0314] in the Second Terrace north of Up Exe show 0.4 to 0.5 m of brown silty clay with a few stones in the top 0.1 m, on 1.5 m of gravel consisting of angular to subangular Culm sandstone and vein quartz up to 0.15 m in diameter, locally iron-cemented. A borehole [at about SS 939 015] near Latchmoor Green, Thorverton, showed the Second Terrace deposits to consist of 1.52 m of fine gravel, on 3.05 m of coarse-grained sand, on 3.05 m of gravel and sandy gravel. An exposure [SS 9330 0058] in the Second Terrace near Nether Exe shows 1.2 m of gravel with subrounded to subangular Culm sandstone and quartz up to 0.1 m in diameter, locally iron-cemented in the top 0.7 m and near the base.

Second Terrace deposits locally form extensive spreads in the Clyst Valley. The terrace forms wide flats in the area around Shermoor Farm [SX 995 949] and north-west of Hayes Farm [SX 990 946], near Clyst Honiton. Smaller patches occur north-west [SX 985 940] and south-west [SX 9850 9325; 984 927] of Clyst Honiton; a borehole [SX 9930 9493] in the northern patch proved 1 m of topsoil and brown sandy clay above 1 m of brown gravel with sand. The deposits north-west of Hayes Farm [SX 990 946] give rise to very gravelly soil. Their thickness is unknown, but they have been exploited for sand and gravel [SX 9880 9442]. In the Culm Valley near Kensham House, Bradninch, a borehole [ST 0057 0348] in the Second Terrace showed 0.5 m of soil, on 1.5 m of soft pale brown sandy clay, on 1.0 m of firm brown sandy gravelly clay, on 1.2 m of sand and gravel, resting on Cadbury Breccia.

The **First Terrace**, its surface generally up to about 1 m above the floodplain, is not widely developed in the district; in most valleys it consists of gravel, locally overlain by silty clay, but in the Broadclyst area of the Clyst Valley, it consists mainly of coarse-grained sand. Boreholes on the First Terrace east of Cutton [SX 977 988] proved up to 3.5 m of reddish brown medium-grained sand. The First Terrace in the Exeter city area is represented by only two small deposits, one [SX 909 918] in the St Thomas area, the other [SX 939 903] near Countess Wear; they consist of reddish brown pebbly clay. At Stoke Canon Paper Mill [SX 937 972], the First Terrace occurs between 0.5 and 1.0 m above the floodplain of the River Culm and consists of silt and clay with local pebble beds. North of Topsham, the First Terrace deposits occur in two small patches [SX 967 896; 969 899] composed of coarse-grained sand with a small amount of gravel. In the southernmost of five small patches of First Terrace around Clyst St Mary in the Clyst Valley, a borehole [SX 9687 9090] proved (beneath 0.15 m of brown clayey topsoil) 5.03 m of grey and reddish brown fine to medium gravel with some sand and a little silty clay, on Dawlish Sandstone.

The First Terrace developed along the River Teign is most extensive below Dunsford, with isolated patches concentrated mostly on the right bank of the river farther downstream; the terraces are narrow, and form gently sloping surfaces. The deposits consist predominantly of brown sand and silt with subrounded sandstone gravel.

A higher level of terraces, shown as **undifferentiated terraces** on the map, is present along the left bank of the River Teign between Dunsford [SX 811 890] and Steps Bridge [SX 805 885], and is partly concealed by extensive spreads of head.

Areas of undifferentiated terraces between Silverton and Ellerhayes are poorly exposed. The most extensive runs from just south of Silverton [SS 958 027] to near Poundsland [SS 962 013]; it has a fan-like upper surface falling gently southwards at about 1°. Surface debris indicates that the deposit is predominantly gravel, Culm sandstone pebbles and cobbles being the predominant clast types, with lesser quantities of vein quartz and very scarce Greensand chert. Thicknesses probably do not exceed 2.5 m.

Spreads of gravel form part of the River Otter terrace system at Crook Plantation [SY 052 865] and Baker's Brake [SY 064 870] in the Bicton area. The deposits consist of gravel with quartzite, Greensand chert, and flint pebbles. The terraces have not been numbered because the survey of the Otter Valley is incomplete.

Details of the river terrace deposits are to be found in the Technical Reports which cover the district (p.x).

BURIED CHANNEL DEPOSITS

Durrance (1969) recognised two phases of buried-channel formation at the mouth of the River Exe (in the adjacent Newton Abbot district), which he related to low sea levels, either within the Devensian glacial period, or possibly within the Wolstonian and Devensian glacial periods. The older channels (?Early Devensian) were cut in rock to depths in excess of 52 m below OD, and filled with gravel. The younger (?Late Devensian) set of channels was excavated partly in this gravel fill, and partly in bedrock, to depths of 30 m below OD. Both sets of channels contain buried terraces.

Boreholes and seismic surveys for the crossing of the River Exe by the M5 Motorway at Topsham indicate a main buried channel excavated in Permian rocks to an average depth of 10.2 m below OD (Durrance, 1974). The channels are filled with gravel to a depth of approximately 4.5 m below OD, overlain by Flandrian silty clay. The channels were considered to be of Late Devensian age and correlated with the younger channels of the lower Exe estuary. Only one phase of buried channel formation was recorded at Topsham, showing that the older (?Early Devensian) channels present near the mouth of the Exe estuary do not extend far upstream. The sequence at Topsham is similar to that found in the upper reaches of the River Teign near Newton Abbot (Durrance, 1971), and it is likely that the younger buried channel of the River Exe is the same age as the buried channel of the River Teign (Durrance, 1980).

There is evidence in the upper reaches of the estuary that marine and freshwater deposits interdigitate. For example, Cullingford (1982) recorded an exposure [SX 962 883] near Topsham in a deposit which was suggested from its sedimentary properties, pollen analysis and beetle fauna, to be of freshwater alluvial origin. Pieces of driftwood from levels between 3.0 and 4.1 m below OD gave radiocarbon dates from about 3300 to about 4000 years BP.

MARINE DEPOSITS

Many estuaries in south-west England were drowned by the rise in sea level (the Flandrian Transgression) which followed melting of the Devensian ice sheets. In the southern part of the Exeter district, marine deposits formed during the Flandrian Stage occupy the estuary of the River Exe approximately as far north as Countess Wear Bridge [SX 942 896], and the valley of the River Clyst as far upstream as Newcourt Barton [SX 970 902]. Large areas of former saltmarsh have been reclaimed and are shown on the map as reclaimed Marine Deposits. The deposits classified on the map as undifferentiated Marine Deposits consist largely of intertidal mud flats, bordered by saltmarsh on the west shore of the estuary from Turf [SX 964 861] northwards to Topsham, and on the east shore from Exton to the lower course of the River Clyst. Thomas (1980) noted that in most of the upper Exe estuary, the reclamation of saltmarsh has proceeded to such an extent that there is little area remaining suitable for the highest intertidal mudflat and saltmarsh to develop. Coarser-grained sediment is present on the floors of tidal channels, locally extending on to the edges of mudbanks, and also occurs on high-tide wave beaches, developed mostly at the foot of artificial embankments (Thomas, 1980).

Boreholes between the railway line north of Exminster [SX 9469 8823] and the mudflats [SX 9549 8878] north-west of Topsham showed between 1.90 and 4.25 m of clay (average thickness in 15 boreholes is 2.8 m), resting on between 1.70 and 5.10 m of gravel (average thickness in 15 boreholes is 3.8 m). A borehole [SX 9539 8870] showed the following representative sequence:

	Thickness m
MARINE DEPOSITS	
Clay, very soft, dark grey to black, silty, with roots and stems of reeds	3.10
Clay, soft, grey-brown, silty, with a little peat	1.15
?LATE DEVENSIAN GRAVEL	
Gravel, grey and brown, with rounded to subangular coarse, medium and fine gravel with a little sand and occasional cobbles	3.25
DAWLISH SANDSTONE	
Breccia, reddish brown, sandy	2.10

ALLUVIUM

Alluvium, the deposits of modern river floodplains, is found in all the main river valleys in the district and along their larger tributaries (Figure 37). In most areas, the alluvium can be divided into a lower unit of gravel and an upper unit of clay or silt; the upper unit may locally contain thin gravel beds or peat-rich beds. The gravels of the Rivers Exe, Creedy and Yeo contain pebbles and cobbles predominantly of Culm sandstone. In contrast, the alluvium of the River Culm and its tributary the River Weaver contain, in addition, Greensand chert. The basal gravels of the River Clyst and its tributaries contain predominantly rounded quartzite derived from the Budleigh Salterton Pebble Beds.

In the Exe Valley east of Thorverton, the upper unit of silty sandy clay and clayey sandy silt is generally between 0.4 and 1.2 m thick and rests on gravel generally less than 1 m thick. Near Nether Exe, an exposure [SS 9293 0041] in the bank of the River Exe shows 1.2 m of brown silty sand with a few gravel layers, on

0.4 m of pebble and cobble gravel, mainly of subrounded Culm sandstone pebbles up to 0.25 m across, and some quartz; some flat pebbles show imbrication. South of Stoke Canon and around Cowley Bridge, the lower gravel unit is up to 6 m thick, overlain by 1 to 2 m of silt and clay. A borehole [SX 9222 9623] on the Exe floodplain near Stafford Bridge proved 2.4 m of sandy clay and silt resting on 5.6 m of gravel, and another [SX 9090 9545], farther downstream at Cowley Bridge, penetrated 1.8 m of clay and silt overlying 5.9 m of gravel. In both boreholes, the bedrock is weathered Crackington Formation shale.

The alluvium of the River Exe in the Exeter city area is about 5 m thick on average; the lower gravel unit is up to 4.2 m thick and is overlain by up to 3.5 m of brown silty and sandy clay. A borehole [SX 9106 9355] penetrated (beneath 0.7 m of made ground) 0.7 m of brown sandy silt with traces of gravel, on 4.2 m of silty and sandy gravel with cobbles, resting on Crackington Formation.

In the Culm Valley, the upper unit of alluvial clay and silt is commonly around 1.4 m thick and overlies gravel around 2.5 m thick. A borehole [ST 0110 0352] east of Bradninch penetrated the following succession:

	Thickness m	Depth m
Topsoil, brown, slightly silty, clayey, with a trace of sand	0.50	0.50
ALLUVIUM		
Clay, brown, slightly silty and organic in parts, soft to firm, slightly sandy; some areas of softer grey silty clay; little subangular to subrounded gravel	1.00	1.50
Gravel, dense, subrounded to subangular, with some medium-and coarse-grained sand; clayey top	2.00	3.50

CLYST ST LAWRENCE FORMATION

Peaty deposits up to 1 m thick overlying gravel are developed in the alluvium of the River Weaver north from Tye Farm [ST 030 035].

The alluvium of the Rivers Creedy and Yeo is similar to that of the River Exe, consisting of silt and clay overlying gravel. The silt and clay commonly contain thin beds and lenses of sand and gravel; organic-rich layers may be present. The gravels are usually poorly sorted, with abundant well-rounded, flattened pebbles and cobbles, mostly of Culm sandstone, with some vein quartz. A typical section [SX 8593 9913] in the bank of the River Yeo west of Downes Mill Farm shows 3.2 m of brown silty clay, becoming gravel-rich towards the base and passing down into 0.2 m of coarse gravel.

The Clyst Valley alluvium consists of silty sandy clay generally between 1.5 and 3 m thick, overlying gravel, which is commonly less than 1 m thick, but which can range up to 2.7 m. Clasts of rounded quartzite are common in the basal gravel unit. Broadclyst Moor, a large boggy area in a tributary valley of the Clyst, is underlain by peaty clay that rests on gravel [around SX 975 980]; because of the high clay content, it has been mapped as alluvium rather than peat. A typical section in the Clyst Valley alluvium north of Clyst Honiton was penetrated in a borehole [SX 9858 9413], as follows:

	Thickness m	Depth m
Topsoil	0.25	0.25
ALLUVIUM		
Clay, firm, reddish brown with traces of carbonaceous matter and roots	1.50	1.75

	Thickness m	Depth m
Clay, sandy, soft, reddish brown with traces of gravel	0.75	2.50
Gravel, sandy, angular and subangular	1.70	4.20
DAWLISH SANDSTONE		
Sand, fine- and medium-grained, silty, with traces of fine gravel	2.55	6.75

About 1.1 km downstream, another borehole [SX 9835 9314] proved 2.45 m of pale brown silty clay resting on 2.55 m of brown sand and gravel. A borehole [SX 9715 9081] near Clyst St Mary proved 1.6 m of brown clay overlying 2.36 m of gravel in a sandy silty clay matrix.

Details of the alluvium are given in the Technical Reports which cover the district (p.x).

ALLUVIAL FAN DEPOSITS

Small alluvial fans have been mapped on the sides of the Exe Valley near Newton St Cyres [SX 8620 9905; 8770 9867], near Chitterley and Bidwell [SS 9390 0498; 9367 0312], at Brampford Speke [SX 9267 9781], near Cowley [SX 9057 9615], and in the Exeter city area [e.g. SX 9059 9477; 9106 9472]. Most of the deposits consist of gravel with silt and clay. A borehole [SX 9285 9772] at Brampford Speke showed the deposit to consist of 1.2 m of gravel with silt and clay, resting on weathered Dawlish Sandstone. An excavation [SX 9059 9477] at Exwick Barton showed 0.2 m of soil resting on 0.7 m of yellowish brown clay with beds of gravel containing subrounded Culm sandstone pebbles. A ditch [SX 9106 9472] in the University grounds showed 0.65 m of brown clay with irregular beds of subangular sandstone gravel, apparently disturbed by frost heaving.

PEAT

A small tract of peat [SX 995 960] occurs about 1 km south-east of Dog Village, Broadclyst. It varies between 0.3 m and greater than 1.2 m in thickness, and rests on sand and gravel. Another small patch has been mapped [SX 989 943] near Hayes Farm, Clyst Honiton. On the Dartmoor Granite, marshy peat with scattered granite boulders is present just south of Westcott [SX 790 870].

MADE GROUND AND WORKED GROUND

Made ground consists of artificial material deposited mainly on the original ground surface. In addition to the embankments for railway lines and major roads, an area of low-lying ground [SX 910 897, 919 889] near Shillingford Abbot, an area [SX 9120 8635] near Kennford, an area [SX 943 885] near Exminster, and some large valleys [SX 918 881, 924 883 and 919 875] around Peamore House were built up or filled during the construction of the M5 and A30 roads. Probably most of the fill was derived from the cuts in the Heavitree Breccia and Dawlish Sandstone. At Silverton Mill, large mounds [SS 979 011 to 981 011] are probably composed mainly of fly ash. Within the Exeter urban area, there are extensive tracts of made ground (Scrivener, 1984). Some of these are obscured by building developments, while others, such as the former refuse tip in the Mincinglake Valley and the areas of fill

Table 19 Landfill waste disposal sites in the district.

Site	Grid reference	Formation	Waste type	Date of licence
West Wheatley Farm	SX 890 918	CkF	I	1990
West Town Farm	SX 854 929	AnSh	I + Ind	1984
Mark's Farm	SX 884 910	AnSh	I	1984
Hazeldene	SX 881 910	AnSh	I	1983
Ide railway cutting	SX 895 901	AlBr	I	1978
Kenbury Wood*	SX 921 870	HvBr	I + Ind	1986
Springdale*	SX 866 923	CkF	I + Ind	1987
West Wheatley Farm*	SX 886 920	CkF	I	1990
Whitestone	SX 871 930	AnSh	I	1980
Punchbowl*	SS 792 007	CrBr/NSBr	H	1978
Hill Barton Farm*	SY 003 911	EMS	I	1989
Hayes Farm	SX 991 943	DaS	I + Ind	1983
Canterbury House Farm	SY 044 893	LMu	I + Ind	1989
Stoke Hill*	SX 938 967	CkF	I + Ind + Paper-making waste	1979
Exwick Barton*	SX 903 947	CkF	I	1990
East Devon Hunt Kennels	SX 990 913	EMS	I + Ind	1982
Sturridge Wood	SX 832 034	BF	I	1984
Bawdenhayes*	SX 825 035	BF	I	1987
Digby Drive	SX 959 916	DaS	I + Ind	1984
Aylesbeare Common	SY 054 901	BSP	I + Ind	1982
Strete Ralegh	SY 053 953	Ayb	I + Ind	1985
Budlake Quarry	SS 980 001	EVR	I + Ind	1978
Heathfield Farm	SX 993 892	EMS	H	1972
Mincinglake	SX 937 944	CkF	H	No licence
Lower Sutton Quarry*	SS 789 046	BF	I + Ind	1983
Holwell Barton Farm	SX 802 995	NSBr	Milk waste	1984

* Sites with a licence in 1994. Key to abbreviations for rock units: CkF – Crackington Formation; AnSh – Ashton Shale Member (Crackington Formation); BF – Bude Formation; EVR – Exeter Volcanic Rocks; AlBr – Alphington Breccia; CrBr – Crediton Breccia; HvBr – Heavitree Breccia; NSBr – Newton St Cyres Breccia; DaS – Dawlish Sandstone; EMS – Exmouth Mudstone and Sandstone; LMu – Littleham Mudstone; Ayb – Aylesbeare Mudstone Group, undivided; BSP – Budleigh Salterton Pebble Beds. Key to abbreviations for types of waste: I – Inert; Ind – Industrial; H – Household.

Information from Devon County Council and Environment Agency.

associated with the flood prevention works in the Exe alluvial plain, have been landscaped to form recreation areas.

Worked ground consists of areas remaining after mineral extraction. Small areas (for example, marlpits) are not shown on the map, but larger areas remaining after gravel and sand working from the Budleigh Salterton Pebble Beds and Otter Sandstone at Beggars Roost Quarry [SY 062 941] and Rockbeare Hill Quarry [SY 062 945] are indicated. These pits are currently being infilled with silt produced during the processing of sand and gravel.

Table 19 gives details of landfill sites licensed for waste disposal by Devon County Council. A total of 26 sites lie within the district, of which 12 were open in 1992.

LANDSLIP

Many slopes on the Crackington Formation and, to a lesser extent, the Bude Formation show evidence of small-scale active landslipping, described in Grainger (1983) and Grainger and Harris (1986). Landslipping is particularly prevalent on the weak, kaolinite-illite shales of the Ashton Shale Member of the Crackington Formation. Very few landslips have been recorded on the outcrop of the Permo-Triassic formations. The geotechnical aspects of landslips on the Crackington and Bude formations are described briefly elsewhere in this memoir (Chapter 11);

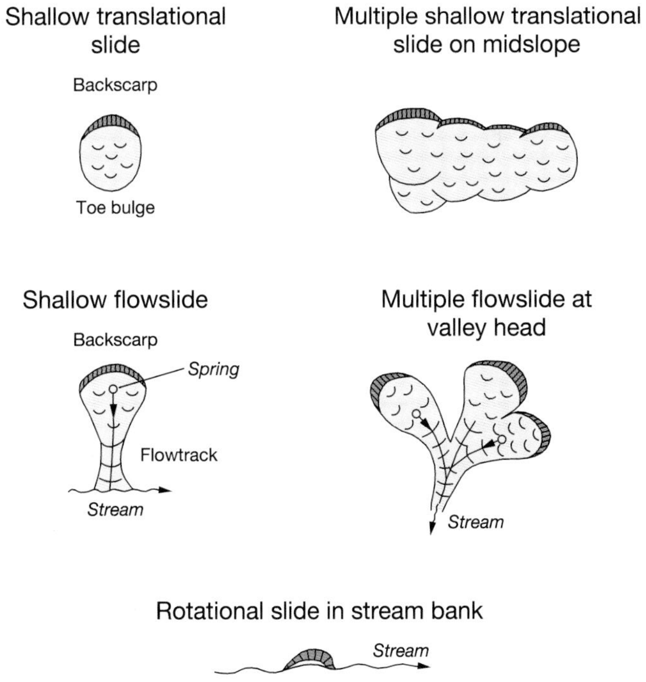

Shallow translational slide

Backscarp

Toe bulge

Multiple shallow translational slide on midslope

Shallow flowslide

Backscarp

Spring

Flowtrack

Stream

Multiple flowslide at valley head

Stream

Rotational slide in stream bank

Stream

Figure 38 Classification of landslip types on the Crackington Formation of the district (after Grainger and Harris, 1986, fig. 9).

details of landslips are given in the Technical Reports for the district.

Grainger and Harris (1986) showed that the active landslips on the Crackington Formation range from shallow, very slow, translational earth slides, to shallow, very slow to extremely slow earth flows (classification of Varnes, 1978). The landslipped material consists of completely disturbed head, residual soil, and weathered bedrock. The translational slides are generally fairly uniform in size and are 10 to 20 m wide, 15 to 40 m long, and about 2 to 4 m thick. However, individual slides may amalgamate to form multiple slides which may be up to 100 m wide and 75 m long. A common feature of slides is the presence of small backscarps. Many shallow translational slides grade downslope into earthflows; a flow track 5 to 10 m wide carries material down to the base of the slope. These combined forms were called flowslides by Grainger and Harris (1986). Small rotational slides occur at the bottoms of slopes adjacent to streams; they have steep backscarps owing to the fact that most of the slide debris is rapidly eroded and transported by the stream. The different types of landslip present on Crackington Formation slopes are summarised diagrammatically in Figure 38; photographs of typical landslip features are shown in Plate 13.

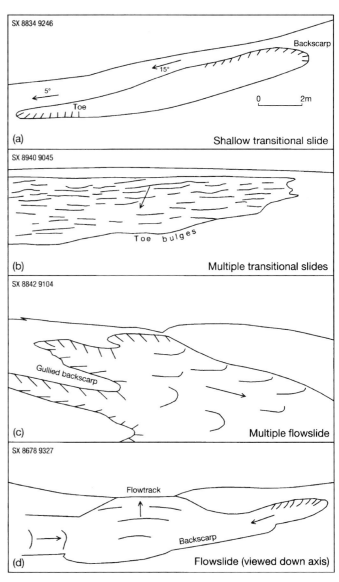

Plate 13 Examples of typical features of active landslips on the Crackington Formation west of Exeter: (a) shallow translational slide [SX 8834 9246] near Cutteridge Gate, Whitestone; (b) multiple translational slides [SX 8940 9045] near West Town Farm, Ide; (c) multiple flowslide [SX 8842 9104] near Mark's Farm, Longdown; (d) flowslide (viewed down axis) [SX 8678 9327] near Whitestone. See text for explanation. Reproduced, with permission, from Grainger (1983, plate 20).

ELEVEN

Economic geology

SOILS AND LAND USE

An account of the soils of the Exeter district is given by Clayden (1971), and the distribution of soil types is shown on the soil map published at the 1:63 360 scale by the Soil Survey (Sheets 325 and 339). More generalised soil associations are shown on the 1:250 000 scale Soil Survey map for south-west England (Sheet 5, with accompanying legend booklet). Part of the present district is described in an account (with accompanying map) of the soils of the Middle Teign Valley by Clayden (1964). The main characteristics of the soils, and their relationship to the geology of the district, are summarised below.

Soils on Carboniferous and igneous rocks

The Dartmoor Granite is characterised by well-drained gritty loams. Small areas of less well-drained loams with peaty surface layers are present in valley bottoms.

The Combe Shale and Teign Chert are characterised by well-drained loamy and gravelly soils with abundant blocky stones derived from the cherts. Dolerites in the Lower Carboniferous sequence, and the lavas within the Permian rocks, give rise to well-drained brown earths consisting of variable thicknesses of stony loams, generally shallow and locally rocky.

Soils on the Bude Formation are mainly well-drained, loamy or silty, and of moderate depth; small areas of clayey soils are locally present. The Trusham Shale and Crackington Formation are characterised predominantly by well- or moderately well-drained brown clay loams. Clayden (1971, p.84) noted that the extent of such soils on the Crackington Formation has been underestimated by many writers, who have considered that the entire outcrop is occupied by poorly drained clay soils on a shale subsoil. The well-drained soils occur mainly on steeper slopes, and thus their distribution is related to the degree of dissection of the landscape. Less well-drained, gleyed brown earths and poorly drained silty clay loams occur mainly on gentle slopes.

The soils developed on the Crackington Formation were considered by Clayden (1964, 1971) to form a hydrologic sequence, in which the formation of a particular soil was closely related to site conditions. However, in the Exeter district, Grainger (1984) also demonstrated some stratigraphical control of soil types. Thus, an east–west-trending belt of clayey soils between Exeter and Tedburn St Mary coincides approximately with the outcrop of the Ashton Shale Member of the Crackington Formation. The differences in soils reflect differences in clay mineralogy: brown earths have formed on steeper slopes of stronger shales dominated by illite and chlorite, while clayier soils have formed on shallower slopes underlain by weaker shales of illite and illite-chlorite.

Soils on the Permo-Triassic rocks

Permian breccias of the district carry mainly well-drained gravelly loams which are in places stony and shallow; steep slopes are locally present. Clayden (1971, p.47) noted that soils on the Cadbury Breccia are likely to be more acid than those on the mixed breccias farther south because they contain more sandstone. Areas of clayey soils around Webberton Cross [SX 879 874], Dunchideock, reflect a more clayey facies of the Alphington Breccia.

Developments of well-drained brown earths of loamy and silty texture approximate to the outcrops of the Thorverton Sandstone, Shute Sandstone, Yellowford Formation, Clyst St Lawrence Formation, and the silts and sands of the Higher Comberoy Formation.

Much of the Dawlish Sandstone and Otter Sandstone outcrops, together with thicker sandstone beds at the top of the Exmouth Mudstone and Sandstone near Woodbury, are occupied by brown earths consisting of sandy loam and loamy sand; small pockets of humus-iron podzols occur locally. Soils on the Otter Sandstone commonly contain stones derived from river terrace deposits.

Most of the outcrop of the Aylesbeare Mudstone Group (excluding the Clyst St Lawrence Formation) is overlain by a thin surface layer containing rounded metaquartzite pebbles derived from the Budleigh Salterton Pebble Beds. This surface layer gives rise to imperfectly or poorly drained gleyed brown earths or, particularly adjacent to the outcrop of the Budleigh Salterton Pebble Beds, to more strongly gleyed soils with a brown or yellowish brown horizon which is usually mottled.

Areas of soils mapped by the Soil Survey as narrow belts along the foot of the Pebble Beds scarp, and on the Littleham Mudstone inliers of Venn Ottery Common, Aylesbeare Common, Hawkerland, and Colaton Raleigh Common were described by Clayden (1971) as 'a somewhat heterogeneous group of strongly gleyed soils in loamy and sometimes gravelly Head over a fine-textured substratum'.

The Budleigh Salterton Pebble Beds outcrop is occupied by soils classed as humus-iron podzols. They are well drained, very acid, very stony, sandy soils with a bleached subsurface horizon. At the base there is commonly a very thin irregular layer of iron-pan resting on in-situ Pebble Beds.

Soils on drift deposits

Soils developed on river terrace deposits consist mainly of loam overlying gravel. They are most extensively developed along the valley of the River Exe between Stoke

Canon and Up Exe. A stony phase is distinguished on terraces near Thorverton and south of Silverton. Soils associated with river terrace deposits between Weaver Bridge and Middle Bolealler, and near Langford, are mainly gleyed brown earths. River terrace deposits in the northeast of the district, between Shutelake and Higher Weaver, give rise to humic gley soils, with loamy upper horizons, over gravel.

Marine deposits of the Exe and Clyst valleys are occupied by clayey groundwater gley soils. Alluvium of the district carries a variety of alluvial soil types including well-drained brown soils on the silty alluvium of the River Exe, gleyed brown soils on reddish brown clayey alluvium of the River Clyst, and gleyed brown soils on alluvium derived mainly from Culm shales. The soils along the higher reaches of the Clyst and its tributaries are in more mixed parent materials and are more variable. Areas of soils flanking the Haldon Hills are developed in flinty and cherty head.

Land use

The district is predominantly agricultural. Its land use has been described by Clayden (1971) and by Applegarth and Cornish (1982). Dairy farming is the most widespread type of farming; it is particularly important on the areas underlain by Aylesbeare Mudstone. Most of the land in this area is classed as grade 3. Mixed farming is more important on land underlain by the Exeter Group, and about half the land is given over to arable crops. Much of the land is grade 1 or 2, especially the well-drained brown earth soils around Thorverton, Silverton and Killerton. The ground underlain by Culm rocks is largely characterised by dairying and the rearing of livestock, combined with cereal growing. Most of the land is classed as grade 3 or 4.

Cider was formerly produced from orchards in the district; the most favoured sites were the south-facing slopes of river terraces. Forestry is locally important, particularly on the outcrop of Culm rocks, for example south of Newton St Cyres, and in the Ashclyst Forest near Broadclyst.

Much of the Budleigh Salterton Pebble Beds outcrop south of West Hill is unenclosed common land; the commons in part owe their survival to the natural infertility and low available water capacity of most of the soils, combined with their stoniness, which makes them difficult to cultivate. The commons are classed as Areas of Outstanding Natural Beauty, and are much used for recreation.

CONSTRUCTION MATERIALS

Sand and gravel

The chief potential resources of sand and gravel in the district lie within the Permo-Triassic formations, especially the Budleigh Salterton Pebble Beds, Dawlish Sandstone and Otter Sandstone; other resources are present in the Quaternary deposits, mainly river terrace deposits and alluvium.

The Budleigh Salterton Pebble Beds are an important resource of sand and gravel in east Devon, and a number of large pits are sited along the outcrop between Blackhill Quarry, near Woodbury, and Whiteball Quarry, near Wellington. One small working pit, at Venn Ottery [SY 065 913 to 067 910], and disused pits at Rockbeare [SY 061 947] and Beggars Roost [SY 062 941], are present in the district. Rockbeare Quarry is now (1995) the site of a processing plant, a block-manufacturing works, and sand and gravel stockpiles. Material from the Foxenholes Quarry [SY 077 948] (just east of the district) and from Venn Ottery Quarry is transported to this site for processing (Vincent and Nicholas, 1982). About 250 000 tonnes per year of sand and gravel are extracted from Blackhill Quarry [SY 031 852] which lies just south of the district. It was estimated that where the Pebble Beds average 30 m in thickness in the Blackhill area, they are likely to yield between 105 000 and 125 000 tons of sand and gravel per acre (Report of Planning Enquiry, 1968, paragraph 36). The deposit contains 10 to 13 per cent by weight of silt and clay; about 50 per cent of the processed material is sand, and 40 per cent is gravel. The sand and crushed pebbles and cobbles are used for surface chippings, concrete aggregate, high-grade concrete sand, and sand for tile manufacture.

The Otter Sandstone is a potential sand resource; the lower 8 m were formerly worked in Beggars Roost Quarry [SY 061 941]. Grain-size analyses from the district and the adjacent area (Edwards, 1989 a,b; 1990) indicate that the sands are predominantly fine to medium grained, with fines content (finer than 0.0625 mm) between 2.9 and 21.6 per cent (average 8.3 per cent).

The Dawlish Sandstone is a potentially large source of building sand. However, the fine grain size, and the local interbeds of breccia (in the south) and mudstones (in the north), partly diminish its economic potential. The only working quarry in the district, Bishop's Court Quarry [SX 963 914], produces ungraded sand for building use. Grading figures for the Dawlish Sandstone of the area between Clyst Honiton and south of Exminster are included in Bristow and Williams (1984) and Bristow (1984 a).

The alluvial gravels and low-level river terrace deposits of the River Exe between Brampford Speke and the northern edge of the district, and in the Culm and Clyst valleys, have high contents of Culm sandstone and vein quartz pebbles, and are a gravel resource. River terrace deposits at higher levels in the district are also potential resources, but are generally small in area and thin. Boreholes in the Stoke Canon and Rewe area have proved up to 3 m of gravel (Scrivener, 1983), and at Nether Exe have proved 2.44 to 7.62 m (average 4.6 m) (Edwards, 1987 a).

The extensive Fourth Terrace of the River Culm between Weaver Bridge [ST 014 034] and Tye Farm [ST 030 035] is possibly up to 5 m thick, but probably mostly less. River terrace deposits between Shutelake and Higher Weaver are between 1 and 3 m thick; there are disused workings (less than 1 m deep) [ST 0455 0472] near Higher Weaver Cross.

In the Clyst Valley, the alluvial gravel is 1.0 to 2.7 m thick (average 1.95 m); grading figures are given by Bristow and Williams (1984, table 3). In the Culm Valley near Rewe, the alluvial gravels are 4.3 m and 4.9 m thick in two boreholes [SX 9454 9883; 9551 9976] (Scrivener, 1983; Bristow, 1983). Farther north, between Hele and Westcott, boreholes indicate that the gravel is 2.0 to 3.1 m thick (average 2.45 m). The gravel is overlain by 1.3 to 1.5 m of clay and silt (average 1.4 m). The alluvial gravels of the River Exe within the urban area have a mean thickness of 2.24 m (Scrivener, 1984), whereas downstream in the Exminster area the mean thickness is 3.6 m (Bristow, 1984a).

The Cadbury Breccia contains a high proportion of fines and does not seem to have been dug, except in a pit [ST 0158 0173] at Roach Copse, west of Clyst Hydon. The Heavitree Breccia, where poorly cemented, could be a possible source of gravel. Generally, however, it is too clayey (Bristow, 1984 a, table 5) for use as an aggregate, but it could be used for road base or for local tracks.

The Upper Greensand of the Haldon Hills was formerly worked for sand, and several pits are present around the outcrop just south of the district [e.g. SX 878 857], but there is no record of workings within the Exeter district.

Roadstone

Vincent and Nicholas (1982) noted that the igneous rocks of Devon provide aggregates which are suitable for road surfacing because their wearing and polishing properties are generally good. The igneous rocks are also suitable for most other aggregate applications. Quarrying of the igneous rocks is generally more difficult and more costly than the quarrying of limestone, because the deposits are smaller and more irregular, the overburden is thick and irregular in many cases, and blasting and crushing are generally more difficult owing to the harder nature of the rocks.

The dolerite sills intruded into the Lower Carboniferous sequence of the middle Teign Valley have been extensively worked for concrete aggregate and roadstone, but only one quarry (Crockham Quarry [SX 8490 8085], near Trusham, outside the district) is currently active. The dolerites of the district have not been much quarried; a small disused working [SX 8512 8646] is present near Christendown Clump, Doddiscombsleigh, and dolerite within the aureole of the Dartmoor Granite was worked for roadstone in a quarry [SX 822 866] near Bridford. Spilitised basalt and associated cherts were worked at Stone Copse Quarry [SX 824 860]. The Ryecroft Sill was extensively worked at Ryecroft Quarry [SX 8432 8475], just outside the district. The basalts at School Wood Quarry [SX 875 871], Dunchideock, were also worked for roadstone and general aggregate, possibly into the 1960s.

The sandstones in the Crackington and Bude formations are a potential source of aggregate, and their properties as skid-resistant road-surfacing material have been investigated by Hawkes and Hosking (1972) and by Cox et al. (1986). Uniformly high Polished Stone Values (PSV) (over 60) and low Aggregate Abrasion Values (AAV) (less than 10) have been demonstrated for typical samples of these sandstones, though the material strength may vary with the state of weathering. The volume ratio of shale to sandstone (commonly greater than 3:1) in the Crackington Formation of the present district would hinder efficient extraction of suitable material. The Bude Formation sandstones are worked for aggregate at several localities in Devon, but not within the district, where the presence of interbedded siltstone with unsuitable PSV and AAV values would make working difficult. Roadstone and aggregate were worked in the Bude Formation at Lower Linscombe [SS 7891 0456] northwest of Newbuildings, East Henstill [SS 8157 0411], Bawdenhayes [SS 8253 0346], and Land Quarry [SS 8320 0330] near Sturridge.

Building stone

Building stones were formerly quarried and used locally from several different formations, most important of which were the Exeter Volcanic Rocks and the Heavitree Breccia. There is no active working in the district.

Dolerite and chert around Doddiscombsleigh have locally been used for building stone. Thin sandstones in the Crackington Formation have been used for rough walls and for the foundations of buildings constructed of cob. The sites from which such material was obtained were commonly pits adjacent to the construction. In the case of cob buildings, the sandstone used in the footings was separated from the weathered shale and clay used to form the cob.

Sandstones in the Bude Formation have been worked for building stone, mostly from small pits near farms. A larger quarry at Ramspit [SS 8625 0317], north of Shobrooke, worked grey and purple-stained turbiditic sandstones; similar material is common in local buildings and in lanes and farm tracks.

The Knowle Sandstone is locally well enough cemented to have been used as a building stone.

The Heavitree Breccia was worked as a building and walling stone for several hundred years in Exeter and throughout the district (Plate 14a). The breccia is only weakly to moderately

Plate 14a Exeter city wall, Western Way, Exeter [SX 9205 9223]. The basal courses are constructed of rough blocks of reddish brown Heavitree Breccia, overlain by squared blocks of purple basalt lava (Photo. Mr A J J Goode).

Plate 14b   Detail of wall at The Quay [SX 9215 9205], Exeter. The stone is purple basalt showing the dolomite veining characteristic of 'Pocombe Stone', derived from quarries on the west side of Exeter. (Photo. Mr A J J Goode).

Plate 14c   Broadclyst Church [SX 9818 9726]. 'Killerton Stone', used in the construction of the part of the church wall shown, is minette (biotite-lamprophyre lava), probably quarried from the Killerton Park area. The hammer is 0.3 m long (A 15322).

cemented, but Berger (1811) noted that it hardened on exposure. It is susceptible to frost and was commonly faced with cob or protected by plaster or a concrete render. Its former widespread exploitation is evident from the presence of many old quarries in the district. The Newton St Cyres Breccia is less well cemented than the Heavitree Breccia, but it is common in buildings and walls in the Newton St Cyres area.

Moderately well-cemented Dawlish Sandstone has been used in buildings and walls, especially in the Broadclyst area, where it was commonly used in conjunction with lava.

The sandstones of the Exmouth Mudstone and Sandstone are generally only weakly to moderately cemented. They have been used in walls and buildings in the Woodbury area. Brighouse (1981, p.237) reported that the north aisle of St Swithin's Church, Woodbury, added in 1530–33, is mainly built of a local soft sandstone which has not weathered well. Local sandstone was also used in the construction of Parsonage House [SY 0126 8769], and was probably quarried from a pit [SY 012 881] 400 m north of the house. Local sandstone was used for the foundations of some cob houses.

The Otter Sandstone is normally rather weakly cemented, but better-cemented horizons, such as intraformational conglomerates, have been locally used for building stone.

The Exeter Volcanic Rocks have for many centuries been used as building stones in the district, and there are many disused quarries. The Romans constructed the city wall from the basalt of Rougemont, upon which the garrison of Isca Dumnoniorum was founded (Plate 14a). Gale (1991) has sug-

Plate 14d Well-rounded pebbles and cobbles, mainly of quartzite, from the Budleigh Salterton Pebble Beds (locally known as 'popples') have been widely used in the construction of walls around Woodbury (A 15324).

Plate 14e Bridford Barytes Mine Quarry [SX 8296 8656]. The middle part of the Teign Chert is exposed in the quarry, where it contains various manganiferous intervals. A 3 m-thick barytes lode (not shown here) is present in the same quarry (A 15265).

gested that vesicular lavas were used in preference to other lava types because they were more resistant to weathering, probably because they dried out more quickly. In more recent times, the principal sources of stone for construction in Exeter city were the quarries west of Exeter, the substantial working at Pocombe [SX 898 915] for example. This provided a characteristic dolomite-veined lava, the 'Pocombe Stone' (Plate 14b). The small existing demand for local basalt to repair old walls and buildings is met from stocks of previously used material.

In the Thorverton area, lavas from Raddon Quarry [SS 9096 0202 to 9102 0201] have been widely used in church and other buildings and for walling, probably from the 12th to the 19th century (Hoskins, 1972, p.262); the stone is locally termed 'Thorverton Stone' or 'Raddon Stone'.

Minette from quarries in Killerton Park, the largest and most recently worked of which [SS 975 005] is on the east side of the outcrop, has been used extensively in buildings in the Broadclyst area. Ashlar blocks of minette have been used in the outer walls of the chapel [SS 9868 0034] at Killerton Park, but the surface of the stone shows a tendency to flake. Killerton Stone has also been used in Broadclyst Church (Plate 14c).

The basalts of the Posbury and Uton areas south-west of Crediton were worked for building stone, known locally as 'Posbury Stone'. The largest quarry [SX 815 978] is at Posbury Clump; smaller workings extend along a north-facing escarpment from a small quarry [SX 8137 9816] north of Posbury Clump, as far as Uton Quarry [SX 8235 9854]. Much of the

Church of the Holy Cross at Crediton is built of Posbury Stone, as well as many other buildings in the area.

Stones in the superficial gravels, such as Greensand cherts, flints, and pebbles from the Pebble Beds, have been used for walls, and parts of buildings. Culm sandstone cobbles and pebbles from the Cadbury Breccia have locally been incorporated in walls and buildings.

Metaquartzite pebbles and cobbles from the pebble beds — locally known as 'popples' or 'Budleigh Buns' (Gale, 1992) — occur abundantly scattered over the Aylesbeare Mudstone outcrop, and have been widely used for walls (Plate 14d), floors and, occasionally, whole buildings. Brighouse (1981) considered that 'Nowhere else in England has the humble cobble-stone reached greater imaginative heights than in Woodbury.....the village still abounds in good popple walls and the first prize must certainly go to the front garden wall at Webbers'. The larger stones were commonly used in the foundations of cob houses, and they can occasionally be seen protruding beneath a coating of pitch.

Brick clay

The Crackington Formation is worked for brick clay at Pinhoe [SX 955 947], Exeter. The shales and sandstones are used to produce facing, engineering and paving bricks. In 1984, about 100 000 tons of shale and sandstone were extracted and crushed for a production of 35 million bricks (information from the quarry manager, Mr G Thompson). A quarry [SX 8690 9638] in the Ashton Shale at Lower Western Copse, near Newton St Cyres, may have been a trial for brick clay. The Lower Carboniferous shale formations are possible resources of brick clay.

The Whipton Formation, and poorly cemented finer-grained parts of the Alphington Breccia, were formerly extensively worked for brickmaking in the Exeter city area, particularly around Newtown [SX 932 930] (Scrivener, 1984).

There are numerous small pits scattered across the outcrop of the Aylesbeare Mudstone. Many were dug for marl, but some were for brick clay. Ussher (1902, p.46) noted a brick pit [ST 0403 0169] east of Clyst Hydon and several pits near Strete Ralegh [SY 053 953]. The most recently worked pit [SY 0537 9527] is about 4 to 6 m deep with small exposures of reddish brown silty clay and mudstone. Another brickpit [SY 0525 9518] was situated just west of Brickyard Road. North-east of Clyst St Mary, small ponds between Holbrook Farm [SX 994 920] and Cat Copse [SX 989 911] probably mark the sites of old workings for brick clay.

Clayey formations of the district were used in the past as cob (a mixture of clay and straw) for domestic and farm buildings. Some of the many pits on the Aylesbeare Mudstone outcrop may have provided clay for this purpose.

Marl and lime

The outcrop of the Aylesbeare Mudstone in the district contains about 900 shallow pits, ranging in size from a few metres across to about 0.5 hectares in area. Some were dug for brick clay, or clay for making cob, but many were probably dug to obtain calcareous clay ('marl') to spread on the non-calcareous soils of the district. Analyses of four samples of Aylesbeare Mudstone clays show contents of calcium carbonate from 3.9 to 12.6 per cent (mean 7.1 per cent)(Bristow, 1984 b; Edwards, 1984b). Ussher (1913) noted that the practice of marling had been discontinued by that date.

Shapter (1864, p.184) noted that the Heavitree Breccia north-west of Exeter is locally sufficiently calcareous to have been worked for lime, but no figure for the carbonate content is quoted. Two analyses for calcium carbonate from the Exminster district showed the Heavitree Breccia to contain between 1 and 5 per cent carbonate (Bristow, 1984 a); it is not known whether these figures are representative. In the Exminster area, the former Limekiln Lane Cottages [SX 9382 8878] and Kiln Close Plantation [SX 9300 8827] are close to old quarries in the Heavitree Breccia. It is possible, however, that the name Limekiln Lane Cottages is derived from a lime-burning industry based on Chalk brought up river by boat to the Matford area.

METALLIFEROUS MINERALS

Baryte

Deposits of baryte were worked at Bridford Barytes Mine [SX 830 865] between 1858 and 1958, when the mine closed. Baryte was also recorded during the development of the Birch Ellers Mine [SX 8264 8701] in the middle part of the 19th century, and at Lawrence Mine [SX 8132 8844]. The baryte mineralisation is located along the Teign Valley lode system, a predominantly north–south-trending 'crosscourse'; this narrow complex of veins, 12 km long, extends south from Dunsford into the adjacent Newton Abbot district, running approximately 1 km from, and more or less parallel with, the margin of the Dartmoor Granite. For much of its length, the lode system coincides with a strike-slip fault, the Teign Valley Fault (Figure 33), that defines the eastern margin of the granite aureole (Barton, 1991). However, in the Exeter district, the production of baryte has been from orebodies trending between north-west and north-north-west, parallel with the strike of the host lithologies within the Lower Carboniferous succession.

The mineral assemblage present in the baryte lodes is typical of low-temperature 'crosscourse' mineralisation in south-west England; studies (Shepherd and Scrivener, 1987; Scrivener et al., 1990) in the Tamar Valley have demonstrated that this mineralisation was generated by the movement of basinal brines rich in sodium and calcium chlorides, during the development of the Permo-Triassic basin. A feature of the Teign Valley lodes, not common elsewhere in south-west England, is the abundance of baryte; this may reflect the interaction of the north–south fracture system with stratabound dispersions of barium which are present within the Teign Chert (Barton, 1991). The regional distribution of barium in soil and stream samples shows a close correlation with the crop of the Lower Carboniferous strata.

The Bridford Barytes Mine [SX 830 865] is situated on the largest known baryte deposit in south-west England, with a total recorded production of 430 000 tonnes of high-grade baryte. The mine was developed to a depth of about 150 m, but most of the production came from workings down to 115 m, below which level sulphides become more abundant. The history of the mine and its production is given by Dines (1956) and by Beer and Ball (1977), the latter drawing on a report by Schmitz (1973) on baryte working in the Teign Valley. The baryte occurs in a number of tabular north-north-west-trending lodes interbedded with the Teign Chert. The principal productive zone is 100 to 200 m wide and contains five major lodes, each about 2 m wide (locally up to 15 m) and inclined at 80 to 85° towards the east. The No. 1 lode has a strike length of about 180 m and is more than 10 m thick for part of its length. It can be seen at surface in a quarry [SX 8296 8656], where it consists of 3 m of ferruginous weathering, white, coarsely crystalline to sugary-textured baryte with minor quartz; approximately 2 m of metalliferous banded cherts form the immediate footwall of the lode (Barton, 1991) (Plate 14e). At depth, pockets and stringers of sphalerite, pyrite, galena and tetrahedrite are present in quartz-baryte lodes; baryte is scarce below the 150 m level.

Individual lodes contain baryte with a little quartz; brecciated inclusions of chert and mudstone are present in places. Minor quantities of sulphides including pyrite, galena, tetrahedrite and sphalerite are present; sulphide abundance increases with depth.

The distribution of barium in stream sediments, panned concentrates and soil samples in the Teign Valley was described by Beer and Ball (1977), who recorded a panned concentrate anomaly of 4.79 per cent Ba north-west of Dunsford [SX 807 896], and 0.031 per cent in Sowton Brook [SX 835 883]. Subsequent soil-sampling traverses showed low to moderate values of barium in these areas, and the authors concluded that baryte mineralisation does not persist north of the Birch Ellers workings; baryte mineralisation is also absent south of Bridford Mine. Anomalously high abundances of barium in stream sediments and panned concentrates were recorded in ground east of Christow and south of Woodah Farm [SX 848 866], coincident with the Lower Carboniferous outcrop.

Lead, zinc and copper

NEWTON ST CYRES AREA

Mining for lead was carried out in the 19th century at a site [SX 8788 9664] between Tinpit Hill and Scrawthorn Plantation, south of Newton St Cyres. There are no production figures. Dines (1956) recorded that the lode 'coursing a little south of east and underlying 20° N' was 0.5 to 1.0 m wide and contained galena, blende, baryte, quartz and clay. Recent geophysical surveys (Scrivener et al., 1985) using the resistivity method, could not trace an east–west-trending structure, but anomalies probably representing fractures were demonstrated trending north-west and north-north-west.

Evidence of mining activity remaining at the present day consists of two partly filled shafts [SX 8788 9664 and 8793 9745] and a collapsed adit [SX 8794 9695] to the north. On the western side of Tinpit Hill, in Common Down Plantation, are two air shafts [SX 8762 9664 and 8772 9662], now collapsed, and the site of a former adit mouth. Dumps near a collapsed shaft [SX 8788 9664] yield fragments of galena and sphalerite, with minor baryte and pyrite. Quartz and siderite form the gangue. The host rocks are shales and sandstones of the Crackington Formation. Typically, the galena from this locality

is very fine-grained and of sheared appearance and the sphalerite dark brown to black. This is in contrast to the aspect of these minerals from the low-temperature cross-course mineralisation of the Teign Valley where unstrained coarsely crystalline galena and pale red or brown sphalerite are the rule. It is possible that the lead-zinc mineralisation of the Newton St Cyres area resulted from the movement of relatively high-temperature fluids during the regional metamorphism and tectonic activity of the Variscan Orogeny. Deposits of this kind have been reported from the Wadebridge area of north Cornwall by Clayton et al. (1990).

TEIGN VALLEY

Small lead and zinc deposits, with associated copper, occur within a north-north-west-trending zone along, or up to 350 m from, the Teign Valley Fault. The mineralised zone extends for about 4.5 km between Dunsford and Christow, and for approximately a further 8 km south of the district (Dines, 1956). Although production data for old workings are not available, the size of the deposits in the district suggests that they were little more than prospects.

Lawrence Mine [SX 8132 8844] and Anna Maria began as separate concerns in 1846 or 1847, and were worked together for a short time after 1852. Lawrence Mine contains seven east-west copper lodes that are intersected by a group of lead lodes trending at 122° (Dines, 1956). The lead lodes carry quartz with baryte, sphalerite and small amounts of galena, and have an overall width of 6.1 to 9.1 m. Three of the copper lodes, 0.9 to 3.4 m wide, were worked in small surface pits that contain chalcopyrite, pyrite and quartz below the gossan. Wheal Anna Maria [SX 8083 8843] (also known as Dunsford Mine) is situated on the north bank of the River Teign, 900 m south-south-west of Dunsford Church. According to Dines (1956), the mine contains four east–west copper lodes, 1.2 to 2.1 m wide, which form gossans near the surface, and consist in large part of quartz and pyrite at depth.

Birch Ellers Mine [SX 8264 8701] (also known as Birch Allers Mine or Birch Mine) was opened for lead and was active between 1850 and 1855. There is no record of production. Two vertical shafts were sunk to 40-fathom and 50-fathom Levels; Dines (1956) reported that the lode is about 1 m wide at the 40-fathom Level and contains galena, sphalerite and baryte. According to Schmitz (1973), a substantial baryte ore-body was encountered in 1854 in the southernmost developments of the mine, but has not been worked. The dumps are of fragments of shale and chert, with some galena, sphalerite, pyrite and baryte.

Manganese

CREDITON TROUGH

Manganese ores were formerly worked from Permian breccia and sandstone host rocks in the southern part of the Crediton Trough, close to the faulted boundary between the Carboniferous and the Permian rocks. Mining activity took place between 1788 and 1849, with some later, unproductive, prospecting from about 1875 (Beer and Scrivener, 1982). The ores were dug from opencast pits and shallow underground workings, after which they were transported to Exeter for concentration; they were sold for use in glassmaking and in the preparation of bleach.

Berger (1811) made direct observations of a manganese mining operation at Upton Pyne, noting that the ore was worked from a bed of indeterminate thickness dipping towards the north at about 30°. The ore occurred beneath an apparently calcareous or siliceous hardground, the 'saalbande' (sic), above which were red sandstone and breccia. De la Beche (1839), writing after the mine at Upton Pyne was closed, considered that the ores were associated with fractures which splayed off the fault separating the Carboniferous and Permian rocks. Dewey and Bromehead (1916) stated that the manganese minerals in the Upton Pyne area form a cement in the sediment host, swelling out locally into small pockets of ore. Scrivener et al. (1985) considered that the manganese ores were worked from small stratiform deposits and that the ore-forming fluids were channelled along minor faults crossing the boundary fault between the Permian and Carboniferous rocks.

A borehole [SX 9108 9773] at Upton Pyne, sited close to the disused and flooded Pound Living Mine, known locally as the Black Pit, proved 51.16 m of Newton St Cyres Breccia in faulted contact with reddened shale and sandstone of the Crackington Formation. Below 41 m depth are a series of carbonate-cemented hardgrounds, and below 46 m is a 2 m zone of pockets and stringers of manganese oxides. The breccia below 47.5 m is fractured and disturbed.

The traces of mineralisation found in the Upton Pyne Borehole confirm the accounts of Berger (1811) and Dewey and Bromehead (1916). Although no examples of in-situ manganese mineralisation can be examined at the present day, material scattered around former workings and prospects confirm that the ores extracted were nodular replacements of the Permian breccia or sandstone host rocks by manganese oxide and carbonate minerals. Specimens of ore commonly comprise earthy, crystalline and colloform manganese oxide minerals with patches of crystalline and colloform pink and brown carbonates. In many specimens, the carbonate patches give rise to a cellular texture. Relict clasts from the sedimentary breccia host rock are very common, and include quartz and potassium-feldspar grains, and lithic fragments. The oxide minerals commonly present in the ores are psilomelane (in part cryptomelane), manganite and pyrolusite, while carbonates are represented by rhodochrosite and mixed $MnCO_3/CaCO_3$ phases (iron-rich carbonate phases have not been recorded). Minor amounts of baryte and quartz are present.

Two specimens of ore (E 65577 and 65578), collected from dump material at the site of the Pound Living Mine, were examined by Mr R Clayton using reflected and transmitted light microscopy, together with electron probe analysis of individual mineral phases; this study revealed a complex paragenetic history for the ore development, summarised as follows:

1. nucleation and growth of colloform, banded rhodochrosite and cryptomelane/psilomelane in botryoidal masses
2. growth of equigranular aggregates of pyrolusite
3. minor fracturing and infill forming veinlets of manganite with pyrolusite and minor carbonate; the formation of the pyrolusite is to some extent at the expense of manganite, which it replaces with a bulk volume reduction to give a cellular texture
4. growth of pyrolusite-polianite needles and prismatic crystals in the cellular structures developed in 3 above
5. patchy growth of aggregates of rhombic carbonate crystals, partly replacing earlier formed oxide phases
6. late stage oxidation due to weathering.

The mineralogy and field relations of the manganese deposits at the southern margin of the Crediton Trough suggest that they were deposited from low-temperature hydrothermal fluids in alternately oxidising (oxide-forming) and reducing (carbonate-forming) environments. Two specimens of manganese ore from dump material at Upton Pyne [SX 9112 9979] and one of ore from Newton St Cyres [SX 8798 9744] were submitted for microthermometric analysis of fluid inclusions in the crystalline carbonate phases. Dr T J Shepherd reports that most of the inclusions examined were monophase, containing liquid only, which confirms a low temperature of formation for the carbonate. Measurement of salinity indicate considerable variation in salinity and of fluid chemistry within individual specimens. The carbonate from Newton St Cyres showed salinities ranging from 1.2 to 14.3 equivalent weight per cent NaCl, while those from Upton Pyne range from 0.3 to 22.9 equivalent weight per cent NaCl. The measurement of initial melting temperatures from the Newton St Cyres material demonstrates that the fluid in the inclusions is essentially H_2O-NaCl, and that some of the Upton Pyne inclusions contain divalent salts, most probably $CaCl_2$, in addition to H_2O-NaCl. While these results do not permit any conclusions to be drawn concerning the deposition of the oxide minerals, it would seem likely that the carbonates were precipitated from the mixing of at least two components, one of which was a calcium chloride-rich brine, similar in composition to the basinal brines responsible for base and precious metal mineralisation in the Variscan basement of southwest England (Scrivener et al., 1985; Shepherd and Scrivener, 1987).

Abundant nodular masses of pyrolusite, haematite and quartz were noted west of Newton St Cyres in the bed [SX 8560 9756] of the Small Brook just north of the Permian–Carboniferous boundary fault. At 700 m east, disturbed ground [SX 8632 9753] and fragments of manganese oxide-cemented breccia in field brash mark the site of a former trial. East of the Shuttern Brook, at West Town, an overgrown pit [SX 8798 9744] near New Barn Cross is surrounded by ground in which manganese oxide-cemented breccia is common. The main site of former mining at Newton St Cyres is in a field [SX 886 975] north of Hayne Farm, known locally as 'Black Pit Field'. Though no trace remains of the excavations, there are abundant fragments of manganese oxide-cemented breccia and of ore scattered across the field. The ore consists of rounded masses, up to 1 cm in diameter, of rhodochrosite in a matrix of manganite. Traces of the host breccia clasts remain in some specimens. Much of the ore is altered to earthy pyrolusite along joints and cracks. A collapsed adit [SX 8906 9762] near Ford Farm was driven north-west apparently as a trial for manganese. The dumps include specimens of breccia with small strings and pods of pyrolusite, but there are no other traces of mineralisation. North of the River Creedy, a partly filled pit [SX 9022 9771] near Langford was the site of a trial for manganese ore. A flooded and partly filled drift, the 'Black Pit' [SX 9112 9779] at Upton Pyne, is the only trace of the former Pound Living Mine, the dumps of which have been ploughed over and distributed over the field to the north of the working. The brash includes nodular fragments of manganese oxide and carbonate ore, earthy pyrolusite-rich material, and sandstone and breccia partly replaced by manganese minerals. Fragments of manganese oxide and carbonate ore, with the characteristic cellular texture, were noted around an old pit [SX 9462 9764] near Huxham Barton. Ussher (1902) noted, without giving details, the presence of manganese at Ratsloe [SX 958 976].

TEIGN VALLEY

Patchy impersistent manganese ores are present within the Teign Chert; in common with manganese deposits elsewhere in the Teign Valley (Morton, 1958), the present occurrences are within the basal part of the Teign Chert, below the *Posidonia* beds. They are oriented parallel with bedding within the cherts (Plate 14e), and are associated with brecciation, sulphide mineralisation and quartz veining.

The manganese deposits were worked in the Christow, Ashton and Doddiscombsleigh areas, mainly between about 1810 and 1875. There is no record of output from the mines; Dines (1956, p.742) stated that the quantity of ore raised is unlikely to have been large, since all the workings are shallow. Beer and Scrivener (1982) noted that the ores were initially used to decolourise glass, later as a reagent in the manufacture of bleach, and finally in the manufacture of steel. Selwood et al. (1984, p.170) noted that true lode structures are absent, and that the ore occurs as beds and nodules interbedded with the cherts and as impregnations in the chert bands. The beds are traceable laterally for up to about 300 m, but, in the absence of good exposure, it is uncertain whether they pinch out or are continuous between workings.

The spatial association of manganese with spilitic basalt, sulphide mineralisation and hydrothermal brecciation of cherts has suggested to a number of authors an origin in a syngenetic exhalative-type system (Morton, 1958; Selwood et al., 1984; Scrivener et al., 1989; Barton, 1991). A model in which metals that include manganese were deposited near the sea floor, during cooling of vigorously convecting fluids driven by the heat of subvolcanic albite-dolerite sills, is consistent with field relations from the present district. The Teign Chert appears to have acted as an impermeable cap-rock; manganese deposition occurred below the cap-rock or along hydraulic fractures that cut the cap-rock (Barton, 1991).

In Scanniclift Copse [SX 8442 8624 to 8447 8636], manganese wad occurs in brecciated Teign Chert over a strike length of about 150 m, the site of the former Teign or Scanniclift Copse Mine. Remains of these workings comprise small surface excavations along the length of the deposit, together with a single adit [SX 8448 8628]. A strike extension of the same deposit occurs to the north-east [SX 8462 8650 to 8473 8655]. A further small working is present near Down Lane [SX 8496 8626].

There is field evidence of manganiferous beds trending at about 023° in the north-west corner [SX 8530 8607] of Mistleigh Copse; Morton (1958, p.51) noted a flooded small adit at this locality. Morton also recorded possible extensions to the manganese beds: '900 yds S.W. of Doddiscombsleigh church, in a wood; subsidence of ground indicates a collapsed underground tunnel' [possibly at SX 8521 8597]; and '1,000 yds. S.W. of Doddiscombsleigh church in a field; a large hole, 30 ft. deep, which appeared in a field in 1956, indicates collapsed underground workings' [possibly at SX 8513 8591].

In Harehill Plantation, old workings for manganese trend at about 145° along a trench several metres deep [SX 8582 8613]. Manganese-impregnated cherts are visible in small exposures. Trial pits are present near the main workings. Morton (1958, p.51) noted 'two small test adits' at this locality. The workings have also been described by Dines (1956, p.748), who wrote that 'Harehill Plantation Mine, 400 yds. S of Doddiscombsleigh church, appears to be little more than a trial. There is a shallow shaft on the northern boundary of the plantation, an adit, driven S.E., with portal about 30 yds. N.W. of the shaft, and an overgrown openwork in the plantation. The dumps consist of hard black, brittle cherts, some red and jaspery chert, and mudstones, occasionally impregnated and coated with wad and pyrolusite.'

Vanadium and uranium

Dark radioactive nodules generally less than 75 mm in diameter (but up to 150 mm), surrounded by greyish green haloes, occur within the Aylesbeare Mudstone, particularly near the base of the Littleham Mudstone Formation (p.79). They were first recorded (Carter, 1931) from the coastal exposures at Littleham Cove [SY 040 803]. They have attracted considerable attention because they are enriched in vanadium, copper, uranium, nickel, cobalt, arsenic and silver (Carter, 1931; Perutz, 1939; Harrison, 1975; Durrance and George, 1976; Durrance et al., 1980; Nancarrow, 1985; Bateson, 1987; Bateson and Johnson, 1992).

Minerals recorded from the vanadiferous nodules include native copper and silver; the copper minerals algodonite, bornite, chalcocite, chalcopyrite, covellite, digenite, djurleite and malachite; the nickel minerals maucherite, niccolite, rammelsbergite and skutterudite; the cobalt minerals langisite, modderite and safflorite; the uranium minerals coffinite and metatyuyamunite; vanadian mica, and montroseite; the arsenic mineral freirinite; the lead minerals clausthalite, galena and molybdomenite; and the zinc mineral sphalerite (Harrison, 1975; Durrance and George, 1976; Nancarrow, 1985).

Carter (1931) used autoradiograph techniques to establish that the nodules were radioactive; he recorded 0.07 per cent U_3O_8 from one nodule. Perutz (1939) recorded uranium contents of 0.3 per cent and 0.5 per cent by radiometric and chemical techniques respectively; Durrance et al. (1980) thought that the chemically determined value was likely to be nearest the correct value. Harrison (1975) noted that the concretions radiometrically assay between 0.006 and 12.3 per cent U_3O_8, most values lying between 0.1 and 0.5 per cent U_3O_8. Carter (1931) recorded 13.96 per cent V_2O_5 from the dark part of a nodule.

The chemistry of 13 nodules, including two from near Aylesbeare within the Exeter district, was examined by Nancarrow (1985), who suggested a classification into two types. Type 1 nodules, all from Littleham Cove, show marked enrichment in U, As, Se, Cu and Co relative to Type 2 nodules, which occur at diverse localities along the Littleham Mudstone outcrop. Detailed chemical data for nodules are given by Nancarrow (1985), and by Bateson and Johnson (1992). Values of vanadium obtained from the nodules vary between 0.03 and 6.7 per cent, with the majority of values above 1 per cent (Bateson and Johnson, 1992).

Samples of Aylesbeare Mudstone analysed by Bateson and Johnson (1992, table 1) contain on average 146 parts per million (ppm) vanadium, close to the mean value for shales, 120 to 130 ppm (Levinson, 1974). Nor is there enrichment of uranium in the host-rock samples: 72 per cent of the samples gave values less than 5 ppm, close to the average value for shale, 4 ppm (Levinson, 1974).

ENGINEERING GEOLOGY

This summary of the engineering geology of the Exeter district is based on geotechnical data which may not be representative of the absolute range of properties of individual lithostratigraphical units. However, they enable the lithostratigraphical units to be characterised generally in terms of their geotechnical properties and likely behaviour in engineering operations, and to be grouped into engineering units, which are shown on the engineering geology map (Figure 39). A full account is given by Forster (1993).

Engineering geology classification

The geological materials present in the Exeter area may be conveniently considered in the following engineering groups:

1. Strong rocks
2. Interbedded strong rocks and mudstones
3. Mudstones, shales and clays
   a) Overconsolidated
   b) Normally consolidated
4. Granular rocks and deposits
   a) Fine-grained
      Well-cemented
      Weakly cemented
   b) Coarse-grained
      Well-cemented
      Weakly cemented
      Uncemented

This classification forms the basis for the engineering geology map (Figure 39).

STRONG ROCKS

Granite

Fresh granite is a very strong rock, as indicated by uniaxial compressive strength values for fresh granite from quarries in Cornwall and Devon (Anon., 1972) which are in the range 100 megaPascals (MPa) to 200 MPa. However, weathering processes have produced, in many areas, a deeply weathered profile from fresh granite to a residual soil which may include core stones of relatively fresh granite in a matrix of coarse granular material (Dearman et al., 1976). Foundation conditions are likely to depend on the state of weathering. Fresh granite generally will offer excellent bearing capacity, and even moderately weathered material should offer reasonable bearing capacity. Highly weathered to residual soils offer a lower bearing capacity, and structures may be subject to differential settlement owing to the presence of core stones, or to uneven fresh bedrock topography. The ease of excavation of granite depends on its weathering grade: fresh or slightly weathered grades require blasting; residual soil may be dug by hand. The stability of excavations is likely to be controlled by the orientation and intensity of jointing in fresh and less weathered material, and high steep faces may be stable. Highly weathered material is unlikely to be stable in high steep faces.

Lavas

The Exeter Volcanic Rocks (p.96) are commonly deeply weathered or altered. However, fresh rock may be strong to very strong, with medium to wide discontinuity spacing. Highly weathered material commonly is reduced to an angular gravel in a firm to very stiff silty clay matrix of intermediate plasticity, and may extend to a depth of 2 m. The highly weathered or fractured lavas generally can be excavated with heavy digging equipment, but ripping is likely to be required in the harder, less fractured, parts. The use of pneumatic rock breakers or blasting may be required where the basalt is unweathered and the discontinuities are medium to widely spaced and tight. There may be considerable variation in strength and discontinuity spacing within the lava rock mass. Therefore, where basalt is present at outcrop, or its presence is suspected at depth, it is recommended that it should be investigated at an early stage of a construction project. Geophysical methods, such as magnetic anomaly surveys, may be of value in locating lavas.

Dolerite and metamorphosed Combe Shale

The dolerites (p.24) occur as sills mostly less than 50 m thick (but locally over 100 m), and comprise massive, hard dolerite which weathers spheroidally. The upper and middle parts of the Combe Shale (p.7) have been locally baked by dolerite intrusions, converting them to hard, splintery rock (adinole and hornfels) for a few metres width adjacent to the intrusion. Exceptionally, up to 15 m may be affected, as at a point [SX 8573 8596] south of Harehill Plantation, Doddiscombsleigh. Blasting or pneumatic tools are likely to be required to excavate these materials, and cut-face stability will be controlled by the orientation and spacing of discontinuities.

Sandstones in Bude Formation

Mappable units of massive fine- to medium-grained sandstones occur within the Bude Formation and have been separately identified as strong rock on the engineering geology map (Figure 39).

INTERBEDDED STRONG ROCKS AND MUDSTONES

Teign Chert

The thinner bedded and weathered material of the Teign Chert (p.8) probably can be excavated by digging, but ripping, the use of pneumatic tools or blasting is likely to be required in fresh more thickly bedded rock. The chert will be abrasive to digging equipment. Foundation conditions are difficult to predict in a formation composed of material of contrasting properties, and must be determined in situ.

Crackington Formation (undivided part)

The Crackington Formation (p.10) consists of folded, weak to very weak, shaly, heavily overconsolidated mudstone with interbeds of moderately strong to strong sandstone in a ratio of between 2:1 and 4:1. It weathers to generally stiff (but ranging from soft to hard), fissured clay with sandstone fragments. It is commonly covered by head, and may be weathered to a considerable depth. Plasticity data show it to be an inorganic clay of low to intermediate plasticity, with material occasionally in the high or very high plasticity classes.

The geotechnical properties and natural slope instability of the Crackington Formation were studied by Grainger (1983, 1984) and Grainger and Harris (1986). Many slopes show topographical features due to active or recent shallow translational slides, earthflows or creep (Plate 13). Slips are commonly 10 to 20 m wide, 15 to 40 m long and 2 to 4 m thick, and locally merge to form areas of complex slipping. The instability is largely confined to the periglacially remoulded material in the near-surface zone. Some parts of the outcrop are more prone to instability than others, and this has been attributed to differences in the clay mineralogy of the parent rock. Grainger (1984) recognised four clay-mineral assemblages: 1. illite; 2. illite plus 5 to 50 per cent kaolinite; 3. illite plus minor amounts of kaolinite and chlorite; and 4. illite plus 5 to 50 per cent chlorite. Grainger and Harris (1986) quoted a range for the residual angle of friction, $ø'_r$ for landslide material of 15 to 25°. These values and the Skempton DeLory equation (Skempton, 1957) give a maximum stable slope angle, with the water table at the surface, in the range 7.5 to 13°. However, Grainger and Harris (1986) noted that a perched water table, held up by impermeable remoulded material, may develop in the near-surface zone. Substantial amounts of water may also be present as an underflow below the impermeable zone, and may create artesian pressures below the lower part of hillslopes which would enable movement to take place at angles less than those predicted by the Skempton DeLory equation. Clay gouge on faulted surfaces may create aquicludes, resulting in localised concentrations of groundwater which could cause slope instability.

The stability of artificial faces in the folded and fractured Crackington Formation is difficult to predict. Each face must be considered individually and its stability assessed in the light of the interaction between the cut face and the pattern and frequency of discontinuities, water inflow, faulting and folding. Excavation may be accomplished by digging in the weathered material and ripping in the less weathered shale and sandstone. The excavated material is generally suitable as fill. Although in places pyritic, the Crackington Formation generally conforms to class 1 conditions for sulphate attack on buried concrete (Anon, 1981).

Bude Formation (excluding mapped sandstones)

The Bude Formation (p.22) consists mainly of shales, claystones and siltstones with stronger units of thick-bedded sandstones (see above). On weathering, the Bude Formation does not weaken as much as the Crackington Formation, and the greater proportion of sandy material renders the resulting clays less plastic. Sherrell (1971) quoted a range for the residual angle of friction for remoulded weathered mudstone of the Bude Formation, at the Nag's Head Landslip [ST 022 051], Westcott, of 23 to 26°. Natural slopes rarely exceed 14°, and most slopes appear stable. However, landslips have taken place on a few steeper slopes of 16 to 20°.

MUDSTONES, SHALES AND CLAYS

Overconsolidated

Hyner Shale and Trusham Shale (undifferentiated); Combe Shale

The Hyner Shale and Trusham Shale (p.7) lie within the metamorphic aureole of the Dartmoor Granite and consist of grey banded hornfels. The Combe Shale (p.7) consists of siliceous mudstones, locally with silty laminae. No geotechnical data are available for these formations.

Ashton Shale

The Ashton Shale Member of the Crackington Formation (p.11) consists of dark grey to black shale with a few thin interbeds of sandstone and siltstone. The non-chloritic (illite-kaolinite) shales give rise to a weathered material of low shear strength, locally prone to slope instability.

Upton Formation

The Upton Formation (p.59) consists of gravelly clay and in places clayey gravel. The gravel clasts are angular to subrounded Culm sandstone fragments, and the clay is composed of angular shale fragments weathered to a clay with a remnant platy texture. It is likely that the geotechnical properties of the formation will be comparable to those of weathered Crackington Formation clays.

Yellowford Formation

The Yellowford Formation (p.70) consists of silty and sandy claystone, clayey sandy siltstone, and silty clayey fine-grained sandstone. Beds of clayey fine-grained breccia are present locally. Near-surface weathering gives rise to

Figure 39 Engineering geology map of the district.

Figure 39 *continued.*

Engineering classification	Lithostratigraphical unit	Foundations	Excavations	Slope stability	Weathering
Strong rocks	Granite / Lavas / Dolerite and metamorphosed Combe Shale / Sandstones in Bude Formation	Excellent bearing capacity on fresh rock, but granite, lavas and dolerite may be deeply weathered to a more compressible residual soil of lower bearing capacity. Chemical attack on buried concrete is unlikely.	Blasting, hydraulic or pneumatic tools needed in fresh rock. Ripping or digging in more weathered material. Fresh core-stones may remain in completely decomposed soil.	Shallow slips may occur in weathered material if undercut. Block slides and wedge failures in cut faces if discontinuities dip out of face. Rock falls and/or topples possible on steep faces.	Deeply weathered to residual soil. Lavas may also have been chemically altered at time of formation.
Interbedded strong rocks and mudstones	Teigs Chert / Crackington Formation (undivided part) / Bude Formation (excluding mapped sandstones)	Generally good bearing capacity for fresh material. Possible differential settlement across sandstone/shale interfaces. Higher plasticity clays from weathered shale may give shrink/swell problems. Chemical attack on buried concrete is unlikely.	Blasting, hydraulic or pneumatic tools needed for thicker sandstones. Thinner sandstones interbedded with shale are rippable or diggable. Chert may cause high rate of wear on machinery.	Shallow slips where shale dominant, weathered material is undercut or water table is high. Falls, block slides and wedge failures in cuttings in sandstone-dominant lithologies.	Mudstones soften to clay, and thinner sandstones are reduced to angular rubble.
Mudstones, shales and clays — Fine-grained deposits — overconsolidated	Hyner Shale and Trusham Shale (undifferentiated) / Combe Shale / Ashton Shale / Upton Formation / Yellowford Formation / Bussell's Member / Poltimore Mudstone / Aylesbeare Mudstone	Generally good bearing capacity if soft superficial material removed. Weathered material of high plasticity may give shrink/swell problems. Chemical attack on buried concrete is unlikely.	Excavations by digging in weathered material, ripping in fresher, harder mudstone. May heave in bottom of excavations due to stress relief.	Shallow slipping in weathered Ashton Shale.	May be deeply weathered and in a softened remoulded condition in the near-surface zone.
Fine-grained deposits — normally consolidated	Alluvium (excluding gravels) / Marine deposits	Low and possibly uneven bearing capacity, high and differential settlement. Possible chemical attack on buried concrete in Marine Deposits.	Easily dug, but may need dewatering and extra support to maintain stability. Running sands may be present.	No natural slope instability.	Surface desiccation may cause a higher strength layer in the near-surface zone.
Granular rocks and deposits — Coarse-grained deposits — well-cemented	Knowle Sandstone	Good bearing capacity. Chemical attack on buried concrete unlikely.	Diggable in weathered material; ripping required in fresh rock.	No known stability problems.	May be weathered to sand in near-surface zone.
Coarse-grained deposits — weakly cemented	Higher Comberoy Formation / Thorverton Sandstone / Creedy Park Sandstone / Shute Sandstone / Whipton Formation / Monkerton Formation / Dawlish Sandstone / Clast St Lawrence Formation / Otter Sandstone	Generally good bearing capacity if near-surface weathered material is removed. Chemical attack on buried concrete unlikely. Fine silty sands may be frost susceptible.	Usually diggable, with ripping in the more cemented parts. Running sand conditions may occur below the water table.	No known natural instability. Cut slopes stable if protected from surface run-off, and if erosion by seepage from perched water tables is controlled.	May be deeply weathered to sand. Loose at the surface, becoming denser with increasing depth.
Coarse-grained deposits — well-cemented	Newton St Cyres Breccia / Heavitree Breccia	Very good bearing capacity. Chemical attack on buried concrete unlikely.	Rippable and diggable in weathered material. Strongest material may need blasting.	No natural instability known. Cut slopes stand well at high angles.	Moderately weathered near surface.
Coarse-grained deposits — weakly cemented	Cadbury Breccia / Bow Breccia / Crediton Breccia / Alphington Breccia / Budleigh Salterton Pebble Beds	Generally good bearing capacity in fresh material. Weathered material may be uncemented and loose near the surface. Settlement low and rapid. Chemical attack on buried concrete unlikely. Silty lithologies may be frost susceptible.	Diggable in weathered and weakly cemented material. Rippable in fresh and better cemented material. Running sand may occur in the weathering zone below the water table.	No natural slope instability known. Cut slopes stable at steep angles if protected from surface runoff and if seepage from perched water tables is controlled.	May be deeply weathered to clayey silty soil with cobbles.
Coarse-grained deposits — un-cemented	Alluvial Gravel / River Terrace Deposits / Head (not shown on map)	Generally good bearing capacity. Settlement low and rapid. Chemical attack on buried concrete unlikely.	Diggable; may meet high water inflow below the water table.	No natural slope instability known, except relict slip surfaces may be present in head leading to potential slope instability.	Surface zone in loose condition in places.

clay, silt, fine-grained sand, and gravelly clay. Natural slopes do not usually exceed 4°. There are no geotechnical data available for the formation.

Bussell's Member and Poltimore Mudstone

These members of the Dawlish Sandstone Formation consist of clayey fine-grained sandstone and mudstone (p.52). No geotechnical data are available.

Aylesbeare Mudstone

The Aylesbeare Mudstone (p.77) consists mainly of weakly calcareous siltstone and silty claystone, with lesser proportions of sandy siltstone. Also present are minor amounts of claystone and silty sandstone and a few mappable lenticular sandstone beds. The moisture content of the Aylesbeare Mudstone has a near-surface range of about 15 to 30 per cent, and decreases to a range of 10 to 30 per cent at a depth of 5 m, suggesting that weathering and stress relief extend to at least that depth. Weathered material has been observed at depths of 9 m and, in exceptional circumstances, 20 m (for example, in a borehole [SY 0655 9711] near Larkbeare). The Aylesbeare Mudstone is composed of inorganic clay of low or intermediate plasticity, but with some high plasticity material. Undrained cohesion values show a wide spread from about 10 kiloPascals (kPa) to over 300 kPa and a general increase in strength with depth. Consolidation data indicate low compressibility, which takes place at a slow rate. Sulphate contents indicate class 1 conditions (Anon., 1981) for sulphate attack on buried concrete. However, reference has been made, in some site investigations, to the possibility of high sulphate values of class 2 and 3 conditions being present. Natural slopes are generally less that 13°, and appear to be stable, except where they are influenced by a perched water table at the base of the overlying Budleigh Salterton Pebble Beds. Steeper slopes in stream cut ravines, disused marl pits (25°) and in road cuttings (30–45°) appear to be stable, although perched water tables in silty or sandy beds impair stability. Excavations of the weathered material should be possible by digging, but ripping may be necessary in the harder mudstones. Some inflow of water into excavations through water-bearing sands may take place. The silty and fine-grained sandy parts of the Aylesbeare Mudstone may be susceptible to frost heave.

Normally consolidated

Alluvium (excluding gravels)

Alluvium (p.137) is composed mainly of soft to stiff, normally consolidated, silty or sandy clay, but includes some silt, sand, and peat. Standard penetration test values range from 10 to 43, but show little correlation with depth. Moisture contents range from 75 to 80 per cent in the near-surface zone, and decrease generally with depth. Plasticity ranges from sandy inorganic clay of low plasticity to inorganic clay of very high plasticity. Undrained cohesion decreases from ground surface to 1 m depth and indicates the presence of a desiccated crust about 1 m thick in some areas. Consolidation data indicate that the alluvium has medium to high compressibility with a medium to slow rate of consolidation. Generally class 1 conditions (Anon., 1981) for sulphate attack prevail, with class 2 or 3 conditions locally present. Alluvium is easily excavated by normal digging methods, but excavations are likely to require support and precautions against running sand may be needed. Excessive water inflow may be met if underlying water-bearing gravels are penetrated. Very weak, soft alluvium may be met below a stronger desiccated crust. The lower material would hinder the movement of vehicles if the desiccated crust were removed or disrupted. The alluvium locally contains minor ribbon-shaped or lenticular bodies of peat or sand which could cause differential settlement of structures founded across their edge.

Marine deposits

The marine deposits (p.137) consist mainly of very soft to firm, normally consolidated silty clay or organic silty clay with a small proportion of sandy lithologies. Standard penetration test ('N') values are generally low, indicating soft (cohesive) or loose (non cohesive) conditions. The median moisture content is 36 per cent, with a wide range reflecting the range of lithologies present; in particular those with a high organic (peat) content give rise to material of very high moisture content. The marine deposits are mainly inorganic clay of intermediate to high plasticity with some inorganic sandy clay of low plasticity, organic or silty clays of high plasticity and some clays, silts and organic material of extremely high plasticity. Undrained cohesion values are generally in the range 1 to 60 kPa (very soft to firm), but higher values up to 100 kPa are found in the top metre indicating that a desiccated crust is present in some areas. Compressibility is generally high, with consolidation taking place at a medium to slow rate. Sulphate content indicates class 1 conditions (Anon., 1981) for sulphate attack on buried concrete, although if saline groundwater is present, sulphate and chloride attack on buried concrete would be possible. The marine deposits are easily excavated with normal digging machinery, but it generally will be necessary to stabilise the sides. Running sand or excessive inflow of water through sand or gravel lenses may be encountered. The movement of vehicles may be difficult where a desiccated crust has been disrupted or removed, or is absent. Foundations, such as rafts to spread the load, or piles to carry the load down through the alluvium to stronger material, may be necessary even for light structures. Lenses or pockets of material such as peat or sand, or relicts of ancient channel infill surrounded by soft silty clay, may cause severe differential settlement, causing damage to buildings, services or roads founded on them.

GRANULAR ROCKS AND DEPOSITS

Fine-grained deposits; well-cemented

Knowle Sandstone

The Knowle Sandstone (p.63) comprises moderately to well-cemented, well-bedded sandstone, with scattered interbeds of breccia. No geotechnical data are available.

Fine-grained deposits; weakly cemented

Higher Comberoy Formation

The Higher Comberoy Formation (p.62) comprises clayey siltstone, silty claystone, clayey very fine-grained sandstone, fine-grained sandstone, and very fine-grained breccia; these lithologies weather to silt, clay, sand and gravel. No geotechnical data are available.

Thorverton Sandstone

The Thorverton Sandstone (p.65) consists mainly of fine- to very fine-grained sandstone, in places with thin beds and partings of clay, and also thin breccia beds; basaltic and lamprophyric lavas occur at various levels. The sandstone is generally weakly cemented and weathers to sand, but some well-cemented beds are present. Natural slopes reach maximum angles of 12°, and near-vertical road cuttings up to 3 m high, in the better-cemented material, have stood for at least 100 years [e.g. at SS 9092 0255]. No geotechnical data are available for the formation.

Creedy Park Sandstone

The Creedy Park Sandstone (p.67) consists mainly of clayey, silty, very weak, fine-grained sandstone, weathering at outcrop to clayey sand. Interbeds of siltstone, claystone and breccia are present throughout. Standard penetration test data show the weathered sandstone to be medium dense near the surface, increasing to dense or very dense at a depth of 5 m.

Shute Sandstone

The Shute Sandstone (p.74) consists mainly of weakly cemented silty sandstone and sandy siltstone, both commonly weathering to silty sand and sandy silt; lenses of breccia and very sandy silty clay occur at a few localities. Natural slopes reach a maximum of 13° where they are capped by river terrace deposits, and appear to be stable. No geotechnical data are available.

Whipton Formation

The Whipton Formation (p.40) consists predominantly of weakly cemented, silty, fine-grained sandstone, with local developments of siltstone, claystone and breccia; these lithologies weather to sand, silt, clay and clayey gravel. Standard penetration test data show it to be loose near the surface and to become very dense below about 5 m depth. The more cohesive material is of stiff to hard consistency and of low to intermediate plasticity. Natural slopes of less than 4° are present on the outcrop, except where active stream erosion leads to localised near-vertical slopes; these are commonly unstable, as in the stream section [SX 9417 9342] south of Poltimore Bridge. A railway cutting near St James's Park Halt [SX 9264 9348], which was cut at an angle of 30°, has remained stable, but is in a relatively well-cemented breccia. Stable artificial slopes of 40 to 50° in weathered Whipton Formation are present west of St Thomas in Dunsford Road [SX 904 914], Hamber Lane [SX 903 911] and Little Johns Cross Hill [SX 902 911].

Monkerton Formation

The Monkerton Formation (p.50) consists mainly of weak, silty or clayey, fine-grained (but locally medium- or coarse-grained) sandstone, locally with silty or sandy heavily over-consolidated mudstone. It weathers to a medium dense to very dense silty clayey sand or firm to very stiff sandy or silty clay. Thin breccia is locally present. The silty or sandy clays are of low to intermediate plasticity; low compressibility is likely, with rapid consolidation. Natural slopes are usually less than 7°.

Dawlish Sandstone

Most of the Dawlish Sandstone (p.51) consists of weakly cemented fine- to medium-grained sandstone, weathering to medium dense to dense sand which becomes very dense with depth. Silty and clayey sandstone and sand are also present, and there are intercalated thin lenses and beds of claystone, clayey siltstone and fine breccia. Thicker units predominantly of mudstone (Poltimore Mudstone and Bussell's Member) have been mapped locally. Standard penetration test data show a general increase in 'N' value with depth, but with a broad scatter, showing that the transition between sand and weakly cemented sandstone is gradual and difficult to determine. Bulk density data indicate a range of values near to the surface of 1.50 megagrams per cubic metre (Mg/m^3) to 1.95 Mg/m^3, but at a depth of 4 m and below, increases to 2.05 to 2.25 Mg/m^3. A change from weathered to unweathered material is thus indicated at a depth of about 4 m. Plasticity data for the clayey and silty parts of the formation show a wide range from inorganic sandy silts of low plasticity to inorganic clays of high plasticity. Natural slopes in the Dawlish Sandstone do not exceed 13°, but many vertical, or near-vertical, artificial faces have remained stable for many years. The 15 m-high face at Bishop's Court Quarry [SX 963 914], Clyst St Mary, is an example. Local erosion and instability take place where water drains over the face, or perched water tables cause seepage on the face. Perched water tables may be due to the presence of thin silt or clay beds within the sandstone. Recent road cuttings have been made at slopes of between 30 and 40° (1:1.5) with a cover of grass or other vegetation as a protection against erosion by surface run-off. Steeply dipping open joints up to 100 mm wide were seen during the construction of the M5 Motorway. Excavation is possible by digging in the weathered zone and ripping in the denser, more cemented material. Where the silt or clay fraction is high and drainage is poor, the movement of vehicles may be impeded in wet weather. Excavations in weathered sandstone below the water table may meet running sand conditions and require control measures such as close boarding or dewatering.

Clyst St Lawrence Formation

The Clyst St Lawrence Formation (p.85) varies from clayey silt to sand, becoming weakly cemented with depth. Natural slopes do not generally exceed 8°, and appear to be stable. There are no geotechnical data.

Otter Sandstone

The Otter Sandstone (p.92) consists mainly of very dense, silty, fine to medium-grained sand, locally sufficiently well cemented to be termed sandstone. Thin units of hard intraformational conglomerate are present at various levels in the formation, and lenticular beds of mudstone occur locally, but are not of mappable extent. Calcretes are present at some horizons. Lenticular units of pebble beds have been mapped at two localities. The Otter Sandstone forms steep stable natural slopes of up to 45°. Vertical faces up to 12 m high have been cut by the River Otter east of the district. Road cuttings, up to 15 m deep, have been successfully excavated at angles of 30 to 40°. Older road cuts with near-vertical faces appear stable up to heights of 3 m, but higher faces show isolated instability due to loose blocks and overhangs, the result of weathering and erosion. Excavation is possible by digging, and by ripping in well-cemented areas; below the water table, running sand conditions may require control. Thin marly or clayey layers may act as minor aquicludes, giving rise to perched water tables which may cause slope or face instability in slopes and cuttings.

Coarse-grained deposits; well-cemented

Newton St Cyres Breccia

The Newton St Cyres Breccia (p.72) is generally fine to medium, weak to moderately weak breccia with a matrix of silty clayey sand; clasts rarely exceed 50 mm. Some weakly cemented, silty, clayey, medium- to fine-grained sand is present locally. Standard penetration test data indicate increasing density from surface to 5 m depth. In less-weathered material, below 5 m depth, unconfined compressive strength values, based on point load index tests, show strength increases from about 2 MPa at 9 m to 10 MPa at 13 m depth. Natural slopes up to 16° on Newton St Cyres Breccia appear stable and it forms stable, near-vertical slopes in excavations, as in a railway cutting [SX 8588 9906], in a road cutting [SX 8790 9804] on the A377 at Newton St Cyres, and in 'Cromwell's Cutting' [SS 8165 0020], Pitt Hill, Crediton.

Heavitree Breccia

The Heavitree Breccia (p.47) is dominantly a moderately weak to moderately strong breccia composed of well-cemented, fine to coarse gravel-sized material, commonly matrix supported, with clasts set in a matrix of poorly-sorted clay-rich, fine- to coarse-grained sandstone. The breccia weathers to medium dense to dense gravelly clayey sand or soft to stiff gravelly sandy clay. Standard penetration test 'N' values show a general increase with depth, but material may show significant weathering to depths of 8 or 9 m below the ground surface. Plasticity data for the weathered matrix indicate that it is a sandy or silty inorganic clay of low plasticity. The undrained cohesion of the weathered matrix ranges from 5 to 105 kPa (soft to stiff) over a depth range of 0 to 8 m. The unconfined compressive strength of the unweathered breccia is estimated at between 5 and 50 MPa, depending on the degree of cementation. Rock quality designation values show an increase in rock quality with depth, reflecting the effect of weathering. In the Shillingford St George area, open joints up to 100 mm in width occur in a quarry at Shillingford Wood [SX 9065 8746] (Bristow, 1984). Methods used in the excavation of the Heavitree Breccia will vary according to its degree of cementation, but it should be possible to excavate even the strongest material with heavy ripping plant, although occasionally blasting may be necessary. Natural slopes are typically up to 30° for scarp slopes and 5 to 10° for dip slopes. Artificially cut vertical faces up to 35 m high have remained stable for many years, but less steep faces may suffer erosion by the action of surface water runoff. The breccia is locally sufficiently well cemented for use as a building stone, and has been widely used in the Exeter area (p.143).

Coarse-grained deposits; weakly cemented

Cadbury Breccia

The Cadbury Breccia (p.59) is composed of unbedded to very roughly bedded breccia consisting predominantly of sandstone gravel and cobbles in a matrix of very poorly sorted gravelly clayey sandy silt. The sandstone clasts are commonly red-stained, and generally do not exceed 0.3 m in diameter near the base of the deposit, decreasing to about 0.1 m at the top. A spring line may be present at the base of the formation, where it meets less permeable rocks of the Bude or Crackington formations. Natural slopes in the Cadbury Breccia are generally less than 17° and appear to be stable. Near-vertical cuttings — such as the 3 m-high cutting [SS 9041 0309] near Raddon Hill Farm — may remain stable for many years.

Bow Breccia

The Bow Breccia (p.62) consists of fine to coarse breccia with a silty sand or sandy silt matrix, and subordinate beds and lenses of finer-grained beds, mostly sandstone. There are no geotechnical data.

Crediton Breccia

The Crediton Breccia (p.68) is a very weak to moderately strong breccia, with clasts mainly less than 40 mm diameter in a poorly sorted silt, sand and clay matrix; lenses or thin beds of reddish brown silty clay are locally present. Standard penetration test data generally show medium to very dense material near to the surface and indicate an increase in strength with depth to at least 7 m below ground surface. Plasticity data for the clayey silt matrix indicate intermediate plasticity. The Crediton Breccia forms natural slopes up to 7°, which appear to be stable.

Alphington Breccia

The Alphington Breccia (p.45) is a low-strength, weakly cemented breccia rich in Culm shale and sandstone fragments and igneous clasts, which weathers to soft to stiff, sandy clay. Natural slopes are usually below 8°, but a few slopes up to 15° occur. No natural instability has been observed. Artificial cuts are usually made at angles of about 25° (1:2). Clast size is usually less than 40 mm diameter, but may range up to 100 mm diameter; boulders of quartz-porphyry, where locally present, may make excavation by digging more difficult.

Budleigh Salterton Pebble Beds

The Budleigh Salterton Pebble Beds (p.87) comprise horizontally bedded, medium dense to very dense, silty sandy gravel with well-rounded cobbles and boulders, mainly of quartzite. Subordinate lenticular beds of gravelly and silty sand occur within the deposit. Cone penetration test data may be inaccurate due to the cone tip impacting on large cobbles, so giving anomalously high 'N' values. However, these data indicate generally dense to very dense material. Particle size distribution data indicate small percentages of silt and clay, but over 50 per cent of the deposit is of gravel size or greater. Plasticity data for the finer material that forms the matrix between the pebbles indicate an inorganic silty, sandy clay of low to intermediate plasticity. The Pebble Beds do not form natural slopes greater than 12°, except where cut by rivers, when slopes up to 25° may form. Artificial slopes, such as quarry faces at Blackhill Quarry [SY 031 852] (just south of the district), may stand near to vertical at a height of 10 m. The stability of cut slopes may be impaired by perched water tables. Excavation should be possible by heavy excavating equipment, although where the Pebble Beds are worked for aggregate, techniques intermediate between hard rock quarrying and sand and gravel extraction are used. Excavations below the water table may suffer rapid inflow of water and, in the sandier lithologies, running sand conditions. The crushed quartzite has good wear and polishing resistance, and the absence of colloidal silica makes it suitable for use as aggregate in concrete (p.143). The Pebble Beds may be used as backfill in trenches and excavations below the water table, but will need the addition of cohesive soil for use in embankments.

Coarse-grained deposits; uncemented

Alluvial gravel

The alluvium (p.137) is commonly underlain by up to 7 m of gravel, which is medium dense to very dense and generally consists of sandy, rounded to subangular gravel with a low proportion of clay or silt. It normally should be possible to excavate the gravel using normal digging equipment, but excavations may need support to maintain side stability, and some provision also may be necessary for dealing with rapid influx of groundwater into excavations below the water table. The gravel should provide good bearing capacity where it crops out, or where foundations need to be carried down through soft alluvium to a stronger stratum.

River terrace deposits

The river terrace deposits (p.134) are mainly gravels, sandy gravels and gravelly sands, but there are significant amounts of finer deposits such as silty or sandy clay, and sandy or gravelly silt. The coarser-grained material is generally medium dense to dense and the finer-grained material is of low plasticity. Compressibility is likely to range from very low for the gravels to medium for the silts or clays and to take place rapidly on the gravels and at a medium rate on the silts and clays. However, if compressible organic clay or peat are present, consolidation will be fast for peat, but slow for organic clay. Excavations

should be possible with normal digging equipment, but support may be required in the less dense material. Excavations below the water table may meet very high rates of water inflow. The river terrace deposits should give good foundation conditions, but pockets of material of contrasting properties should be removed to avoid differential settlement.

Head

Head (p.132; Plate 12) is composed of a variable mixture of the bedrock and superficial lithologies present upslope of its position. It is inherently inhomogeneous in lithology and geotechnical properties. In terms of their engineering behaviour, the three types of head recognised in the district may be considered together. The behaviour of head is not accurately predictable, but is related to the source material. It is generally of low strength and may be excavated with digging machinery. It will require support in excavations and, although commonly thin, may be up to 4 m thick at the foot of slopes. It may contain perched water tables or water-bearing sands which will flow into excavations, causing collapse. The solifluction process that was active in the formation of head may have left relict shear surfaces within the deposit, and movement may take place on these if the toe of the slope is excavated, if the slope is loaded, if water is introduced into the slope, or if the drainage of water from the slope is impeded. Engineering works on slopes covered by thick head should include a slope-stability investigation at the site investigation stage. Lateral changes in lithology, and hence in geotechnical properties, may give rise to uneven settlement of a structure, and care must be taken to determine bearing capacity when constructing on head deposits. Where deposits are not excessively thick it may be desirable to remove them and place the foundation on bedrock.

HYDROGEOLOGY AND WATER SUPPLY

The district includes much of the catchment of the River Exe and its tributaries, the rivers Clyst, Culm and Yeo. A small area in the south-east is in the catchment of the River Otter, and part of the south-west of the district lies within the River Teign catchment. The estuary of the River Exe extends into the southern margin of the district, and the lower reaches of this river and the River Clyst are tidal.

Higher ground in the north, west and east of the district reaches elevations of 240 m above OD, with an average annual rainfall of over 1000 mm, while in the valley of the River Exe it is about 800 mm. Potential evaporation is about 430 millimetres per year (mm/a), but the actual evapotranspiration is smaller. The remaining effective precipitation contributes to surface run-off and groundwater recharge. Recharge commonly occurs between November and February.

The more important aquifers are the Otter Sandstone and Budleigh Salterton Pebble Beds (Sherwood Sandstone Group) in the south-east of the district, and the

Dawlish Sandstone Formation (Exeter Group), which crops out in a north-south band between the Crediton Trough and the Exe estuary.

The groundwater abstraction licence data (Table 20) show that most of the geological units are exploited for groundwater, though some are of restricted potential. The majority of the 530 licensed abstractions in the district are for general agricultural use, with some for irrigation only and some for domestic supply. This represents an important resource, with a mean licensed abstraction of 2520 cubic metres per year (m^3/a), excluding unlicensed abstractions of up to 20 m^3/day. There are four public water supply abstractions: one from the Otter Sandstone and Budleigh Salterton Pebble Beds; two from the Dawlish Sandstone Formation; and one from the younger breccias of the Exeter Group. In total, the licensed abstraction for public water supplies is 3 202 450 m^3/a, which represents 64 per cent of the total licensed abstractions. There are ten industrial abstractions which are confined to the generally more productive aquifers and have a total licence of 519 900 m^3/a. These figures represent maximum rates which in general are not fully utilised.

The Hydrogeological Map 'The Permo-Trias and other minor aquifers of south-west England' (Institute of Geological Sciences, 1982) provides a graphical summary of the hydrogeology of the district.

Carboniferous rocks

In the Crackington and Bude formations, groundwater storage and flow are largely restricted to fractures in sandstones. Fracture apertures decrease with depth, and the highest yields are mostly obtained from the uppermost 50 m beneath the water table (Monkhouse, 1985). In the valleys, the water table is usually within 10 m of the ground surface. Borehole yields vary from 0.3 to 0.7 litres per second (l/s) for drawdowns in excess of 12 m. There are 147 licensed abstractions (Table 20); 57 are springs, the remainder boreholes and wells. Some may fail in prolonged drought owing to limited groundwater storage. The groundwater is moderately soft and usually has a carbonate hardness in the range 100 to 200 milligrams per litre (mg/l) and chloride concentration of less than 30 mg/l. Dissolved iron is sometimes present.

Exeter Group: below the Dawlish Sandstone

Fracture flow and modest intergranular permeability induce locally steep hydraulic gradients and a water table which may rise and fall seasonally by up to 10 m. The rocks in the lower part of the Exeter Group (Upton Formation up to the Thorverton Sandstone) are locally moderately well cemented. Groundwater storage and flow are restricted to fractures which are usually poorly developed, and transmissivity lies in the range 0 to 10 m^2/day (Davey,

Table 20 Licensed abstraction data of the Environment Agency.

Aquifer	Public water supply		Industrial abstraction		Mean abstraction	Maximum abstraction	Total licensed abstractions		
	m^3/a	No. of licences	m^3/a	No. of licences			m^3/a	No. of licences	
								Wells	Springs
Alluvium	—	—	—	—	1649	8710	34 635	21	0
River terrace deposits	—	—	27 273	2	5007	20 455	40 052	8	0
Otter Sandstone and Budleigh Salterton Pebble Beds	945 340	1	11 364	1	7870*	24 000*	984 689	3	3
Budleigh Salterton Pebble Beds	—	—	23 864	2	3633	18 182	61 756	10	7
Aylesbeare Mudstone	—	—	63 363	2	3019	61 999	377 412	107	18
Dawlish Sandstone	1 916 160	2	—	—	3709*	38 187†	2 057 116	33	7
Exeter Group below Dawlish Sandstone	340 950	1	393 204	—	3138†	31 900†	1 198 645	78	73
Igneous, fine-grained	—	—	—	—	2566‡	13 274	28 669	2	5
Granite	—	—	—	—	1663	3318	13 303	3	5
Carboniferous rocks	—	—	830	1	1554	11 116	226 859	90	57
Totals	3 202 450	4	519 898	10	2521	—	5 023 138	355	175

* Excludes public water supply licensed abstractions.
† Excludes public water supply and industrial licensed abstractions.
‡ Excludes one large licensed abstraction.

1981a). Up to 20 per cent of the transmissivity of the Bow Breccia can be attributed to intergranular permeability (Institute of Geological Sciences, 1982). This is a poor aquifer, in which wells can fail in dry summers. The younger Exeter Group sequence in the Crediton Trough (Creedy Park Sandstone up to the Shute Sandstone) is generally less well cemented and commonly rubbly with a clayey matrix. The transmissivity is usually dependent upon fractures, and is in the order of 1 to 100 m^2/day with a storativity of 10^{-4}. The average borehole yield is about 2.5 l/s for a drawdown of 15.0 m. There are currently 151 licensed abstractions (Table 20) from the breccias, totalling 1 198 645 m^3/year; half are springs. The abstractions include a public supply well [SX 827 987] at Uton in the Crediton Trough, and two industrial abstractions [SS 991 027] near Hele, and [SS 837 002] at Crediton. Total carbonate hardness of the groundwater is about 200 mg/l. The concentration of the chloride and sulphate ions is usually between 15 and 25 mg/l, and iron content is low. Nitrate levels can be high.

Igneous rocks

These rocks are aquicludes which have low intergranular porosity and permeability. Borehole drilling in these rocks is speculative, and its success is dependent on penetrating weathered or fractured rock. Groundwater quality from the granite is typically soft, with a pH of 5 or 6.

Exeter Group: Dawlish Sandstone

The most productive aquifer in the district is the Dawlish Sandstone Formation within the Exeter Group. The aquifer is partly confined beneath the Aylesbeare Mudstone Group. The hydraulic gradient is generally about 1 in 200 and the seasonal fluctuation in piezometric level is in the order of 5 m in the unconfined lower part of the aquifer and 1 to 2 m in the remainder of the outcrop. The lower part of the aquifer has an average permeability of about 0.6 metres per day (m/d) and the upper part of only 5 × 10^{-3} m/d. The permeability declines with decreasing particle size. Medium- to coarse-grained well-cemented sandstone has a permeability of about 0.3 m/d; moderately cemented sandstone of the same grade has a permeability of about 1.3 m/d, whereas that for a fine- to medium-grained sandstone is of the order 10^{-2} m/d (Davey, 1981a). Both intergranular and fracture flow are important, and Davey (1981) calculated that intergranular permeability is 10 to 40 per cent of field transmissivity in the Starved Oak Cross No. 2 Borehole [SX 9130 9879] near Brampford Speke. Transmissivity is usually in the range 100 to 300 m^2/d. In the Crediton Trough and the Clyst Valley, typical yields for a 250 mm diameter borehole are 10 to 12 l/s; although 40 l/s was obtained from a borehole at Stoke Canon for a drawdown of 30 m after 6 days. An average abstraction rate is 6 l/s for a drawdown of 7 m, which gives a specific capacity of 0.88 l/s/m. The total licensed abstraction for the aquifer is 2 057 000 m^3/a from 40 licensed sources, of which seven are springs (Table 20). The boreholes include two public supply sources, the Fortescue Farm Borehole [SX 9287 9938]

and the Burrow Farm Borehole [SX 9408 9958], which offer a combined licensed abstraction of 1 916 000 m^3/a. The majority of licenses are for agricultural and domestic uses; they have a mean licensed abstraction of 3700 m^3/y. Groundwater quality is good, with bicarbonate hardness of about 200 mg/l. Chloride and sulphate concentrations generally fall within the range 15 to 25 mg/l.

Aylesbeare Mudstone Group

The Aylesbeare Mudstone Group is regarded as an aquiclude, but small supplies are derived from interbedded sandstone horizons. The permeability varies considerably, from 9 × 10^{-3} m/d in mudstones to 1.6 m/d where there is a substantial thickness of sandstone (Davey, 1981). A 100 m-deep borehole near Woodbury penetrated some sandstone and siltstone beds up to 1 m thick and had a transmissivity of 40 to 45 m^2/d (Davey, 1989). Yields are typically about 1 l/s with a drawdown of 12 m, although discharges up to 23 l/s have been recorded. The group has a total of 125 licensed abstractions totalling 377 400 m^3/a (Table 20). There are two large industrial abstractions amounting to 63 300 m^3/a, but the majority are agricultural and domestic with a mean abstraction of 3020 m^3/a. The quality of the groundwater is variable. Sodium and chloride concentrations may be high; iron may exceed 0.2 mg/l. Sulphur and calcium are present in the range 15 to 50 mg/l and 30 to 90 mg/l respectively, but can be as high as 1000 mg/l and 560 mg/l respectively, due to the presence of evaporites in the rocks.

Budleigh Salterton Pebble Beds and Otter Sandstone

The Budleigh Salterton Pebble Beds and the overlying Otter Sandstone are hydraulically interconnected and together form an important aquifer in the east and south of the district. The resource is utilised by several public water supplies in the Otter Valley. The hydraulic gradient attains a maximum of about 1 in 25 to the south-east, with a seasonal fluctuation of the water table in the order of 1 m. Springs develop from the Pebble Beds on to the underlying Aylesbeare Mudstone Group (Plate 8a) and also on to the dip slope of the Otter Sandstone, because of its generally lower permeability. Streams are generally in hydraulic connection with the water table and are gaining, but recent pumping in the Otter Valley has lowered the water table and caused some streams to lose.

Flow is predominantly intergranular, although fracture flow also occurs. Boreholes usually penetrate both formations, and permeability lies in the range 1 to 5 m/d for the whole aquifer. The permeability of the Pebble Beds is particularly variable and can be several times higher than the overlying sand. The strata form a multilayered aquifer with a leaky confined response to pumping tests. Specific yield is of the order of 0.1 and storativity is in the range 10^{-3} to 10^{-4}. Many boreholes respond to fluctuations in barometric pressure, showing the confining effect of the marl bands in the Pebble Beds and the cemented horizons in the sandstone.

Borehole yields depend on the saturated thickness and are mostly in the range 30 to 45 l/s, for a drawdown of 25 to 46 m. Due to the unconsolidated nature of the aquifer, wells must be carefully designed and constructed to limit the ingress of sand. There are only 23 licensed abstractions in the district, including a public water supply of 945 340 m^3/a and three industrial abstractions with a total licence for 1046 000 m^3/a. Excluding the public water supply source, the Pebble Beds have a mean abstraction licence of 3600 m^3/a and the combined Otter Sandstone and Budleigh Salterton Pebble Beds have a mean licence of 7900 m^3/a.

The quality of the groundwater in the Otter Sandstone at outcrop is excellent, moderately hard, calcium-bicarbonate-rich water, typically with a total dissolved solids concentration of 450 mg/l and pH of 7.5. The groundwater from the Budleigh Salterton Pebble Beds is a sodium, potassium, chloride water which is acidic, with a pH of 5 and low total dissolved solids (100 mg/l). These chemical differences imply that little or no mixing of the groundwaters occurs although the two formations are hydraulically connected. Where mixing does take place, for example due to pumping, complex conditions result, and groundwater quality deteriorates. If groundwater is pumped from the Pebble Beds via the overlying Otter Sandstone, pumped water can have iron concentrations up to 10 mg/l due to solution of iron from the Otter Sandstone.

Quaternary (river terrace deposits)

The river terrace deposits are of limited thickness and shallow depth, and tend to be very permeable. They are easily polluted, and can be in hydraulic continuity with rivers and local bedrock aquifers. Small agricultural and domestic water supplies are produced from wells in the coarser horizons and two industrial abstractions draw from the river terrace deposits at Crediton on the River Yeo.

Waste disposal at landfill sites

Of the 26 licensed landfill sites in the district, nine were open in 1994 (Table 19). Nearly all the licenses include inert waste and most also include general industrial waste. Ten sites are in the Carboniferous outcrop mostly in the Crackington Formation. The site at the Punchbowl [SS 792 007] includes household waste and the site at Stoke Hill [SX 938 967] includes paper-making waste. Due to the generally low permeability of the Carboniferous strata, these sites appear to pose no current threat to groundwater quality. There are four sites in the younger sequence of breccias of the Exeter Group. Two sites, now closed, are sited on the Dawlish Sandstone, but they were licensed to accept only inert and general industrial waste and, therefore, are likely to pose little risk to groundwater quality. Nevertheless, the high permeability and extensive use of this groundwater source should in general make the Dawlish Sandstone unacceptable for landfill. The outcrop of the Aylesbeare Mudstone Group includes five waste sites. Only one, sited on the outcrop of the Exmouth Mudstone and Sandstone at Heathfield Farm [SX 993 892] was licensed to accept household waste. The generally low permeability of these strata limits the risk of groundwater pollution. One waste site, now closed, was licensed for inert and general industrial waste in the Budleigh Salterton Pebble Beds at Aylesbeare Common [SY 054 901]. This is located west of the Otter Valley groundwater divide, on a narrow outcrop overlying the Littleham Mudstone, and therefore is unlikely to represent a risk to the main groundwater resource to the west.

REFERENCES

Most of the references listed below are held in the Library of the British Geological Survey at Keyworth, Nottingham. Copies of the references can be purchased subject to current copyright legislation.

ANON. 1972. *Natural stone directory. Quarries in Britain and Ireland currently producing in block and slab form.* (London: Stone Industries.)

ANON. 1981. Concrete in sulphate bearing soils and ground waters. *Building Research Establishment Digest*, No. 250.

APPLEGARTH, S M, and CORNISH, L. 1982. *Agricultural land classification Sheet 176 Exeter.* (Bristol: Ministry of Agriculture, Fisheries and Food).

AULT, L A, EDWARDS, R A, and SCRIVENER, R C. 1990. Preliminary report on the application of resistate geochemistry to the New Red Sandstone of the Exeter district. *British Geological Survey Technical Report*, WI/90/1.

AULT, L A, HASLAM, H W, EDWARDS, R A, and SCRIVENER, R C. 1993. Geochemistry of Exeter Group (New Red Sandstone: Permian) rocks of the Exeter district (1:50 000 sheet 325). *British Geological Survey Technical Report*, WI/93/16.

BARTON, C M. 1992. Geology of the Bridford and Dunsford district (Devon). 1:10 000 sheet SX 88 NW and part of 1:10 000 sheet SX 78 NE. *British Geological Survey Technical Report*, WA/92/51.

BHATIA, M R, and CROOK, K A W. 1986. Trace element characteristics of greywackes and tectonic setting discrimination of sedimentary basins. *Contributions to Mineralogy and Petrology*, Vol. 92, 181–193.

BATESON, J H. (compiler) 1987. Geochemical and geophysical investigations of Permian (Littleham Mudstone) sediments of part of Devon. *British Geological Survey, Mineral Reconnaissance Programme Report*, No. 89.

BATESON, J H, and JOHNSON, C C. 1992. Reduction and related phenomena in the New Red Sandstone of south-west England. *British Geological Survey Technical Report*, WP/92/1.

BEER, K E, and SCRIVENER, R C. 1982. Metalliferous mineralisation. 117–147 in *The geology of Devon*. DURRANCE, E M, and LAMING, D J C (editors). (Exeter: University of Exeter.)

BEER, K E, and BALL, T K. 1977. Mineral investigations in the Teign Valley, Devon. Part 1 — Barytes. *Mineral Reconnaissance Programme, Report of the Institute of Geological Sciences*, No. 12.

BENTON, M J. 1990. The species of *Rhynchosaurus*, a rhynchosaur (Reptilia, Diapsida) from the Middle Triassic of England. *Philosophical Transactions of the Royal Society of London*, B, Vol. 328, 213–306.

BENTON, M J, WARRINGTON, G, NEWELL, A J, and SPENCER, P S. 1994. A review of the British Middle Triassic tetrapod assemblages. 131–160 in *In the shadow of the dinosaurs: early Mesozoic tetrapods*. FRASER, N C, and SUES, H-D (editors). (New York: Cambridge University Press.)

BERGER, J F. 1811. Observations on the physical structure of Devonshire and Cornwall. *Transactions of the Geological Society of London*, Series 1, Vol. 1, 93–184.

BRAMMALL, A, and HARWOOD, H F. 1923. The Dartmoor Granite: its mineralogy, structure and petrology. *Mineralogical Magazine*, Vol. 20, 39–53.

BRIGHOUSE, U W. 1981. *Woodbury — a view from the Beacon.* (Callington: Penwell.)

BRISTOW, C R. 1983. *Geology of Sheet SX 99 NE (Broadclyst, Devon). Geological report for DoE: land use planning.* (Exeter: Institute of Geological Sciences.)

BRISTOW, C R. 1984a. *Geological notes and local details for 1:10 000 sheets: SX 98 NW (Exminster, Devon).* (Exeter: British Geological Survey.)

BRISTOW, C R. 1984b. *Geological notes and local details for 1:10 000 sheets: Sheet SY 09 NW (Whimple, Devon).* (Exeter: British Geological Survey.)

BRISTOW, C R, and WILLIAMS, B J. 1984. *Geological notes and local details for 1:10 00 sheets: Sheet SX 99 SE (Pinhoe and Clyst St Mary, Devon).* (Exeter: British Geological Survey.)

BRISTOW, C R, EDWARDS, R A, SCRIVENER, R C, and WILLIAMS, B J. 1985. *Geology of Exeter and its environs. Geological Report for DoE* (Exeter: British Geological Survey.)

BRISTOW, C R, and SCRIVENER, R C. 1984. The stratigraphy and structure of the Lower New Red Sandstone of the Exeter district. *Proceedings of the Ussher Society*, Vol. 6, 68–74.

BUTCHER, N E, and HODSON, F. 1960. A review of the Carboniferous goniatite zones in Devon and Cornwall. *Palaeontology*, Vol. 3, 75–81.

BUTCHER, N E, and KING, A F. 1967. Exeter, Okehampton and Bude. 38–43 *in* Excursion A1, B1. Bristol–Mendip area and South-West England. 6th International Congress of Carboniferous Stratigraphy and Geology.

CAMPBELL-SMITH, W. 1963. Description of the igneous rocks represented among pebbles from the Bunter Pebble Beds of the Midlands of England. *Bulletin of the British Museum of Natural History (Mineralogy)*, Vol. 2, 1–17.

CARTER, G E L. 1931. An occurrence of vanadiferous nodules in the Permian beds of south Devon. *Mineralogical Magazine*, Vol. 22, 609–613.

CHADWICK, R A. 1985. Permian, Mesozoic and Cenozoic structural evolution of England and Wales in relation to the principles of extension and inversion tectonics. 9–25 in *Atlas of onshore sedimentary basins in England and Wales: post-Carboniferous tectonics and stratigraphy*. WHITTAKER, A (editor). (Glasgow and London: Blackie.)

CHESHER, J A. 1968. The succession and structure of the Middle Teign Valley. *Proceedings of the Ussher Society*, Vol. 2, 15–17.

CHESLEY, J T, HALLIDAY, A N, SNEE, L W, MEZGER, K, SHEPHERD, T J, and SCRIVENER, R C. 1993. Thermochronology of the Cornubian batholith in southwest England: Implications for pluton emplacement and protracted hydrothermal mineralization. *Geochimica et Cosmochimica Acta*, Vol. 57, 1817–1835.

CLAOUÉ-LONG, J C, ZHANG ZICHAO, MA GUOGAN, and DU SHAOHUA. 1991. The age of the Permian–Triassic boundary. *Earth and Planetary Science Letters*, Vol. 105, 182–190.

CLARKE, R F A. 1965. Keuper miospores from Worcestershire, England. *Palaeontology*, Vol. 8, 294–321.

CLARY, T. 1988. Geometry and architecture of a Triassic sheet sandstone (Abstract). *British Sedimentological Research Group Meeting*, March, 1988.

CLAYDEN, A W. 1908a. Note on the discovery of footprints in the 'Lower Sandstones' of the Exeter district. *Report and Transactions of the Devonshire Association for the Advancement of Science, Literature and Art*, Vol. 40, 172–173.

CLAYDEN, A W. 1908b. On the occurrence of footprints in the Lower Sandstones of the Exeter district. *Quarterly Journal of the Geological Society of London*, Vol. 64, 496–500.

CLAYDEN, B. 1964. Soils of the Middle Teign Valley district of Devon. *Soil Survey of Great Britain. Bulletin*, No. 1.

CLAYDEN, B. 1971. Soils of the Exeter district. *Memoir of the Soil Survey of Great Britain*, Sheets 325 and 339.

CLAYTON, G, COQUEL, R, DOUBINGER, J, GUEINN, K J, LLOBOZIZK, S, OWENS, B, and STREEL, M. 1977. Carboniferous miospores of western Europe: illustration and zonation. *Mededelingen Rijks Geologische Dienst*, Vol. 29, 1–71.

CLAYTON, R E, SCRIVENER, R C, and STANLEY, C J. 1990. Mineralogical and preliminary fluid inclusion studies of lead-antimony mineralisation in north Cornwall. *Proceedings of the Ussher Society*, Vol. 7, 258–262.

COCKS, L R M. 1989. Lower and Upper Devonian brachiopods from the Budleigh Salterton Pebble Bed, Devon. *Bulletin of the British Museum of Natural History (Geology)*, Vol. 45, 21–37.

COCKS, L R M. 1993. Triassic pebbles, derived fossils and the Ordovician to Devonian palaeogeography of Europe. *Journal of the Geological Society of London*, Vol. 150, 219–226.

COCKS, L R M, and LOCKLEY, M G. 1981. Reassessment of the Ordovician brachiopods of the Budleigh Salterton Pebble Bed, Devon. *Bulletin of the British Museum of Natural History (Geology)*, Vol. 35, 111–124.

COLLINS, F G. 1911. Notes on the Culm of South Devon: Part I — Exeter district. With a report on the plant-remains, by E A Newell Arber; and notes on Carboniferous cephalopoda from the neighbourhood of Exeter, by G C Crick. *Quarterly Journal of the Geological Society of London*, Vol. 67, 393–414.

COLLINSON, J D. 1986. Deserts. 95–112 in *Sedimentary environments and facies*, Second edition. READING, H G (editor). (Oxford: Blackwell Scientific Publications.)

CONYBEARE, W D, and PHILLIPS, W J. 1822. *Outlines of the geology of England and Wales*. (London: William Phillips.)

COPE, J W C, INGHAM, J K, and RAWSON, P F. 1992 (editors). Atlas of palaeogeography and lithofacies. *Geological Society Memoir*, No. 13.

CORNWELL, J D. 1967. The palaeomagnetism of the Exeter lavas, Devonshire. *Geophysical Journal of the Royal Astronomical Society*, Vol. 12, 181–196.

CORNWELL, J D. 1991. Interpretation of magnetic anomalies in the western part of the Exeter district. *British Geological Survey, Regional Geophysics Group, Project Note*, PN/91/9.

CORNWELL, J D, EDWARDS, R A, ROYLES, C P, and SELF, S J. 1990. Magnetic evidence for the nature and extent of the Exeter lavas. *Proceedings of the Ussher Society*, Vol.7, 242–245.

CORNWELL, J D, ROYLES, C P, and SELF, S J. 1992a. Geophysical evidence for the form of New Red Sandstone basins in the Exeter district. *British Geological Survey Technical Report*, WK/92/9.

CORNWELL, J D, ROYLES, C P, and SELF, S J. 1992b. Evidence from magnetic surveys for the nature and extent of the Exeter

Volcanic Rocks. *British Geological Survey Technical Report*, WK/92/12.

COSGROVE, M E. 1972. The geochemistry of the potassium-rich Permian volcanic rocks of Devonshire, England. *Contributions to Mineralogy and Petrology*, Vol. 36, 155–170.

COX, F C, DAVIES, J R, and SCRIVENER, R C. 1986. *The distribution of high grade sandstone for aggregate usage in parts of south west England and South Wales. Report for the Department of the Environment.* (Keyworth: British Geological Survey.)

CREER, K M. 1957. Palaeomagnetic investigations in Great Britain. IV. The natural remanent magnetisation of certain stable rocks from Great Britain. *Philosophical Transactions of the Royal Society*, A, Vol. 250, 111–129.

CULLINGFORD, R A. 1982. The Quaternary. 249–290 in *The geology of Devon*. DURRANCE, E M, and LAMING, D J C (editors). (Exeter: University of Exeter.)

DANGERFIELD, J, and HAWKES, J R. 1969. Unroofing of the Dartmoor Granite and possible consequences with regard to mineralization. *Proceedings of the Ussher Society*, Vol. 2, 122–131.

DANGERFIELD, J, and HAWKES, J R. 1981. The Variscan granites of south-west England: additional information. *Proceedings of the Ussher Society*, Vol. 5, 116–120.

DARBYSHIRE, D P F, and SHEPHERD, T J. 1985. Chronology of granite magmatism and associated mineralization, SW England. *Journal of the Geological Society of London*, Vol. 142, 1159–1177.

DARBYSHIRE, D P F, and SHEPHERD, T J. 1987. Chronology of magmatism in south-west England: the minor intrusions. *Proceedings of the Ussher Society*, Vol. 6, 431–438.

DAVEY, J C. 1981a. The hydrogeology of the Permian aquifers in central and east Devonshire. Unpublished PhD thesis, University of Bristol.

DAVEY, J C. 1981b. Geophysical studies across the Permian outcrop in central and east Devonshire. *Proceedings of the Ussher Society*, Vol. 5, 354–361.

DAVEY, J C. 1982. Aquifer characteristics of the Permian deposits in central Devonshire. *Proceedings of the Ussher Society*, Vol. 5, 354–361.

DAVEY, J C. 1989. Groundwater in the Aylesbeare Group. *Proceedings of the Ussher Society*, Vol. 7, 191.

DAVIDSON, T. 1866–71. *A monograph of the British fossil Brachiopoda*. Part VII. The Silurian Brachiopoda. (London: Palaeontographical Society.)

DAVIDSON, T. 1870. Notes on the brachiopoda hitherto obtained from the Pebble-bed of Budleigh Salterton, near Exmouth, Devonshire. *Quarterly Journal of the Geological Society of London*, Vol. 26, 70–90.

DAVIDSON, T. 1880. On the species of brachiopoda that characterise the 'Grès Armoricain' of Brittany together with a few observations on the Budleigh Salterton 'Pebbles'. *Geological Magazine*, Vol. 7, 337–343.

DAVIDSON, T. 1881. *Monograph of the British fossil Brachiopoda*. Vol. IV, Part IV. Devonian and Silurian Brachiopoda that occur in the Triassic Pebble Bed of Budleigh Salterton in Devonshire. (London: Palaeontographical Society.)

DE LA BECHE, H T. 1835. Note on the trappean rocks associated with the (New) Red Sandstone of Devonshire. *Proceedings of the Geological Society of London*, Vol. 2, 196–198.

DE LA BECHE, H T. 1839. *Report on the geology of Cornwall, Devon, and West Somerset*. (London: Longman, Orme, Brown, Green and Longmans.)

DEARMAN, W R. 1963. Wrench-faulting in Cornwall and south Devon. *Proceedings of the Geologists' Association*, Vol. 74, 265–287.

DEARMAN, W R, BAYNES, F J, and IRFAN, Y. 1976. Practical aspects of periglacial effects on weathered granite. *Proceedings of the Ussher Society*, Vol. 3, 373–381.

DEWEY, H, and BROMEHEAD, C N. 1916. Tungsten and manganese ores. *Memoir of the Geological Survey of Great Britain.*

DINES, H G. 1956. The metalliferous mining region of South West England. *Memoir of the Geological Survey of Great Britain.*

DOWNES, W. 1881. On the occurrence of Upper Devonian fossils in the component fragments of the Trias near Tiverton. *Transactions of the Devonshire Association*, Vol. 19, 293–297.

DURRANCE, E M. 1969. The buried channels of the Exe. *Geological Magazine*, Vol. 106, 174–189.

DURRANCE, E M. 1971. The buried channels of the Teign estuary. *Proceedings of the Ussher Society*, Vol. 2, 299–306.

DURRANCE, E M. 1974. Gradients of buried channels in Devon. *Proceedings of the Ussher Society*, Vol. 3, 111–119.

DURRANCE, E M. 1980. A review of the geology of the Exe Estuary. 41–71 *in* Essays on the Exe Estuary. BOALCH, G T (editor). *The Devonshire Association for the Advancement of Science, Literature and Art, Special Volume*, No. 2.

DURRANCE, E M. 1985. A possible major Variscan thrust along the southern margin of the Bude Formation, south-west England. *Proceedings of the Ussher Society*, Vol. 6, 173–179.

DURRANCE, E M, and GEORGE, M C. 1976. Metatyuyamunite from the uraniferous-vanadiferous nodules in the Permian marls and sandstones of Budleigh Salterton, Devon. *Proceedings of the Ussher Society*, Vol. 3, 435–440.

DURRANCE, E M, MEADS, R E, BALLARD, R R B, and WALSH, J N. 1978. Oxidation state of iron in the Littleham Mudstone Formation of the New Red Sandstone Series (Permian–Triassic) of southeast Devon, England. *Bulletin of the Geological Society of America*, Vol. 89, 1231–1240.

DURRANCE, E M, MEADS, R E, BRINDLEY, R K, and STARK, A G W. 1980. Radioactive disequilibrium in uranium-bearing nodules from the New Red Sandstone (Permian–Triassic) of Budleigh Salterton, Devon. *Proceedings of the Ussher Society*, Vol. 5, 81–88.

EDMONDS, E A, WRIGHT, J E, BEER, K E, HAWKES, J R, WILLIAMS, M, FRESHNEY, E C, and FENNING, P J. 1968. The geology of the country around Okehampton. *Memoir of the Geological Survey of Great Britain*, Sheet 324 (England and Wales).

EDWARDS, R A. 1984a. *Geological notes and local details for 1:10 000 sheets: parts of Sheets SS 80 SE and SS 90 SW (Thorverton, Devon).* (Exeter: British Geological Survey.)

EDWARDS, R A. 1984b. *Geological notes and local details for 1:10 000 sheets: Sheet SY 09 SW (Aylesbeare, Devon).* (Exeter: British Geological Survey.)

EDWARDS, R A. 1984c. *Geological notes and local details for 1:10 000 sheets: Sheet SY 08 NW (Woodbury, Devon).* (Exeter: British Geological Survey.)

EDWARDS, R A. 1987a. *Geological notes and local details for 1:10 000 sheets: Sheet SS 90 SW (Thorverton, Devon).* (Exeter: British Geological Survey.)

EDWARDS, R A. 1987b. *Geological notes and local details for 1:10 000 sheets: Sheet SS 90 SE (Silverton and Bradninch, Devon).* (Exeter: British Geological Survey.)

EDWARDS, R A. 1988. Geology of the Clyst Hydon district (Devon). 1:10 000 Sheet ST 00 SW. *British Geological Survey Technical Report*, WA/88/4.

EDWARDS, R A. 1989a. Geology of the Talaton district (Devon). 1:10 000 Sheet SY 09 NE. *British Geological Survey Technical Report*, WA/89/1.

EDWARDS, R A. 1989b. Geology of the Plymtree and Payhembury district (Devon). 1:10 000 Sheet ST 00 SE. *British Geological Survey Technical Report*, WA/89/2.

EDWARDS, R A. 1990. Geology of the West Hill district (Devon). 1:10 000 Sheet SY 09 SE. *British Geological Survey Technical Report*, WA/90/17.

EDWARDS, R A. 1991. Geology of the Dunchideock district (Devon). 1:10 000 Sheet SX 88 NE. *British Geological Survey Technical Report*, WA/91/7.

EDWARDS, R A. 1998. Geology of the Ide, Longdown, Holcombe Burnell and Great Fulford areas (Devon). Parts of 1:10 000 sheets SX89SE and SW and parts of sheets SX79SE. *British Geological Survey Technical Report*, WA/98/11.

EDWARDS, R A, and SMITH, S A. 1989. Budleigh Salterton Pebble Beds and Otter Sandstone boundary in East Devon. *British Geological Survey Technical Report*, WA/89/97.

EDWARDS, R A, WARRINGTON, G, SCRIVENER, R C, JONES, N S, HASLAM, H W, and AULT, L. 1997. The Exeter Group, South Devon, England: a contribution to the early post-Variscan stratigraphy of NW Europe. *Geological Magazine*, Vol. 134, 177–197.

EXLEY, C S, STONE, M, and FLOYD, P A. 1983. Composition and petrogenesis of the Cornubian granite batholith and post-orogenic volcanic rocks in Southwest England. 152–177 in *The Variscan fold belt of the British Isles.* HANCOCK, P L (editor). (Bristol: Adam Hilger Ltd.)

FAURE, G. 1986. *Principles of isotope geology* 2nd edition. (New York: John Wiley and Sons.)

FOLK, R L. 1968. *Petrology of sedimentary rocks.* (Austin: Hemphills.)

FOLK, R L, and WARD, W C. 1957. Brazos River bar, a study in the significance of grain-size parameters. *Journal of Sedimentary Petrology*, Vol. 27, 3–27.

FORSTER, A. 1993. The engineering geology of the Exeter district. *British Geological Survey Technical Report*, WN/91/16.

FORSTER, S C, and WARRINGTON, G. 1985. Geochronology of the Carboniferous, Permian and Triassic. 99–113 *in* The chronology of the geological record. SNELLING, N J (editor). *Memoir of the Geological Society of London*, No. 10.

FORTEY, N J. 1991. The Exeter Volcanic Rocks: petrology and mineralogy. *British Geological Survey Technical Report*, WG/91/35.

FORTEY, N J. 1992. The Exeter Volcanic Rocks: geochemistry. *British Geological Survey Technical Report*, WG/92/7.

FRESHNEY, E C, EDMONDS, E A, TAYLOR, R T, and WILLIAMS, B J. 1979. Geology of the country around Bude and Bradworthy. *Memoir of the Geological Survey of Great Britain*, Sheets 307 and 308 (England and Wales).

GALE, A W. 1991. The building stones of Exeter. *Transactions of the Devonshire Association*, Vol. 123, 272–273.

GALE, A W. 1992. *The building stones of Devon.* (Exeter: Devonshire Association.)

GIBBS, A D. 1984. Structural evolution of extensional basin margins. *Journal of the Geological Society of London*, Vol. 141, 609–620.

GLENNIE, K W. 1972. Permian Rotliegendes of Northwest Europe interpreted in light of modern desert sedimentation studies. *Bulletin of the American Association of Petroleum Geologists*, Vol. 56, 1048–1071.

GLOPPEN, T G, and STEEL, R J. 1981. The deposits, internal structure and geometry in six alluvial fan–fan delta bodies (Devonian–Norway) — a study in the significance of bedding sequences in conglomerates. 49–69 in Recent and ancient nonmarine depositional environments: models for exploration. ETHRIDGE, F G, and FLORES, R M (editors). *Special Publication of the Society of Economic Paleontologists and Mineralogists*, No. 31.

GRAINGER, P. 1983. Aspects of the engineering geology of mudrocks, with reference to the Crackington Formation of SW England. Unpublished PhD thesis, University of Exeter.

GRAINGER, P. 1984. The influence of clay mineralogy and diagenesis of Upper Carboniferous shales on soil formation in parts of Devon. *Journal of Soil Science*, Vol. 35, 599–606.

GRAINGER, P, and GEORGE, M C. 1978. Clay mineral studies of Crackington Formation shales near Exeter. *Proceedings of the Ussher Society*, Vol. 4, 145–155.

GRAINGER, P, and HARRIS, J. 1986. Weathering and slope stability on Carboniferous mudrocks in south-west England. *Quarterly Journal of Engineering Geology*, Vol. 19, 155–173.

GRAINGER, P, and WITTE, G. 1981. Clay mineral assemblages of Namurian shales in Devon and Cornwall. *Proceedings of the Ussher Society*, Vol. 5, 168–178.

GREGORY, K J. 1969. Geomorphology. 27–42 in *Exeter and its region*. BARLOW, F (editor). (Exeter: University of Exeter.)

GREGORY, R G, and DURRANCE, E M. 1987. The Otterton Trough: implications of groundwater helium data. *Proceedings of the Ussher Society*, Vol. 6, 542–547.

GRIMMER, S C, and FLOYD, P A. 1986. Geochemical features of Permian rift volcanism — A comparison of Cornubian and Oslo basic volcanics. *Proceedings of the Ussher Society*, Vol. 6, 352–359.

HALLIDAY, A N. 1980. The timing of early and main stage ore mineralisation in southwest Cornwall. *Economic Geology*, Vol. 75, 752–759.

HAMBLIN, R J O. 1973a. The Haldon Gravels of south Devon. *Proceedings of the Geologists' Association*, Vol. 84, 459–476.

HAMBLIN, R J O. 1973b. The clay mineralogy of the Haldon Gravels. *Clay Mineralogy*, Vol. 10, 87–97.

HAMBLIN, R J O, and WOOD, C J. 1976. The Cretaceous (Albian–Cenomanian) stratigraphy of the Haldon Hills, south Devon, England. *Newsletters in Stratigraphy*, Vol. 4, 135–149.

HARRIS, N B W, PEARCE, J A, and TINDLE, A G. 1986. Geochemical characteristics of collision-zone magmatism. 67–81 in Collision tectonics. COWARD, M P, and RIES, A C (editors). *Special Publication of the Geological Society of London*, No.19.

HARRISON, R K. 1975. Concretionary concentrations of the rarer elements in Permo-Triassic red beds of south-west England. *Bulletin of the Geological Survey of Great Britain*, No. 52, 1–26.

HARROD, T R, CATT, J A, and WEIR, A H. 1973. Loess in Devon. *Proceedings of the Ussher Society*, Vol. 2, 554–564.

HASLAM, H W. 1990. The geochemistry of samples from the Bude and Crackington formations north of Exeter. *British Geological Survey Technical Report*, WP/90/3.

HASLAM, H W, and SCRIVENER, R C. 1991. A geochemical study of turbidites in the Bude and Crackington formations north of Exeter. *Proceedings of the Ussher Society*, Vol. 7, 421–423.

HAUBOLD, H. 1973. Die Tetrapodenfährten aus dem Perm Europas. *Freiberger Forschungshefte*, C, Vol. 285, 5–55.

HAWKES, J R. 1982. The Dartmoor granite and later volcanic rocks. 85–116 in The geology of Devon. DURRANCE, E M, and LAMING, D J C (editors.) (Exeter: Exeter University Press.)

HAWKES, J R, and HOSKING, J R. 1972. British arenaceous rocks for skid-resistant road surfacings. *Transport and Road Research Laboratory Report*, No. LR488.

HENSON, M R. 1970. The Triassic rocks of south Devon. *Proceedings of the Ussher Society*, Vol. 2, 172–177.

HENSON, M R. 1971. The Permo-Triassic rocks of south Devon. Unpublished PhD thesis, University of Exeter.

HENSON, M R. 1972. The form of the Permo-Triassic basin in south-east Devon. *Proceedings of the Ussher Society*, Vol. 3, 447–457.

HENSON, M R. 1973. Clay minerals from the Lower New Red Sandstone of South Devon. *Proceedings of the Geologists' Association*, Vol. 82, 429–445.

HESS, J C, and LIPPOLT, H J. 1986. ^{40}Ar/^{39}Ar ages of tonstein and tuff sanidines: new calibration points for the improvement of the Upper Carboniferous time scale. *Chemical Geology (Isotope Geoscience Section)*, Vol. 59, 143–154.

HIGGS, R. 1984. Possible wave-influenced sedimentary structures in the Bude Formation (Westphalian), north Cornwall and north Devon. *Proceedings of the Ussher Society*, Vol. 5, 477–478.

HIGGS, R. 1986. 'Lake Bude' (early Westphalian, SW England): storm-dominated siliciclastic shelf sedimentation in an equatorial lake. *Proceedings of the Ussher Society*, Vol. 6, 417–418.

HIGGS, R. 1991. The Bude Formation (Lower Westphalian), SW England: siliciclastic shelf sedimentation in a large equatorial lake. *Sedimentology*, Vol. 38, 445–469.

HINDE, G J, and FOX, H. 1895. On a well-marked horizon of radiolarian rocks in the Lower Culm Measures of Devon, Cornwall and West Somerset. *Quarterly Journal of the Geological Society of London*, Vol. 51, 609–667.

HOBSON, B. 1892. On the basalts and andesites of Devonshire known as 'felspathic traps'. *Quarterly Journal of the Geological Society of London*, Vol. 48, 496–507.

HOFMANN, B A. 1991. Mineralogy and chemistry of reduction spheroids in Red Beds. *Mineralogy and Petrology*, Vol. 44, 107–124.

HOLLOWAY, S. 1985. The Permian. 26–30 in *Atlas of onshore sedimentary basins in England and Wales: post-carboniferous tectonics and stratigraphy*. WHITTAKER, A (editor.) (Glasgow and London: Blackie.)

HOSKINS, W G. 1972. *Devon*. (Newton Abbot: David & Charles.)

HOUSE, M R, and SELWOOD, E B. 1964. Palaeozoic palaeontology in Devon and Cornwall. 45–86 in *Present views of some aspects of the geology of Cornwall and Devon*. HOSKING, K F G, and SHRIMPTON, G J (editors). (Penzance: Royal Geological Society of Cornwall.)

HULL, E. 1892. A comparison of the red rocks of the south Devon coast with those of the Midland and Western counties. *Quarterly Journal of the Geological Society of London*, Vol. 48, 60–67.

HUTCHINS, P F. 1958. Devonian limestone pebbles in Central Devon. *Geological Magazine*, Vol. 95, 119–124.

HUTCHINS, P F. 1963. The Lower New Red Sandstone of the Crediton Valley. *Geological Magazine*, Vol. 100, 107–128.

HUXLEY, T H. 1869. On Hyperodapedon. *Quarterly Journal of the Geological Society of London*, Vol. 25, 138–152.

INSTITUTE OF GEOLOGICAL SCIENCES. 1982. *Hydrogeological map of the Permo-Trias and other minor aquifers of south-west England*. (London: Institute of Geological Sciences.)

IRVINE, T N, and BARAGER, W R A. 1971. A guide to the chemical classification of the common volcanic rocks. *Canadian Journal of Earth Science*, Vol. 8, 523–548.

IRVING, A. 1888. The red-rock series of the Devon coast section. *Quarterly Journal of the Geological Society of London*, Vol. 44, 149–163.

IRVING, A. 1892. Supplementary note to the paper on the red rocks of the Devon coast section. *Quarterly Journal of the Geological Society of London*, Vol. 48, 68–80.

JACKSON, P C. 1984. The influence of palaeoceanography on the Dinantian of S.W. England and the Rheinisches Schiefergebirge. *European Dinantian Environments, Abstracts, first meeting — Manchester, April 11–13, 1984*. 37–39, Department of Earth Sciences, Open University, Milton Keynes.

JACKSON, P C. 1985. Sedimentology, stratigraphy and palaeoceanography of some Lower Carboniferous hemipelagic sequences. Unpublished PhD thesis, University of Oxford.

JACKSON, P C. 1992. The Dinantian stratigraphy of north-west Devon. *Proceedings of the Yorkshire Geological Society*, Vol. 48, 447–460.

JOHNSTON-LAVIS, H J. 1876. On the Triassic strata which are exposed in the cliff-sections near Sidmouth, and a note on the occurrence of an ossiferous zone containing bones of a labyrinthodon. *Quarterly Journal of the Geological Society of London*, Vol. 32, 274–277.

JONES, A P, and SMITH, J V. 1985. Phlogopite and associated minerals from Permian minettes in Devon, south England. *Bulletin of the Geological Society of Finland*, Vol. 57, 89–102.

JONES, N S. 1992a. Sedimentology and depositional history of the Crediton Trough area, S.W. England. *British Geological Survey Technical Report*, WH/92/112R.

JONES, N S. 1992b. Sedimentology of the Permo-Triassic of the Exeter area, S.W. England. *British Geological Survey Technical Report*, WH/92/122R.

KEMP, S J. 1992. X-ray diffraction analyses of samples from the Exeter Volcanic Rocks. *British Geological Survey, Mineralogy and Petrology Group Brief Report*.

KIDSON, C. 1962. Denudation chronology of the River Exe. *Transactions of the Institute of British Geographers*, Vol. 31, 43–66.

KNILL, D. 1969. The Permian igneous rocks of Devon. *Bulletin of the Geological Survey of Great Britain*, No. 29, 115–138.

KRUMBEIN, W C. 1934. Size frequency distribution of sediments. *Journal of Sedimentary Petrology*, Vol. 4, 65–77.

LAMING, D J C. 1965. Age of the New Red Sandstone in South Devonshire. *Nature, London*, Vol. 207, 624–625.

LAMING, D J C. 1966. Imbrication, palaeocurrents and other sedimentary features in the lower New Red Sandstone, Devonshire, England. *Journal of Sedimentary Petrology*, Vol. 36, 940–959.

LAMING, D J C. 1968. New Red Sandstone stratigraphy in Devon and West Somerset. *Proceedings of the Ussher Society*, Vol. 2, 23–25.

LAMING, D J C. 1970. A guide to the New Red Sandstone of Tor Bay, Petitor and Shaldon. *Report and Transactions of the Devonshire Association for the Advancement of Science, Literature and Art*, Vol. 101, 207–218.

LAMING, D J C. 1982. The New Red Sandstone. 148–178 in *The geology of Devon*. DURRANCE, E M, and LAMING, D J C (editors). (Exeter: University of Exeter.)

LEAT, P T, THOMPSON, R N, MORRISON, M A, HENDRY, G L, and TRAYTHORN, S C. 1987. Geodynamic significance of post-Variscan intrusive and extrusive potassic magmatism in SW England. *Transactions of the Royal Society of Edinburgh: Earth Sciences*, Vol. 77, 349–360.

LEEDER, M R. 1982. Upper Palaeozoic basins of the British Isles — Caledonide inheritance versus Hercynian plate margin processes. *Journal of the Geological Society of London*, Vol. 139, 479–491.

LEEDER, M R. 1987. Tectonic and palaeogeographic models for Lower Carboniferous Europe. 1–20 in European Dinantian environments. MILLER, J, ADAMS, A E, and WRIGHT, V P (editors). *Geological Journal Special Issue*, No. 12.

LEONARD, A J, MOORE, A G, and SELWOOD, E B. 1982. Ventifacts from a deflation surface marking the top of the Budleigh Salterton Pebble Beds, east Devon. *Proceedings of the Ussher Society*, Vol. 5, 333–339.

LEVINSON, A A. 1974. *Introduction to exploration geochemistry*. (Calgary: Applied Publishing Ltd.)

LEVY, A. 1827. On a new mineral substance called Murchisonite. *Philosophical Magazine*, Vol. 1, 448–452.

LOWE, D R. 1982. Sediment gravity flows: II. Depositional models with specific reference to the deposits of high-density turbidity currents. *Journal of Sedimentary Petrology*, Vol. 52, 279–297.

MADER, D. 1985a. *Beiträge zur Genese des germanischen Buntsandsteins*. (Hannover: Sedimo-Verlag.)

MADER, D. 1985b. Braidplain, floodplain and playa lake, alluvial-fan, aeolian and palaeosol facies composing a diversified lithogenetical sequence in the Permian and Triassic of south Devon (England). 15–64 in *Aspects of fluvial sedimentation in the Lower Triassic Buntsandstein of Europe. Lecture notes in Earth Sciences*, Vol. 4. MADER, D (editor). (Berlin: Springer-Verlag.)

MADER, D. 1990. *Palaeoecology of the flora in Buntsandstein and Keuper in the Triassic of middle Europe*. (Stuttgart: Gustav Fischer.)

MADER, D, and LAMING, D J C. 1985. Braidplain and alluvial fan environmental history and climatological evolution controlling the origin and destruction of aeolian dune fields and governing overprinting of sand seas and river plains by calcrete pedogenesis in the Permian and Triassic of south Devon (England). 519–528 in *Aspects of fluvial sedimentation in the Lower Triassic Buntsandstein of Europe*. MADER, D (editor). (Berlin: Springer-Verlag.)

MARTIN, E C. 1908. The New Red (Permian) gravels of the Tiverton district. *Geological Magazine*, Vol. 5, 150–157.

McKEEVER, P J, and HAUBOLD, H. 1996. Reclassification of vertebrate trackways from the Permian of Scotland and related forms from Arizona and Germany. *Journal of Paleontology*, Vol. 70, 1011–1022.

METCALFE, A T. 1884. On further discoveries of vertebrate remains in the Triassic strata of the south coast of Devonshire, between Budleigh Salterton and Sidmouth. *Quarterly Journal of the Geological Society of London*, Vol. 40, 257–262.

MILLER, J A, SHIBATA, K, and MUNRO, M. 1962. The potassium-argon age of the lava at Killerton Park, near Exeter. *Geophysical Journal*, Vol. 6, 394–396.

MILLER, J A, and MOHR, P A. 1964. Potassium-argon measurements on the granites and some associated rocks from southwest England. *Geological Journal*, Vol. 4, 105–126.

MILNER, A R, GARDINER, B G, FRASER, N, and TAYLOR, M A. 1990. Vertebrates from the Middle Triassic Otter Sandstone Formation of Devon. *Palaeontology*, Vol. 33, 873–892.

MILODOWSKI, A E, STRONG, G E, WILSON, K S, ALLEN, D J, HOLLOWAY, S, and BATH, A H. 1986. *Diagenetic influences on the aquifer properties of the Sherwood Sandstone in the Wessex Basin. Investigation of the geothermal potential of the UK.* (Keyworth, Nottingham: British Geological Survey.)

MITCHELL, G F, and ORME, A R. 1967. The Pleistocene deposits of the Isles of Scilly. *Quarterly Journal of the Geological Society of London,* Vol. 123, 59–92.

MITCHELL, G F, PENNY, L F, SHOTTON, F W, and WEST, R G. 1973. A correlation of Quaternary deposits in the British Isles. *Special Report of the Geological Society of London,* No. 4.

MONKHOUSE, R A. 1985. A statistical study of specific capacities in the Upper Carboniferous (Culm Measures) of south-west England and their use in predicting borehole yields. *British Geological Survey Technical Report,* WD/85/1.

MORTON, A C. 1992. Heavy mineral assemblages in the Budleigh Salterton Pebble Beds and the overlying Otter Sandstone from a borehole near Exeter. *British Geological Survey Technical Report,* WH/92/216C.

MORTON, R D. 1958. The stratigraphy, structure and igneous rocks of the Central Teign Valley. Unpublished PhD thesis, University of Nottingham.

MORTON, R D, and SMITH, D G W. 1971. Differentiation and metasomatism within a Carboniferous spilite-keratophyre suite in S.W. England. *Special Paper of the Mineralogical Society of Japan,* No. 1, 127–133.

MURCHISON, R I. 1867. *Siluria.* (London: J. Murray.)

NANCARROW, P H A. 1985. Vanadiferous nodules from the Littleham Marl, near Budleigh Salterton, Devon. *British Geological Survey Mineralogy and Petrology Report,* No. 85/12.

PATON, R G. 1974. Capitosauroid labyrinthodonts from the Trias of England. *Palaeontology,* Vol. 17, 253–289.

PEACHEY, D, ROBERTS, J L, VICKERS, B P, ZALASIEWICZ, J A, and MATHERS, S J. 1985. Resistate geochemistry of sediments — a promising tool for provenance studies. *Modern Geology,* Vol. 9, 145–157.

PEARCE, J A, and CANN, J R. 1973. Tectonic setting of basic volcanic rocks investigated using trace element analysis. *Earth and Planetary Science Letters,* Vol. 19, 290–300.

PEARCE, J A, HARRIS, N B W, and TINDLE, A G. 1984. Trace element discrimination diagrams for the tectonic interpretation of granitic rocks. *Journal of Petrology,* Vol. 25, 956–983.

PENGELLY, W. 1864. On the chronological value of the New Red Sandstone System of Devonshire. *Transactions of the Devonshire Association for the Advancement of Science, Literature and Art,* Vol. 1, Report of Second Meeting (1863), 30–43.

PERUTZ, M. 1939. Radioactive nodules from Devonshire, England. *Mineralogische und Petrographische Mitteilungen,* Vol. 51, 141–161.

PHILLIPS, J. 1841. *Figures and descriptions of the Palaeozoic fossils of Cornwall, Devon and West Somerset.* London.

PICKARD, R. 1949. The geology of Milber Down. *Transactions of the Devonshire Association,* Vol. 81, 217–226.

POLLARD, J E. 1976. A problematic trace fossil from the Tor Bay Breccias of south Devon. *Proceedings of the Geologists' Association,* Vol. 87, 105–108.

POLWHELE, R. 1793–1806. *The history of Devonshire.* (Three volumes, 1793, 1797, 1806). Reprinted 1977 with an introduction by A L Rowse. (Dorking: Kohler and Coombes.)

PURVIS, K, and WRIGHT, V P. 1991. Calcretes related to phreatophytic vegetation from the Middle Triassic Otter Sandstone of South West England. *Sedimentology,* Vol. 38, 539–551.

RAMSBOTTOM, W H C. 1977. Major cycles of transgression and regression (mesothems) in the Namurian. *Proceedings of the Yorkshire Geological Society,* Vol. 41, 261–291.

RAMSBOTTOM, W H C, CALVER, M A, EAGAR, R M C, HODSON, F, HOLLIDAY, D W, STUBBLEFIELD, C J, and WILSON, R B. 1978. A correlation of Silesian rocks in the British Isles. *Special Report of the Geological Society of London,* No. 10.

REID, C, BARROW, G, SHERLOCK, R L, MACALISTER, D A, DEWEY, H, and BROMEHEAD, C N. 1912. The geology of Dartmoor. *Memoir of the Geological Survey of Great Britain,* Sheet 338 (England and Wales).

RIDGWAY, J M. 1974. A problematical trace fossil from the New Red Sandstone of south Devon. *Proceedings of the Geologists' Association,* Vol. 85, 511–517.

RILEY, N J. 1983. Palaeontological report on spot samples from surface localities in the Crackington Formation of IGS One-inch sheet 325. Submitted by Dr J M Thomas (University of Exeter). *Unpublished Report of the British Geological Survey,* No. PDL 83/77.

RILEY, N J. 1984. Palaeontological report on recently collected material from the Crackington Formation of One-inch Sheet 325. *Unpublished Report of the British Geological Survey,* No. PDL 84/47.

RILEY, N J. 1991. Late Visean-Namurian biostratigraphy; various localities, Exeter Sheet 325. *British Geological Survey Technical Report,* WH/91/66R.

RILEY, N J. 1992a. Macrofaunal biostratigraphy from various localities in the Crackington Formation, Exeter Sheet 325, 1991 Field Season. *British Geological Survey Technical Report,* WH/92/18R.

RILEY, N J. 1992b. Biostratigraphy of fossiliferous clasts from the Cadbury Breccia (Late Carboniferous–Early Permian) of the Exeter district. *British Geological Survey Technical Report,* WH/92/22R.

RILEY, N J. 1993. Dinantian (Lower Carboniferous) biostratigraphy and chronostratigraphy in the British Isles. *Journal of the Geological Society of London,* Vol. 150, 427–446.

RUNDLE, C C. 1976. K-Ar ages for lamprophyre dykes from SW England. *Institute of Geological Sciences, Isotope Geology Unit Report,* No. 80/9.

RUNDLE, C C. 1981. K-Ar ages for micas from S.W. England. *Institute of Geological Sciences, Isotope Geology Unit Report,* No. 81/10.

SADLER, P M. 1974. Trilobites from the Gorran Quartzites, Ordovician of south Cornwall. *Palaeontology,* Vol. 17, 71–93.

SCHMITZ, C J. 1973. The Teign valley lead mines. *Northern Cave and Mine Research Society, Sheffield, Occasional Publication,* No. 6.

SCOTESE, C R, BAMBACK, R K, BARTON, C, VAV DER VOO, R, and ZIEGLER, A M. 1979. Palaeozoic base maps. *Journal of Geology,* Vol. 87, 217–277.

SCRIVENER, R C. 1983. *Geology of Sheet SX 99 NW (Brampford Speke, Devon). Geological report for DoE: land use planning.* (Exeter: Institute of Geological Sciences.)

SCRIVENER, R C. 1984. *Geological notes and local details for 1:10 000 sheets: Sheet SX 99 SW (Exeter, Devon).* (Exeter: British Geological Survey.)

SCRIVENER, R C. 1988. Geology of the Newton St Cyres district (Devon). 1:10 000 Sheet SX 89 NE. *British Geological Survey Technical Report,* WA/88/5.

SCRIVENER, R C, DARBYSHIRE, D P F, and SHEPHERD, T J. 1990. Crosscourse mineralisation and basin development in SW England (Abstract). *Proceedings of the Ussher Society*, Vol. 7, 306.

SCRIVENER, R C, COOPER, B V, DURRANCE, E M, and MEHTA, S. 1985. Mineralisation at the southern margin of the Crediton Trough, near Exeter, Devon (Abstract). *Proceedings of the Ussher Society*, Vol. 6, 272.

SCRIVENER, R C, and EDWARDS, R A. 1990. Field excursion to the New Red Sandstone of the eastern Crediton Trough, 3rd January 1990. *Proceedings of the Ussher Society*, Vol. 7, 304–305.

SCRIVENER, R C, and EDWARDS, R A. 1991. Geology of the Shobrooke district (Devon). 1:10 000 Sheet SS 80 SE. *British Geological Survey Technical Report*, WA/90/24.

SCRIVENER, R C, LEAKE, R C, LEVERIDGE, B E, and SHEPHERD, T J. 1989. Volcanic-exhalative mineralisation in the Variscan province of SW England. *Terra Abstracts*, Vol. 1, 125.

SCRIVENOR, J B. 1948. The New Red Sandstone of Devonshire. *Geological Magazine*, Vol. 85, 317–332.

SEDGWICK, A, and MURCHISON, R I. 1840. On the physical structure of Devonshire, and on the subdivisions and geological relations of its older stratified deposits. *Transactions of the Geological Society of London*, Vol. 2, 633–704.

SELWOOD, E B, EDWARDS, R A, SIMPSON, S, CHESHER, J A, HAMBLIN, R J O, HENSON, M R, RIDDOLLS, B W, and WATERS, R A. 1984. Geology of the country around Newton Abbot. *Memoir of the British Geological Survey*, Sheet 339 (England and Wales).

SELWOOD, E B, and McCOURT, S. 1973. The Bridford Thrust. *Proceedings of the Ussher Society*, Vol. 2, 529–535.

SHAPTER, T. 1842. *The climate of the south of Devon; and its influence upon health: with short accounts of Exeter, Torquay, Babbicombe, Teignmouth, Dawlish, Exmouth, Budleigh-Salterton, Sidmouth, &c.* Second edition 1862. (London: John Churchill.)

SHEPHERD, T J, and SCRIVENER, R C. 1987. Role of basinal brines in the genesis of polymetallic vein deposits, Kit Hill-Gunnislake area, SW England. *Proceedings of the Ussher Society*, Vol. 6, 491–497.

SHERRELL, F W. 1970. Some aspects of the Triassic aquifer in east Devon and west Somerset. *Quarterly Journal of Engineering Geology*, Vol. 2, 255–286.

SHERRELL, F W. 1971. The Nag's Head Landslips, Cullompton By-Pass, Devon. *Quarterly Journal of Engineering Geology*, Vol. 4, 37–73.

SMITH, D B, BRUNSTROM, R G W, MANNING, P I, SIMPSON, S, and SHOTTON, F W. 1974. A correlation of Permian rocks in the British Isles. *Special Report of the Geological Society of London*, No. 5.

SMITH, S A. 1989. The sedimentology of the Budleigh Salterton Pebble Beds: a preliminary report. *British Geological Survey Technical Report*, WH/89/198C.

SMITH, S A. 1990. The sedimentology and accretionary styles of an ancient gravel-bed stream: the Budleigh Salterton Pebble Beds (Lower Triassic), southwest England. *Sedimentary Geology*, Vol. 67, 199–219.

SMITH, S A, and EDWARDS, R A. 1991. Regional sedimentological variations in Lower Triassic fluvial conglomerates (Budleigh Salterton Pebble Beds), southwest England: some implications for palaeogeography and basin evolution. *Geological Journal*, Vol. 26, 65–83.

SPENCER, P S, and ISAAC, K P. 1983. Triassic vertebrates from the Otter Sandstone Formation of Devon, England. *Proceedings of the Geologists' Association*, Vol. 94, 267–269.

STEIGER, R H, and JÄGER, E. 1977. Subcommission on geochronology: convention on the use of decay constants in geo- and cosmochronology. *Earth and Planetary Science Letters*, Vol. 36, 359–362.

STRECKEISEN, A. 1976. To each plutonic rock its proper name. *Earth Science Reviews*, Vol. 12, 1–33.

STRONG, G E, and MILODOWSKI, A E. 1987. Aspects of the diagenesis of the Sherwood Sandstones of the Wessex Basin and their influence on reservoir characteristics. 325–37 *in* Diagenesis of sedimentary sequences. MARSHALL, J D (editor). *Special Publication of the Geological Society of London*, No. 36.

STRONG, G E, and SMITH, S A. 1989. Petrology of Budleigh Salterton Pebble Beds samples from the Exeter area, Devon. *British Geological Survey Technical Report*, WH/89/264R.

TANDY, B C. 1973. A radiometric and geochemical reconnaissance of the Permian outcrop and adjacent areas in south-west England. *Radioactive and Metalliferous Minerals Unit Report*, No. 315. Institute of Geological Sciences.

TANDY, B C. 1974. New radioactive nodule and reduction feature occurrences in the Littleham–Larkbeare area of Devon. *Radioactive and Metalliferous Minerals Unit Report*, No. 316. Institute of Geological Sciences.

TASLER, R, CADKOVA, Z, DVORAK, J, FEDIUK, F, CHALOUPSKY, J, JETEL, J, KAISEROVA-KALIBOVA, M, PROUZA, V, SCHOVANKOVA-HRDLOCKOVA, D, STREDA, J, STRIDA, M, and SETLIK, J. 1979. *Geologie ceske casti vnitrosudetske panve.* [Geology of the Bohemian part of the Intra-Sudetic Basin.] (Praha: Ustredni ustav geologicky.)

THOMAS, H H. 1902. The mineralogical constitution of the Bunter Pebble-Bed in the West of England. *Quarterly Journal of the Geological Society of London*, Vol. 58, 620–632.

THOMAS, H H. 1909. A contribution to the petrography of the New Red Sandstone in the West of England. *Quarterly Journal of the Geological Society of London*, Vol. 65, 230–245.

THOMAS, J M. 1963. Sedimentation in the Lower Culm Measures round Westleigh, northeast Devon. *Proceedings of the Ussher Society*, Vol. 1, 71–72.

THOMAS, J M. 1980. Sediments and sediment transport in the Exe Estuary. 73–87 *in* Essays on the Exe Estuary. BOALCH, G T (editor). *The Devonshire Association for the Advancement of Science, Literature and Art, Special Volume*, No. 2.

THOMSON, G F, CORNWELL, J D, and COLLINSON, D W. 1991. Magnetic characteristics of some pyrrhotite-bearing rocks in the United Kingdom. *Geoexploration*, Vol. 27, 23–41.

THORPE, R S. 1987. Permian K-rich volcanic rocks of Devon: petrogenesis, tectonic setting and geological significance. *Transactions of the Royal Society of Edinburgh: Earth Sciences*, Vol. 77, 361–366.

THORPE, R S, COSGROVE, M E, and VAN CALSTEREN, P W C. 1986. Rare earth element, Sr- and Nd-isotope evidence for petrogenesis of Permian basaltic and K-rich volcanic rocks from south-west England. *Mineralogical Magazine*, Vol. 50, 481–490.

TIDMARSH, W G. 1932. The Permian lavas of Devon. *Quarterly Journal of the Geological Society of London*, Vol. 86, 712–775.

TUCKER, M E. 1991. *Sedimentary petrology: an introduction to the origin of sedimentary rocks.* (Second edition). (Oxford: Blackwell Scientific Publications.)

TURNER, C. 1975. Der Einfluss grosser Mammalier auf die interglaziale Vegetation. *Quartar-paleontologie*, Vol. 1, 13–19.

USSHER, W A E. 1875. On the subdivisions of the Triassic rocks between the coast of west Somerset and the south coast of Devon. *Geological Magazine*, Vol. 2, 163–168.

USSHER, W A E. 1876. On the Triassic rocks of Somerset and Devon. *Quarterly Journal of the Geological Society of London*, Vol. 32, 367–394.

USSHER, W A E. 1877. A classification of the Triassic rocks of Devon and west Somerset. *Transactions of the Devonshire Association for the Advancement of Science*, Vol. 9, 392–399.

USSHER, W A E. 1892. The Permian in Devonshire. *Geological Magazine*, Vol. 9, 247–250.

USSHER, W A E 1902. The geology of the country around Exeter. *Memoir of the Geological Survey of Great Britain*, Sheet 325.

USSHER, W A E. 1913. The geology of the country around Newton Abbot. *Memoir of the Geological Survey of Great Britain*, Sheet 339.

VARNES, D J. 1978. Slope movement types and processes. 11–33 in Landslides — analysis and control. SCHUSTER, R L, and KRIZEK, R J (editors). *Transactions of the Research Board, National Academy of Science, Washington DC, Special Publication*, No. 176.

VICARY, W. 1865. On the feldspathic traps of Devonshire. *Transactions of the Devonshire Association*, Vol. 1, 43–49.

VICARY, W, and SALTER, J W. 1864. On the Pebble-bed of Budleigh Salterton with a note on the fossils. *Quarterly Journal of the Geological Society of London*, Vol. 20, 283–302.

VINCENT, A, and NICHOLAS, C. 1982. Industrial minerals. 291–305 in *The geology of Devon*. DURRANCE, E M, and LAMING, D J C (editors). (Exeter: University of Exeter.)

WALKER, A D. 1969. The reptile fauna of the 'Lower Keuper' Sandstone. *Geological Magazine*, Vol. 106, 470–476.

WARRINGTON, G, and SCRIVENER, R C. 1988. Late Permian fossils from Devon: regional geological implications. *Proceedings of the Ussher Society*, Vol. 7, 95–96.

WARRINGTON, G, and SCRIVENER, R C. 1990. The Permian of Devon, England. *Review of Palaeobotany and Palynology*, Vol. 66, 263–272.

WARRINGTON, G, AUDLEY-CHARLES, M G, ELLIOTT, R E, EVANS, W B, IVIMEY-COOK, H C, KENT, P E, ROBINSON, P L, SHOTTON, F W, and TAYLOR, F M. 1980. A correlation of Triassic rocks in the British Isles. *Special Report of the Geological Society of London*, No.13.

WENTWORTH, C K. 1922. A scale of grade and class terms for clastic sediments. *Journal of Geology*, Vol. 30, 377–392.

WHITAKER, W. 1869. On the succession of beds in the 'New Red' on the south coast of Devon, and on the locality of a new specimen of *Hyperodapedon*. *Quarterly Journal of the Geological Society of London*, Vol. 25, 152–158.

WILLIAMS, B J. 1983. *Geology of Sheet SX 98 NE (Topsham, Devon). Geological report for DOE: land use planning.* (Exeter: Institute of Geological Sciences.)

WRIGHT, V P, MARRIOTT, S B, and VANSTONE, S D. 1991. A 'reg' palaeosol from the Lower Triassic of south Devon: stratigraphic and palaeoclimatic implications. *Geological Magazine*, Vol. 128, 517–523.

APPENDIX 1

Major- and trace-element analyses of the Exeter Volcanic Rocks of the district

Locality Sample Grid reference Lithology	Dunch LC1979 SX 875 873 basalt	Dunch LC1081 SX 875 873 basalt	Know I LC1083 SX 874 895 basalt	Know I LC1084 SX 874 895 basalt	WT qy LC1092 SX 896 903 basalt	WT cut LC1091 SX 887 904 basalt	Pocm LC1086 SX 900 914 basalt	Killert LC1093 SS 975 005 minette
SiO_2	55.62	53.77	56.64	54.62	56.85	47.92	45.63	46.33
TiO_2	1.27	1.26	1.45	1.46	1.41	1.68	1.72	1.06
Al_2O_3	14.65	14.66	16.48	16.72	16.20	15.97	14.93	13.41
Fe_2O_3	9.70	9.88	8.78	9.44	8.94	11.25	10.52	6.17
MnO	0.10	0.09	0.07	0.14	0.09	0.15	0.17	0.16
MgO	4.38	6.31	3.77	1.61	2.11	3.12	4.55	4.52
CaO	6.35	5.86	5.14	3.91	5.44	4.85	10.16	9.00
Na_2O	3.06	2.79	3.13	2.57	2.97	2.16	2.74	1.52
K_2O	2.04	2.04	2.67	6.28	3.05	4.98	2.43	7.59
P_2O_5	0.38	0.30	0.22	0.22	0.22	0.34	0.33	0.89
Loss on ignition	3.03	3.60	2.22	3.62	3.13	8.02	7.23	8.11
Ba	303	273	324	386	341	369	426	3529
Rb	83	69	89	119	103	48	65	83
Sr	250	248	261	156	233	90	360	2466
Th	5	6	8	7	7	4	4	16
U	2	nd	nd	nd	1	nd	1	nd
Zr	115	114	135	132	128	133	145	427
Nb	11	11	13	13	13	16	19	11
Y	15	17	19	19	18	17	17	20
V	105	108	118	124	113	133	135	86
Cr	333	252	129	106	118	170	179	131
Co	41	37	27	18	36	28	55	24
Ni	123	128	35	22	38	73	174	105
Cu	39	33	15	25	24	20	12	328
Zn	116	175	286	70	204	75	323	296
La	20	20	23	24	23	22	24	140
Ce	41	41	49	50	48	50	51	290
Pr	4.88	4.89	5.89	5.94	5.74	6.19	6.19	34.79
Nd	19.60	19.80	23.30	23.40	23.00	24.00	25.10	130.60
Sm	4.16	4.21	4.93	4.94	4.81	4.88	5.27	19.46
Eu	1.21	1.21	1.40	1.36	1.39	1.58	1.65	4.69
Gd	4.20	4.34	4.98	4.87	4.78	5.03	5.30	10.87
Dy	4.07	4.15	4.70	4.59	4.51	4.40	4.79	5.77
Ho	0.69	0.70	0.80	0.77	0.76	0.76	0.81	0.95
Er	2.01	2.07	2.37	2.27	2.20	1.84	2.33	2.81
Yb	1.70	1.70	1.97	1.91	1.81	1.70	1.84	1.72
Lu	0.25	0.26	0.30	0.28	0.27	0.27	0.28	0.27

Key to abbreviations for localities (grid references are given in the table).

Dunch:	Dunchideock	Radd:	Raddon Quarry
Know I:	Knowle Quarry, Ide	Uton:	Uton Quarry
WT qy:	West Town Quarry, Ide	Posb:	Posbury Clump Quarry
WT cut:	cutting West Town, Ide	Spenc:	Spencecombe Quarry
Pocm:	Pocombe Quarry, Exeter	Mead:	Meadowend Quarry
Killert:	Killerton Park Quarry	Know C:	Knowle Hill Quarry
Columb:	Columbjohn Wood quarries	nd:	not detected

Killert LC1094 SS 975 005 minette	Columb LC1096 SS 962 002 minette	Columb LC1097 SS 964 003 basalt	Radd LC1101 SS 908 010 dolerite	Uton LC1114 SX 815 985 basalt	Posb LC1106 SX 815 978 basalt	Spenc LC1109 SS 794 017 basalt	Mead LC1108 SS 797 024 lamprophyre	Know C LC1110 SS 789 023 lamprophyre
46.51	55.54	46.50	41.15	49.03	54.11	44.05	52.56	54.98
1.18	1.40	1.84	1.57	1.38	1.63	1.51	1.46	1.67
14.36	16.44	17.51	14.59	15.49	14.81	16.35	15.21	17.49
6.14	8.08	12.62	10.99	10.84	9.10	11.82	8.35	9.14
0.15	0.20	0.31	0.47	0.15	0.12	0.18	0.07	0.19
2.16	1.51	4.34	4.40	3.20	2.34	1.41	5.29	0.97
9.66	2.20	2.47	9.27	8.01	4.52	10.00	5.25	2.52
1.46	1.99	1.70	1.93	2.67	4.66	4.09	2.59	4.87
8.84	9.57	6.79	3.59	1.81	2.70	0.45	5.29	4.81
1.00	1.11	0.41	0.27	0.23	0.28	0.24	1.05	1.03
7.13	0.95	6.50	12.58	7.67	6.17	10.30	2.35	2.19
3736	3960	515	289	296	159	212	2825	1804
92	69	56	66	41	24	9	152	128
2343	3004	160	112	228	85	96	1131	621
15	16	5	4	2	5	4	17	17
2	nd	2	nd	1	nd	nd	nd	1
447	511	162	117	127	132	127	447	496
12	14	16	13	12	18	10	23	27
21	26	18	15	17	15	17	23	24
115	76	117	118	149	87	77	111	142
140	155	206	215	167	188	191	233	268
14	16	43	32	37	27	35	30	34
118	76	135	89	117	72	106	106	110
77	49	19	17	63	28	17	11	28
94	70	154	148	106	43	74	126	68
151	153	26	19	19	22	17	70	71
303	333	57	39	41	44	38	144	146
36.81	38.96	6.75	4.65	4.75	5.36	4.63	17.07	17.30
138.40	143.50	26.90	20.00	20.20	21.90	19.90	64.80	65.10
20.45	21.30	5.35	4.32	4.45	4.79	4.52	11.78	11.49
5.00	5.27	1.58	1.36	1.41	1.54	1.37	3.00	2.95
11.58	12.53	5.44	4.56	4.77	4.72	4.83	8.99	8.90
6.33	7.24	5.06	4.39	4.60	4.23	4.70	6.43	6.70
1.05	1.23	0.86	0.73	0.79	0.71	0.79	1.05	1.12
3.06	3.65	2.48	2.11	2.31	2.05	2.30	3.04	3.25
1.99	2.43	1.93	1.73	1.93	1.70	1.85	2.27	2.49
0.30	0.38	0.29	0.25	0.29	0.25	0.27	0.35	0.37

Appendix 1 *continued*

APPENDIX 2

Geological Survey photographs

Copies of the photographs are deposited for reference in the British Geological Survey library at Keyworth, Nottingham NG12 5GG and at the British Geological Survey, St Just, 30 Pennsylvania Road, Exeter, EX4 6BX. National Grid references are given in square brackets; those of general views are of the viewpoints. Dates of photographs are also given. The photographs were taken by J Rhodes (1934, 1945), J M Pulsford (1963), C J Jeffery (1970), H J Evans (1984), and T P Cullen (1991).

A 6385   Pocombe Quarry [SX 899 914]. 1934.

A 6386   Pocombe Quarry [SX 899 914]. 1934.

A 8161   Bridford Barytes Mine Quarry [SX 829 866]. 1945.

A 8162   Bridford Barytes Mine. View in stope above the 380 ft level on No. 4 Vein [SX 830 865]. 1945.

A 8163   Bridford Barytes Mine [SX 830 865]. The 380 ft level on No. 3 Vein. 1945.

A 8164   Bridford Barytes Mine [SX 830 865]. The 280 ft level. 1945.

A 8165   Bridford Barytes Mine [SX 830 865]. Wide stope above the 380 ft level on No. 3 Vein. 1945.

A 8166   Bridford Barytes Mine [SX 830 865]. Another view of stope in A 8165. 1945.

A 8167   Bridford Barytes Mine [SX 830 865]. New Shaft. 1945.

A 8168   Bridford Barytes Mine [SX 830 865]. View of conveyor belt from crusher near shaft, to mill. 1945.

A 8169–71   Bridford Barytes Mine Mill [SX 830 865]. Views of mill. 1945.

A 8172   Bridford Barytes Mine [SX 830 865]. View of treatment tables in mill. 1945.

A 8173–4   Bridford Barytes Mine [SX 830 865]. Views of mill. 1945.

A 10090   Bude Formation in Lower Linscombe Quarry [SS 789 046]. 1963.

A 10109   Knowle village — Spencecombe Lava (olivine basalt) [SS 796 015]. 1963.

A 11575   Barytes veins in Bridford Barytes Mine Quarry [SX 8295 8656]. 1970.

A 15264   Combe Shale in Stone Copse Quarry [SX 8237 8603]. 1991.

A 15265   Teign Chert in Bridford Barytes Mine Quarry [SX 8296 8656]. 1991.   (Plate 14e).

A 15266   Teign Chert and spilitised basalt in Stone Copse Quarry [SX 8243 8600]. 1991.

A 15267   Teign Chert, Mistleigh Copse, Doddiscombsleigh [SX 8543 8599]. 1991.   (Plate 1a).

A 15268   Crackington Formation, Springdale, Whitestone [SX 8654 9208]. 1991.

A 15269   Fold in Crackington Formation, Eastern Cotley Wood, Perridge, Longdown [SX 8639 9070]. 1991.

A 15270   Folds and cleavage in Crackington Formation, Springdale, Whitestone [SX 8654 9208]. 1991.

A 15271   Crackington Formation, Bonhay Road, Exeter SX 9143 9269]. 1991.

A 15272–3   Sole markings on Crackington Formation sandstone, Pinhoe Brickpit, Exeter [SX 9550 9458]. 1991.   (Plate 1c).

A 15274   Crackington Formation, Pinhoe Brickpit, Exeter [SX 9551 9460]. 1991.   (Plate 1b).

A 15275   Bude Formation, East Henstill Quarry, West Sandford [SS 8158 0413]. 1991.   (Plate 1d).

A 15276   Bude Formation, Dowrich Quarry, Windmill Hill, Sandford [SS 8224 0508]. 1991.

A 15278–80   Heltor Rock, Dartmoor [SX 7997 8703]. 1991. (Plate 11a).

A 15281   Megacrystic layering in Dartmoor Granite, Burnicombe Down [SX 8025 8700]. 1991. (Plate 11b).

A 15282   Otter Sandstone at Beggars Roost Quarry, Rockbeare, West Hill [SY 0614 9419]. 1991. (Plate 8c).

A 15283   Budleigh Salterton Pebble Beds and Otter Sandstone, Beggars Roost Quarry, Rockbeare, West Hill [SY 061 941]. 1991.   (Plate 7c).

A 15284–5   Ventifact bed at top of Budleigh Salterton Pebble Beds, Beggars Roost Quarry, Rockbeare, West Hill [SY 0606 9414]. 1991.   (Plate 7b).

A 15286   Otter Sandstone, Stoneyford Farm, Hawkerland [SY 0625 8865]. 1991.   (Plate 8b).

A 15287   Budleigh Salterton Pebble Beds, Beggars Roost Quarry, Rockbeare, West Hill [SY 062 941]. 1991. (Plate 7a).

A 15288   Dawlish Sandstone, Lake Bridge, Brampford Speke [SX 9275 9780]. 1991.   (Plate 5b).

A 15289   Aeolian cross-bedding in Dawlish Sandstone, Clyst St Mary [SX 9706 9101]. 1991.   (Plate 5c).

A 15290   Cross-bedding in Dawlish Sandstone, Exminster [SX 9423 8811]. 1991.

A 15291   Cross-bedding in Dawlish Sandstone, Broadclyst [SX 9855 9746]. 1991.

A 15292–3   Heavitree Breccia, Shillingford Wood Quarry, Shillingford St George [SX 9065 8746]. 1991.

A 15294–5   Heavitree Breccia, Heavitree Quarry, Exeter [SX 9498 9214]. 1991.   (Plate 4d).

A 15296   Heavitree Breccia, A30 Link Road cuttings near Exminster [SX 9292 8797]. 1991.   (Plate 4c).

A 15297   Newton St Cyres Breccia, Cromwell's Cutting, Pitt Hill, Crediton [SS 8165 0020]. 1991.

A 15298   Newton St Cyres Breccia, Newton St Cyres [SX 8795 9804]. 1991.

A 15299   Heavitree Breccia, M5 Motorway cuttings near Exminster [SX 9306 8792]. 1991.   (Plate 4b).

A 15300   Crediton Breccia, Forches Cross, Crediton [SS 8318 0099]. 1991.

A 15301   Alphington Breccia, Mincinglake Brook, Exeter [SX 9422 9326]. 1991.

A 15302   Crediton Breccia, Forches Cross, Crediton [SS 8318 0099]. 1991.   (Plate 6b).

A 15303–4   Alphington Breccia, The Quay, Exeter [SX 9214 9203]. 1991.   (Plate 5a).

A 15305   Knowle Sandstone, Folly Farm, Uton [SX 8188 9846]. 1991.

A 15306   Bow Breccia, Green Lane, West Sandford [SS 8108 0282]. 1991.

A 15307   Cadbury Breccia, Rhode Farm, Bradninch [SS 9853 0440]. 1991.

A 15308–9   Cadbury Breccia, Raddon Hill Farm, Thorverton [SS 9041 0309]. 1984.   (Plate 6a).

A 15310   Olivine-basalt, School Wood Quarry, Dunchideock [SX 8755 8726]. 1991.   (Plate 10a).

A 15311   Base of lava, School Wood Quarry, Dunchideock [SX 8750 8708]. 1991.   (Plate 10c).

A 15312   Basalt and overlying brecciated facies, Pocombe Quarry, Exeter [SX 8991 9140]. 1991.

A 15313–4   Basalt of the Exeter Volcanic Rocks, Pocombe Quarry [SX 8985 9149]. 1991.   (Plate 10b).

A 15315   Olivine-basalt, Posbury Quarry [SX 8151 9783]. 1991.

A 15316   Periglacial disturbances in the Budleigh Salterton Pebble Beds, Venn Ottery Quarry [SY 0655 9124]. 1991.

A 15317   Blanket head and regolith on Creedy Park Sandstone, Aller Barton, Sandford [SS 8095 0212]. 1991.

A 15318   Landslipping on Ashton Shale (Crackington Formation), Osbornes Farm, Old Wheatley [SX 8860 9239].

A 15319   Blanket head and regolith on Crackington Formation, Springdale, Whitestone [SX 8659 9213]. 1991.   (Plate 12).

A 15320   Barytes lode overlying metalliferous Teign Chert, Bridford Barytes Mine Quarry [SX 8296 8656]. 1991.

A 15321   Building stones in Exe Vale Hospital lodge buildings, Exminster [SX 9420 8809]. 1991.

A 15322   Building stones of Broadclyst Church [SX 9818 9726]. 1991.   (Plate 14c).

A 15323   Building stones of Woodbury Church [SY 0094 8720. 1991.

A 15324   Pebble wall in Woodbury village [SY 0115 8735]. 1991.   (Plate 14d).

A 15325   Building stones in wall, School Lane, Thorverton [SS 9254 0208]. 1991.

A 15326   Workings at Pinhoe Brickpit, Pinhoe, Exeter [SX 955 946]. 1991.

A 15327   Alphington Breccia scenery, Shillingford St George [SX 9060 8733]. 1991.

A 15328   Heavitree Breccia escarpment, Shillingford Abbot [SX 9125 8914]. 1991.   (Plate 4a).

A 15329   Scenery south-east of Killerton Park, Broadclyst [SX 9749 0048]. 1991.

A 15330   Scenery north-east of Killerton Park, Broadclyst [SX 9749 0048]. 1991.

A 15331   The Raddon Hills from viewpoint near Exeter Hill Cross, Shobrooke [SS 8780 0195]. 1991.

A 15332   Newton St Cyres Breccia escarpment, from viewpoint at Camp Bridge, Shobrooke [SX 8670 9968]. 1991.

A 15333   Alphington Breccia and Exeter Volcanic Rocks scenery, Dunchideock [SX 8784 8893]. 1991.

A 15334   View of Exeter from near Knowle Quarry [SX 8722 8959]. 1991.

A 15335   Heavitree Breccia escarpment from viewpoint at The Danny, Topsham Road, Exeter [SX 9393 9073]. 1991.

A 15336   Raddon Hills from viewpoint near Yellowford, Thorverton [SS 925 005]. 1991.

A 15337   Sandstone topography (Exmouth Mudstone and Sandstone), Windmill Hill, near Woodbury Salterton [SY 016 901]. 1991.

A 15338   Scenery around Killerton Park, from viewpoint near Yard Downs [SS 9768 0380]. 1991.

A 15339   Scenery to the south-west of Crediton [SX 8196 9922]. 1991.

A 15340   Scenery to the west of Crediton [SX 8163 9984]. 1991.

A 15341   Crediton Trough scenery from Upton Pyne Church [SX 9102 9771]. 1991.

A 15342   Feature of Budleigh Salterton Pebble Beds on Littleham Mudstone [SY 038 859]. 1991. (Plate 8a).

A 15343   Scenery between Woodbury Castle and the Exe Estuary [SY 030 872]. 1991.

FOSSIL INDEX

Latinised names only are listed.

No distinction is made here between a positively determined fossil genus or species and examples doubtfully referred to them i.e. with the qualifications ?, cf. or aff.

Amaliae   17
Aviculopecten losseni (van Koenen, 1879)   10

Bilinguites bilinguis (Salter, 1864)   21
Bilinguites gracilis (Bisat, 1924)   18, 19
Bollandoceras micronotum (Phillips, 1836)   8, 9

Calamites   13, 19
Calamospora   16, 22
Caneyella   13, 16, 18
Caneyella minor (Brown, 1841)   18
Caneyella semisulcata (Hind, 1897)   18
Cenellipsis   8, 10
Cheilichnus bucklandi (Jardine) McKeever and Haubold, 1996   43
Chonetes   10
Chonetes (*Semenewia*)   10
Convolutispora   31
Crassispora kosankei (Potonié and Kremp) Bharadwaj, 1957   16, 31

Densosporites   22
Densosporites anulatus (Loose) Smith and Butterworth, 1967   16
Dunbarella   13, 18, 19
Dunbarella rhythmica (Jackson, 1927)   18

Entogonites grimmeri (Kittl, 1904)   8

Florinites pumicosus (Ibrahim) Schopf, Wilson and Bentall, 1944   31

Glyphioceras   10
Glyphioceras reticulatum Phillips, 1836   10
Glyphioceras spirale Phillips, 1836   9
Goniatites inconstans Phillips, 1836   13
Goniatites sphaericostriatus Bisat, 1924   8, 19
Granulatisporites   22

Hodsonites magistrorum (Hodson, 1957)   18
Homoceras beyrichianum (de Koninck, 1843)   13, 14
Homoceras smithi (Brown, 1841)   18
Homoceras striolatum (Phillips, 1836)   13
Homoceras subglobosum Bisat, 1924   11
Homoceras undulatum (Brown, 1841)   13, 14
Homoceratoides   18

Isohomoceras subglobosum (Bisat, 1924)   18

Klausipollenites schaubergeri (Potonié and Klaus) Jansonius, 1962   43

Lueckisporites virkkiae Potonié and Klaus emend. Clarke, 1965   31, 43
Lusitanites   8, 10, 18
Lycospora   31
Lycospora pusilla (Ibrahim) Schopf, Wilson and Bentall, 1944   16, 22

Neoglyphioceras   8, 10, 18
Neuropteris   13
Neuropteris schlehani Stur, 1877   13

Perisaccus granulosus (Leschik) Clarke, 1965   43
Phillipsia cliffordi Woodward, 1884   9
Posidonia   13, 18, 21
Posidonia becheri (Bronn, 1828)   8, 9, 10
Posidonia membranacea M'Coy, 1851   8, 10, 18

Posidonia minor (Brown, 1841)   18
Posidonia obliquata Brown, 1841   18
Posidonomya becheri Bronn, 1828   10
Potonieisporites novicus Bharadwaj, 1954   31
Protohaploxypinus microcorpus (Schaarschmidt) Clarke, 1965   43

Raistrickia saetosa (Loose) Schopf, Wilson and Bentall, 1944   16
Reticuloceras   18, 21
Reticuloceras bilingue (Salter, 1864)   14
Reticuloceras coreticulatum Bisat and Hudson, 1943   16, 18, 21
Reticuloceras gracile Bisat, 1924   13
Reticuloceras moorei Bisat and Hudson, 1943   13
Reticuloceras nodosum Bisat and Hudson, 1943   13, 16, 18, 19
Reticuloceras pulchellum (Foord, 1903)   16, 18
Reticuloceras regularum Bisat and Hudson, 1943   13
Reticuloceras reticulatum (Phillips, 1836)   16, 18
Reticuloceras stubblefieldi Bisat and Hudson, 1943   18

Sanguinolites ellipticus (Hind, 1897)   7
Seminula ambigua J de C Sowerby, 1824   10
Straparollus   18
Stroboceras sulcatum (Fleming, 1828)   10
Sudeticeras   8, 10, 18

Taeniaesporites ortisei Klaus, 1963   43

Urnatopteris tenella (Brongniart) Kidston, 1884   13

Vallites henkei (H Schmidt, 1925)   18
Vallites striolatus (Phillips, 1836)   12, 14, 16, 18, 19

GENERAL INDEX

^{40}Ar/^{39}Ar   63, 66, 102
A30 Link Road   49, 171
A30 road   11
A38 road cuttings   50
acid igneous rocks   62, 68
acid lava   47, 70, 72
acid tuff   35, 62, 63, 68, 69, 70, 73, 111
actinolite   9, 117, 118
adinole   7, 8, 26, 150
adit   10, 147, 149
adularia   104
aeolian   34, 35, 44, 51, 76, 79, 92, 94
aeolian cross-bedding   52, 171
aeolian dune   3, 34, 53, 54, 56, 79, 104
aeromagnetic anomaly   101
aeromagnetic anomaly map   98, 102, 121
aeromagnetic data   120
agate   110
agglomerate   50, 96, 111
aggregate   143, 157
Aggregate Abrasion Values (AAV)   143
albite dolerite   7, 8, 24, 25, 26, 27, 28, 117, 118, 149
albite hornfels   26
algodonite   149
allanite   91, 117
Aller Barton   67, 133, 172
Allercombe   87
alluvial fan   3, 34, 35, 50, 71, 75, 76, 138
alluvial gravel   153, 157
alluvium   3, 74, 77, 84, 87, 130, 131, 137, 138, 142, 153, 154, 157, 158
Alphin Brook   12, 22
Alphington   45, 96, 100, 101
Alphington Breccia   30, 31, 32, 33, 35, 38, 40, 44, 45, 46, 47, 48, 51, 76, 111, 119, 129, 133, 134, 135, 139, 141, 146, 153, 156, 171, 172
Alpine Orogeny   129
Alportian   13, 14, 15, 17
  H$_{2a}$ Zone   17, 19
  H$_{2b}$ Zone   13, 14, 17, 19
  H$_{2b1}$ Zone   15
amesite   117
ammonoid   5, 6, 8, 11, 13, 14, 17, 18, 19, 21, 33
ammonoid zones   17
amphibians   32, 34, 92
amphibole   25, 26, 27, 28, 118
amygdales   25, 27, 40, 107, 108, 109, 110
analcime   102, 106, 107
anatase   91
andalusite   26, 91
andraditic garnet   91
Anglian Stage   130
angular unconformity   67
Anisian   32, 33, 92
ankerite   12, 96

annelid   56
anthracoceratid   18
anticline   7, 11, 12, 19, 21, 67, 125
  Ashton   124
  Birch Down   7, 124
  Bridford   7, 24, 26, 124
  Bridford Sill   24
  Doddiscombsleigh   7, 9, 26, 27, 124
  Drewsteignton Anticline   120
  Mistleigh Copse   124
  Yendacott   70, 129
apatite   25, 91, 102, 104, 114, 117, 118
aphotic conditions   5
aplite   34, 116
aplogranite   111, 114, 116
aquatic crustaceans   34
aquiclude   151, 156, 159
aquifer   87, 157, 159, 160
Ar/Ar data   113
Areas of Outstanding Natural Beauty   142
Arenig   87
arkoses   77, 89
Armorica   5
Arnsbergian   14, 17
  E$_2$ Zone   14, 19
  E$_{2a}$ Zone   17
  E$_{2b}$ Zone   14, 17
  E$_{2c}$ Zone   14, 17
arsenic   149
arthropods   32, 34, 92
Artinskian   33
Asbian   9
  B$_1$ Zone   8
  B$_2$ Zone   8, 9
Ash Farm   23
Ashclyst Farm   85
Ashclyst Forest   10, 12, 19, 86, 101, 120, 128, 134, 142
Ashlake Cross   110
Ashton   149
Ashton Shale   3, 10, 11, 12, 14, 15, 17, 18, 117, 125, 126, 146, 151, 153, 172
Ashton Shale Member   5, 10, 11, 139, 151
augite   25, 96, 102, 104, 105, 106, 118
augite-biotite   104
augite-quartz   105
aureole   7, 8, 11, 117, 118, 151
  Dartmoor Granite   116
autoradiograph   149
axinite   28, 116, 117
Aylesbeare   1, 77, 79, 83, 84, 85, 86, 87, 133
Aylesbeare Common   87, 139, 141, 160
Aylesbeare Group   30, 77, 85
Aylesbeare Mudstone   3, 32, 34, 52, 56, 77–86, 87, 91, 128, 133, 134, 142, 146, 149, 150, 153, 154, 158
Aylesbeare Mudstone Group   29, 30, 31, 32, 33, 35, 77–86, 87, 128, 139, 141, 159, 160

Babylon Farm   66, 108
back-arc basins   5, 7
Bagshot Beds   119

Bailey   133
Baker's Brake   137
Bampfylde House   109
barium   146, 147
Barley Lane   40, 109
Barnfield Cottages   56
Barnstaple   130
baryte   9, 145, 146, 147, 148, 171
barytes lode   172
basalt   24, 35, 38, 40, 42, 44, 45, 46, 50, 62, 63, 65, 66, 67, 74, 96, 98, 102, 103, 104, 106, 107, 108, 109, 110, 111, 143, 144, 145, 169, 170, 172
basaltic lava   40, 65, 155
basic lava   49
basinal brines   146, 148
Bawdenhayes   22, 139, 143
Beare   59, 60, 61, 107, 109
Beare Farm   109
Beedles Terrace   44
beetle fauna   137
Beggars Roost   89, 142
Belfield House   56
Berrysbridge   66
Bicton   31, 95, 137
Bicton College   92, 95
Bicton Common   87, 91, 92, 93
Bicton Gardens   92
Biddypark Lane   46, 110
Bideford Bay   120
Bideford Formation   17
Bidwell   61, 138
Bidwell Cross   61
biostratigraphy   13, 31
biotite   25, 27, 28, 50, 102, 103, 104, 107, 111, 114, 117, 118
biotite-cordierite hornfels   117
biotite-granite   72, 98, 112, 113, 114, 116
biotite-lamprophyre   65, 102, 144
biotitic-microsyenite   98
bioturbation   32, 33
birch   9, 117, 126
Birch Ellers   147
Birks Halls, University of Exeter   19
bisaccate pollen   31, 44, 67, 68, 72
bivalve   5, 7, 13, 19, 33, 87, 119
Black Pit   148, 149
Black Pit Field   148
Blackall Road   109
Blackhill   89, 134
Blackhorse   58, 136
Blackingstone Rock   114
blanket head   10, 61, 63, 132, 172
blende   147
blue granite   113
Blue Hayes   136
Bob's Close Copse   86
Bodmin   33
Bodmin Granite   34, 112, 113
Bolsovian   22, 32, 33
Bondhouse   Copse 21
Bonhay Road   8, 13, 14, 19, 126, 171
boreholes   41, 158, 159
  Blackhill   83, 84
  Burrow Farm   54, 136, 159
  Bussell's Farm Exploration   52, 56

Bussell's Farm Production   56
Dawlish Sandstone   56
Fortescue Farm   54, 135, 159
Huxham   56
Professor A Stuart   56
Rockbeare   133
Sandy Lane   54
Starved Oak Cross No. 1   54
Starved Oak Cross No. 2   54, 74, 159
Upton Pyne   148
Venn Ottery   79, 80, 83, 84, 85
Withycombe Raleigh Common   83, 84
Yellowford   66, 70, 71
bornite   149
boron-rich fluids   34
Bouguer gravity anomalies   120, 122, 123
Bow   63
Bow Beds   29, 68
Bow Breccia   29, 31, 32, 33, 34, 35, 38, 59, 60, 62, 63, 64, 67, 68, 75, 102, 111, 112, 113, 153, 156, 159, 171
   sedimentary features   63
Bow Conglomerate   29, 62
Bowley   23
Bowley Wood   23
brachiopod   7, 9, 10, 87
Bradley Farm   63
Bradninch   59, 60, 61, 62, 128, 136, 138, 171
braided rivers   3, 91, 93
braidplain   94
Brampford Speke   51, 54, 55, 56, 58, 74, 135, 138, 142, 159, 171
branchiopod crustaceans   32, 92
Brandirons Corner   64, 65, 102
breccia   9, 30, 34
Bremridge Farm   22
brick clay   146
Brickyard Copse   86
Brickyard Road   146
Bridford   20, 24, 26, 27, 28, 96, 98, 113, 114, 117, 118, 124, 125, 126
Bridford Lamprophyre   24, 97, 118
Bridford Sill   7, 9, 24, 25, 26
Bridford Thrust   126
Bridford Wood   8, 11, 12, 117
Bridfordmills   13
Brigantian   10, 17
   P_{1c} Zone   8, 10, 19
   P_{1d} Zone   8, 10, 18
   P_{1d}–P2b Zone   18
   P_{2a} Zone   8
   P_{2b} Zone   10
   P_{2c} Zone   8
Broad Oak   92
Broadclyst   1, 32, 51, 52, 53, 56, 77, 86, 98, 102, 104, 109, 128, 136, 138, 142, 144, 145
Broadclyst Church   144, 145, 172
Broadclyst Moor   138
Brook Hill   62
brookite   91
brown earth   142
bryozoans   7
Bude Formation   3, 5, 7, 10, 15, 17, 22, 33, 35, 59, 63, 75, 85, 120, 122, 124, 125, 126, 132, 133, 134, 139, 141, 143, 151, 153, 156, 171
Budlake   65, 107, 109
Budlake-Columbjohn   96
Budleigh Buns   146
Budleigh Salterton   77, 87, 88, 92
Budleigh Salterton Pebble Beds   1, 3, 30, 31, 32, 33, 34, 77, 87, 88, 89, 90, 91, 92, 93, 94, 95, 128, 129, 130, 133, 134, 137, 139, 141, 142, 145, 153, 154, 157, 158, 159, 160, 171, 172
   sedimentary features of the   91
building stone   143, 144, 156, 172
Buller's Hill Gravel   119
bullion   17
buried channel deposits   130, 137
buried channels   3, 4, 130
Burn River   23
Burnicombe   114
Burnicombe Down   113, 114, 115, 116, 171
Burnwell   20
Burrow   33, 77, 86
Burrow Farm   53, 98, 100, 102
burrow structures   32
Bussell's Farm   52, 56, 57, 136
Bussell's Member   30, 31, 51, 52, 54, 56, 153, 154, 155
Bussell's Mudstone   136

Cadbury   23, 59, 61, 133, 134
Cadbury Beds   29
Cadbury Breccia   29, 31, 32, 33, 35, 38, 59, 60, 61, 62, 63, 65, 68, 70, 75, 85, 107, 111, 128, 136, 141, 142, 146, 153, 156, 171, 172
calc-flintas   117
calcic amphibole   91
calcite   27, 92, 96, 102, 104, 105, 106, 107, 109, 110, 117
calcite veins   109
calcium carbonate   47, 50, 146
calcrete   92, 94, 156
Calthorpe Road   19
Camp Bridge   172
Canns Farm   108
Canterbury House Farm   139
capitosaurids   32, 92
Carboniferous   1, 3, 5, 6, 7, 10–22, 29, 31, 32, 33, 34, 35, 40, 52, 61, 63, 72, 86, 91, 101, 116, 120, 121, 123, 125, 132, 133, 147, 148, 160
   boundary, Permian   32, 96
   hydrogeology   158
   Lower   5, 7, 24, 59, 120, 126, 141, 143, 146, 147
   Upper   5, 17, 128
Carwithen Copse   70
cassiterite   91
Cat Copse   146
Cawsand   34, 113
Cawsand Bay   90
celadonite   109
Cenomanian   119
central area of Exeter   41
chalcedony   10
chalcocite   149
chalcopyrite   20, 147, 149
Chalk   3, 119, 146
channel fills   68, 69
channel scours   91
channels   66, 72
Channons   108
Chapel Downs   68, 69
Chapel Farm   108
Chapel Farm, Raddon   66
chemical analyses   35
Cherry Tree Cottage   10
chert   7, 8, 9, 10, 14, 19, 28, 35, 40, 45, 46, 49, 50, 59, 62, 69, 72, 74, 75, 91, 117, 133, 143, 149
   Greensand chert   133, 134, 135, 136, 137, 146
   jasper   10
   mineralogical compositions   9
   radiolarian   3, 10
   Teign   3, 24, 26, 27, 28
chiastolite hornfelses   117
Chilton   22
Chitterley   23, 136, 138
chlorite   9, 11, 12, 25, 26, 27, 28, 77, 104, 105, 107, 117, 118, 141, 151
Chokierian   11, 12, 14, 15, 17
   H_1 Zone   18
   H_{1a} Zone   11, 12, 14, 17, 18
   H_{1a1-3} Zone   15
   H_{1b} Zone   13, 14, 15, 17, 19
   H_{1b1} Zone   15
   H_{2a1} Zone   18
chonetoid debris   61
Christendown Clump   9, 24, 25, 27, 28, 143
Christendown Clump Tuff   25
Christow   147, 149
Christow Common   117
Christow Lamprophyre   24
Church Bridge   17
Church Farm, Cadbury   61
Church Lane, Shobrooke   70
Church of the Holy Cross at Crediton   146
Clapham   46
clast types   34, 35
clausthalite   149
clay mineralogy   5, 11, 12, 19, 96, 126
cleavage   7, 8, 9, 12, 20, 21, 22, 125, 171
Clifton Hill brickpit   46
climbing ripples   79
clinochlore   27
clinopyroxene   102
clinozoisite   25, 28, 118
Clyst   134, 135, 142
Clyst Honiton   56, 86, 135, 136, 138, 142
Clyst Hydon   19, 31, 59, 60, 85, 142, 146
Clyst Sands   30, 51
Clyst St George   135
Clyst St Lawrence   19, 59, 62, 85
Clyst St Lawrence Formation   30, 31, 32, 77, 85, 86, 138, 141, 153, 155
Clyst St Mary   52, 53, 58, 136, 138, 146, 171
Clyst Valley   135, 136, 138, 142, 159
Clyst William   86, 87, 90, 91

Clystlands   136
cob   143, 144, 146
cobalt   149
Cod Wood   117
coffinite   149
Colaton Raleigh   94
Colaton Raleigh Common   91, 134, 141
College Farm   95
colloidal silica   8, 157
Columbjohn   52, 56, 57, 96, 102, 135
Columbjohn Borehole   52
Columbjohn Wood   98, 107, 109
columnar jointing   27
Combe Shale   1, 3, 5, 7, 24, 26, 27, 117,
    128, 141, 150, 153, 171
Combesatchfield   62, 66, 109
Common Down Plantation   147
conglomerate   89
conifers   34
conodont   5, 18
construction materials   142
Coombe Barton   61, 70
Coombe House   68
Coombland Wood   11
copper   79, 147, 149
Copperwalls Farm   17
Copplestone Down   9, 28
corals   119
cordierite   117
Cornubian   91
    batholith   111
    massif   91
Coryton breccias   30
Cotley Wood   21
Countess Wear   47, 49, 135, 136
Countess Wear Bridge   137
Country House Inn   49
covellite   149
Cowley   11, 12, 14, 17, 135, 138
Cowley Bridge   138
Cowley Bridge Road   135
Cowley Hill   17
Crablake Farm   50, 59
Crackington Formation   3, 5, 8, 10–22,
    33, 40, 44, 52, 59, 63, 76, 77, 85, 86,
    96, 101, 104, 107, 109, 117, 120, 122,
    123, 125, 126, 128, 132, 133, 134,
    138, 139, 140, 141, 143, 146, 147,
    148, 151, 156, 158, 160, 171, 172
    engineering   151–153
Crediton   1, 29, 31, 34, 60, 67, 68, 69,
    72, 73, 102, 103, 107, 112, 124, 129,
    145, 159, 160, 171, 172
Crediton Beds   29, 68, 72
Crediton Breccia   29, 31, 32, 33, 34, 35,
    38, 45, 60, 67, 68, 69, 70, 71, 72, 74,
    75, 76, 107, 112, 113, 133, 134, 139,
    153, 156, 171
Crediton Conglomerates   29, 68, 72
Crediton Trough   1, 3, 4, 10, 22, 29, 30,
    31, 32, 33, 34, 35, 38, 39, 40, 51, 52,
    59, 60, 62, 63, 74, 75, 76, 77, 86, 96,
    98, 99, 100, 107, 111, 113, 120, 122,
    123, 126, 127, 128, 129, 147, 148,
    158, 159, 172
    aeromagnetic map   123
Creedy   134, 135, 136

Creedy Barton   72, 74
Creedy Lakes   67
Creedy Park   67
Creedy Park Sandstone   29, 31, 32, 33,
    34, 35, 38, 63, 67, 68, 75, 133, 153,
    155, 159, 172
Creedy Valley   135
creep   151
Cretaceous   1, 3, 119, 124
crevasse splays   79
crinoid   10, 61
Crockham sill   26
crocodile   93
Cromerian   130
Cromwell's Cutting   72, 73, 112, 156, 171
Crook Plantation   137
Crooklake Brook   11
cross-bedding   42, 53, 54, 56, 58, 63, 66,
    68, 70, 72, 73, 74, 79, 84, 85, 91, 92
    trough   48
cross-bedding azimuths   53
cross-lamination   12, 13, 20, 63, 65, 70,
    92, 117
crosscourse mineralisation   146, 147
Crossmead   109, 111
Crossmead Hall   109
Crownhill Lane   66
crustaceans   33, 53, 56
cryptomelane   148
ctenosauriscid   32, 92
Cullompton Hill   62
Culm   35, 40, 42, 45, 46, 51, 54, 59, 60,
    61, 62, 63, 66, 68, 69, 70, 74, 75, 97,
    104, 109, 111, 120, 123, 128, 134,
    135, 136, 137, 138, 142, 146, 156
Culm Measures   1, 3, 5
Culm Valley   135, 136, 138, 142
Culver   21
Culver Tunnel   21
current lineations   20
Cutteridge Gate   12, 140
Cutton   136

Danes Hill Lane   56
Danes Wood   56
Darnaford   21
Dartmoor   33, 75, 112, 115, 119, 171
Dartmoor Granite   1, 3, 5, 7, 8, 11, 12,
    24, 25, 27, 29, 34, 35, 38, 72, 90, 92,
    94, 96, 97, 111, 112, 113–116, 120,
    121, 124, 125, 128, 138, 141, 143,
    146, 151, 171
Dawlish Sandstone   3, 30, 31, 32, 34, 35,
    38, 43, 50, 51–59, 74, 76, 77, 86, 87,
    104, 134, 136, 137, 138, 139, 141,
    142, 144, 153, 155, 158, 160, 171
Dawlish Sandstone Formation   33, 34,
    35, 154, 158, 159
debris flow   3, 34, 46, 61, 63, 65, 68, 75,
    76
Deep Lane   69, 113
delta   3
Deodara Glen   109
depositional environments   79, 91, 93
desert dunes   3, 34
desiccation cracks   49, 54, 79
Devensian   130, 132, 135, 137

Devon County Council   139
Devonian   1, 3, 5–9, 35, 40, 59, 61, 75,
    91, 120, 125, 126
    Middle   62, 75
    Upper   5, 7
    Carboniferous boundary   7
    Gedinnian   87
dewatering pipes   76
Digby Drive   139
digenite   149
Dilly Bridge   11
Dillybridge Brook   11
dimorphoceratid   18
Dinantian   5
diopside   9, 102, 117, 118
dissolution channels   105, 106
djurleite   149
Doddiscombsleigh   7, 8, 9, 10, 11, 12,
    14, 20, 24, 25, 26, 28, 124, 125, 126,
    128, 143, 149, 150, 171
Doddiscombsleigh church   149
Dog Village   104, 135, 138
dolerite   24, 27, 50, 108, 109, 117, 141,
    143, 150, 153, 170
    Ryecroft Sill   25
dolomite   40, 45, 96, 107, 108, 109, 117,
    144
Down Lane   8, 9, 26, 28
Downes   136
Downes Mill Farm   74, 138
dumortierite   118
dump   147, 148, 149
Dunchideock   1, 38, 33, 40, 46, 63, 96,
    98, 103, 104, 105, 106, 107, 125, 133,
    134, 141, 143, 169, 172
Dunchideock Barton   45, 46, 110
Dunchideock Bridge   104, 110
Dunchideock Church   110
dunes   76
Dunland   1
Dunsford   12, 13, 124, 126, 136, 146, 147
Dunsford Church   147
Dunsford Hill   44
Dunsford Road   155
Dunsmore   71, 107, 108
dysaerobic   5

earth flows   140, 151
East Bowley   23
East Coombe   23
East Ford   17
East Henstill   143
Eastacott   74, 98
Eastern Cotley Wood   21, 171
Easternhill Farm   20
Efford   70
electron probe analysis   105, 148
element analyses   112, 169
elephant   130
Ellerhayes   70, 71, 136
Elston   63
elvan   90, 112
engineering classification   153
engineering geology   150–156
English Channel   90
Environment Agency   158
equisetalean   32, 34, 92

evaporites   159
excavations   153
Exe   54, 70, 134, 142
Exe Breccia   30, 53, 77
Exe estuary   1, 3, 4, 130, 135, 137, 158, 172
Exe Group   30
Exe Vale Hospital   58, 172
Exe Valley   135, 136, 137, 138
Exeter   1, 3, 8, 13, 18, 31, 32, 34, 35, 38, 39, 40, 42, 43, 44, 45, 46, 49, 50, 51, 59, 61, 62, 63, 72, 76, 77, 96, 98, 104, 107, 109, 120, 126, 135, 136, 138, 141, 143, 145, 146, 156, 171, 172
  Airport   56, 87, 136
  Castle   109
Exeter Formation   51
Exeter Group   3, 29, 30, 31, 32, 33, 34, 35–76, 96, 111, 112, 113, 142, 158, 159, 160
  geochemistry   112
  hydrogeology   158, 159
  School   135
Exeter Hill Cross   172
Exeter School   135
Exeter Volcanic Rocks   1, 3, 4, 29, 31, 34, 35, 40, 44, 63, 65, 96–111, 112, 113, 120, 139, 143, 144, 150, 169, 172
Exeter–Barnstaple railway   74
Exminster   47, 50, 52, 53, 58, 128, 137, 138, 142, 146, 171
Exminster Breccia   30
Exminster Hill   47
Exminster Sandstone   30, 51
Exmouth   77
Exmouth Beds   30
Exmouth Formation   30
Exmouth Mudstone   83
Exmouth Mudstone and Sandstone   30, 31, 53, 77, 83, 84, 128, 139, 141, 144, 160, 172
Exton   58, 77, 135, 137
Exwick   11, 18, 135
Exwick Barton   11, 14, 18, 138, 139
Exwick Barton Cottage   18
Exwick Mill   19

facies associations   53
Fair Oak Farm   56
Famennian   5, 7
Farringdon   84
Farringdon Wood   84
Farringdon, Woodbury Salterton   83
fault   19, 20, 126, 128
  Permian   129
  reverse   19
  strike   19
  thrust   20
  Variscan   128
faults   76, 91, 113, 114, 126, 127, 128, 148
  Christow   147
  Dunsford   147
  Hill Barton   50, 129
  Honiton Road   129
  Mincinglake   128
  Pinhoe   50, 129
  Raddon Hills   59, 60, 128

Sticklepath   128
  Teign Valley   126, 128, 146, 147
  Woodah   128
feldspar porphyries   89
feldspar xenocrysts   96
feldspathic conglomerate   87
Feldspathic Traps   96
felsite   34
ferrohastingsite   9, 117
Ferry Road   58
fish   10, 32, 33, 34, 92, 93
Flandrian   130, 132, 137
flint   133, 134, 135, 136, 137, 146
floodplain   1, 134, 136, 138
floods   65
flow foliation   104, 105, 108
flow-banding   111
flowslides   140
fluid inclusions   148
flute cast   13, 19, 20, 21
fluvial channels   48, 76, 79
fluvial facies association   79
fluvial sheetfloods   54, 79
fly ash   138
fold   19, 21, 125, 127, 129, 171
  recumbent   19
Folly Farm   171
Folly Farm, Uton   108
footprints   32, 33, 43, 53
Forches Cross   60, 68, 69, 171
Ford Brook   16
Ford Dunsford   20
Ford Farm   149
Fordland Farm   21
Fordton Cross   68, 69, 72, 74
Fore Street   44
foreland basins   7
Fortescue Farm   54, 135
Forward Green   23
fossil ice-wedge polygons   134
fossil localities   6
fossil plant   13, 29, 43
foundations   153
Foxenholes   89
Frasnian   87
freirinite   149
Fremington   130
Fremington Till   130
Frogmire   67
frost creep   132
frost heaving   134
Fursdon   59, 61
Fursdon Barton   62
Fursdon House   61
Fursdon Lodge   61
Furzy Copse   87

Gahard formations   87
galena   147, 149
gamma-ray log   62
gangue   147
garnet   91
gastropod   13, 18
Gedinnian   87
gelifluction   132
geochemistry   3, 25, 28, 34, 38, 83, 96, 116

geophysical surveys   147
geotechnical data   150
giant ox   130
glaciation   3
glacioeustatic changes   5
gleyed brown earths   141, 142
goethite   92, 105
Goldwell   68
Gondwanan continent   5
Goosemoor   94
Gorran Quartzites   87
gossans   147
graded bedding   13, 22, 28
granite   34, 35, 50, 59, 74, 111, 112, 113–116, 118, 150, 153, 158
  aureole   25, 26, 28, 75, 146
  tors   1
granodiorite   114
granodiorite, Clampitt Down   114
gravel and sand   139
gravity surveys   4, 128
Great Fulford   19
Great Gutton   63, 68
Great Huish Farm   11
Great Knoll   110
Great Leigh   26
Great Pitt Farm   108
Green Lane   65, 171
Greenlands Plantation   108
Greenslinch   60, 61
Grès Armoricain   87
Grès de May   87
greywacke   19
Grindle Brook   84, 85
grit   89
groove casts   13, 15, 20, 21, 22
Ground magnetic profile   101, 102
Ground magnetic surveys   98
groundwater   158, 160
groundwater gley soils   142
Gunstone House   63
gymnosperm pollen   34
gypsiferous mudstones   79
gypsum   79, 84

Hackworthy   11
Hackworthy Corner   17
haematite   19, 77, 92, 96, 98, 102, 103, 104, 105, 107, 110, 114, 148
Haldon Gravels   119
Haldon Hills   1, 119, 133, 142, 143
Halscombe   40, 96, 104, 110
Halsfordwood   12, 14, 17
Hand and Pen   86, 133
Harehill Copse   27
Harehill Plantation   7, 8, 10, 24, 26, 28, 149, 150
Harehill Sill   8, 24, 26, 27
Harford   17
Harpford   87
Harpford Common   92, 94, 134
Hart's Lane, Whipton   50
Hartland Cross   108
hastingsitic   118
Haven Banks   44
Hawkerland   93, 141, 171
Hayes Farm   136, 138, 139

Hayne Barton   136
Hayne Farm   148
Hayne House   60, 71
Hazel Wood   109
Hazeldene   139
head   1, 3, 44, 62, 66, 67, 68, 75, 77, 84, 85, 86, 111, 119, 130, 132–134, 141, 142, 151, 153, 157
Heal-eye Stream   71
Heathfield   75
Heathfield Farm   139, 160
Heavitree   45, 47, 49, 50
Heavitree Breccia   30, 31, 33, 32, 34, 35, 38, 45, 47–50, 52, 53, 58, 72, 76, 98, 112, 129, 134, 138, 139, 142–144, 146, 153, 156, 171, 172
Heavitree Brewery   133
Heavitree Conglomerate   47
heavy minerals   61, 89, 90, 91, 92
Hedgemoor   114
Hele   10, 60, 61, 62, 142, 159
Hele Payne   62
Heltor   8, 114
Heltor Rock   1, 8, 113, 114, 115, 116, 171
hemisphaerica-dichotoma Zone   7
hexactinellid sponge spicules   18
Higher Ashton   111
Higher Barton   24
Higher Comberoy   59
Higher Comberoy Farm   62
Higher Comberoy Formation   29, 31, 32, 62, 141, 153, 155
Higher Coombe Farm   134
Higher Greendale   79, 84
Higher Lowton   117
Higher Lowton Down   7
Higher Metcombe   91
Higher Rew   133
Higher Roach Farm   23
Higher Weaver   136, 142
Higher Weaver Cross   136, 142
Hill Barton   47, 50, 51
Hill Barton Farm   139
Hill Copse   8, 22, 26
Hill Farm   21
hill wash   132
Hillhead   91
Hillside   94
hippopotamus   130
Holbrook Farm   146
Hollis Head   109
Holmead   96
Holocene   89
homoceratid   21
Honiton   130
Honiton Road   56
horn   133
hornblende   118
hornfels   7, 8, 11, 12, 26, 28, 35, 45, 46, 47, 49, 51, 56, 62, 63, 65, 68, 69, 70, 72, 73, 74, 75, 90, 91, 117, 134, 135, 150, 151
horsetail   19
Hospital Lane   51
Hoxnian   130
Hulk Lane   71
humic gley soils   142

Hundred Acre Copse   11
Huxham   52, 107, 109
Huxham Barton   149
hydraulic gradient   159
hydrogeology   4, 157–160
hydrothermal brecciation   7
hydrothermal chert breccias   9
hydrothermal kaolinite   119
Hyner Shale   1, 5, 7, 151, 153

Ice House   104
ice sheet   1, 3, 130, 137
ice-wedge casts   134
iddingsite   102, 103, 105, 107, 110, 111
Ide   40, 45, 46, 104, 107, 110, 139, 140
Ide Brake   21
Idestone   40, 46, 100, 101, 104, 110, 126
Idestone Hill   13
igneous clasts   111
igneous enclaves   114
igneous rock   65, 74, 75, 143, 159
ignimbrite   65, 111
illite   12, 77, 141, 151
   mixed-layer   77
ilmenite   25, 26, 102, 105, 106, 107
imbrication   63, 70, 74, 87, 91
induan   3
inertinite   16, 22
insect   32, 33, 34, 92
interdune   51, 53, 56, 76
interdune deposits   3, 34, 54
interglacial deposit   130
interstadials   130
intertidal mudflat   137
intraformational breccia   66, 74
intraformational conglomerate   92, 93, 144, 156
Ipswichian Stage   130
iron   110, 117, 158, 159
iron oxide   19, 27, 40, 61
iron-oxide   104, 105, 108, 110
iron pan   92, 141
iron pyrites   26, 27
Isca Dumnoniorum   144
isostatic uplift   29
isotope data for acid igneous rocks   113

Jackmoor Cross   136
Jaspilitic   9
joints   11, 22, 23, 74, 104, 108, 109, 110, 116
Jurassic   3, 124

K-feldspar   118
K/Ar radiometric age determination   50, 118
kaolinite   11, 77, 89, 92, 139, 151
Kazanian   32, 33
Kazanian–Tatarian   31
Kelland Brook   11, 17
Kenbury Wood   139
Kennford   45, 47, 50, 98, 138
Kennford Breccias   30
Kensham House   62, 136
Kensham Lodge   85
Kenson Hill   62

Killerton   33, 66, 98, 100, 101, 107, 111, 142
Killerton Gardens   104
Killerton Park   1, 60, 65, 67, 98, 101, 102, 104, 107, 109, 111, 128, 144, 145, 172
Killerton Stone   144, 145
Kiln Close Plantation   146
Kinderscoutian   11, 12, 14, 17, 19
   R_1 Zone   12, 14, 18, 19
   R_{1a} Zone   11, 14, 17, 18, 19, 22
   R_{1a} Zone?   13
   R_{1a1} Zone   15, 18
   R_{1b} Zone   13, 17, 19, 21
   R_{1b2} Zone   15, 18
   R_{1b3} Zone   18
   R_{1b}–R_1 Zone   18
   R_{1c} Zone   13, 17, 19, 21
   R_{1c1-3} Zone   15, 18
   R_{1c4} Zone   15, 18, 21
Kingsand   113
Kingsford   21
Kingston   92, 94
Kingswell   21
Knowle   63, 65, 67, 68, 72, 107, 108, 171
Knowle Hill   33, 96, 98, 104
Knowle Sandstone   29, 30, 31, 32, 33, 35, 38, 40, 45, 59, 62, 63–65, 67, 75, 102, 104, 107, 111, 113, 143, 153, 154, 171
Kungurian   33
kyanite   90

labradorite   27, 103, 105, 107, 117, 118
Ladinian   33
lag conglomerate   93
lake   79
   sabkha-playa   79
Lake Bridge   51, 56, 58, 171
Lake Cottage   20
Lake Farm   20, 134
Lake House   20
Lakeham   111
lamprophyre   29, 35, 50, 62, 96, 98, 101, 102–104, 111, 118, 170
   lava   65, 98, 155
   microsyenites   63, 102, 104
Land Farm   23
Land Quarry Plantation   22
Landevennec   87
landfill waste disposal sites   139
landslip   11, 139, 140, 151, 172
Langford   85, 135, 142, 149
Langford Court   85
langisite   149
Langsettian Stage   22
Langstone Breccias   30
Langstone Point and Exmouth Shrubbery breccias   30
Larkbeare   86, 91, 154
Larkbeare Avenue   86
Latchmoor Green   66, 136
latior Zone   7
lava   3, 40, 45, 46, 50, 52, 63, 65, 69, 70, 101, 110, 111, 141, 144, 150, 153
lava flows   3
Lawrence Castle   1, 45, 111, 119
lead   147, 149

Legars Upton   59
Leigh Cross   20
Leign Farm   12, 113
levees   79
Licensed abstraction data   158, 159
licensed landfill sites   160
lime   146
Limekiln Lane Cottages   50, 146
limonite   92
lineation   21
Little Clyst William   91
Little Houndbeare Farm   87, 134
Little Johns Cross Hill   155
Little Westcott   114, 116
Littleham Beds   30
Littleham Cove   79, 149, 150
Littleham Formation   30
Littleham Mudstone   30, 31, 77, 79, 83,
   84, 92, 93, 129, 139, 141, 150, 160, 172
Littleham Mudstone Formation   149
Littleham Mudstones   30
Livingshayes   62
Llandeilo   87
load casts   12, 13, 15, 17, 20
lode   146, 147, 149
loess   130
Longbrook Street   44, 109
Longdown   14, 16, 20, 21, 126, 134,
   140, 171
Lord Haldon Hotel   134
Lord's Meadow   67
Lower Ashton   8, 24, 26
Lower Byes Plantation   134
Lower Comberoy Farm   109
Lower Creedy   59, 63
Lower Creedy House   63
Lower Heltor   114, 116
Lower Linscombe   22, 143
Lower Marls   30, 77
Lower Sandstone   30, 51
Lower Wear   50
Lower Western Copse   11, 146
Lowley Wood   20
Lowton   7
Luccombs Farm   59
Lyalls   133
Lympstone Common   132
Lynch Plantation   23
Lyndhane   85

M5 Motorway   23, 44, 48, 50, 58, 85,
   137, 155
   cuttings   47, 171
made ground   44, 67, 68, 85, 138
Madison Avenue   46
magnetic anomalies   97, 98, 116, 120
   anomaly map   101
   survey   97, 150
magnetic basement   120, 123
magnetic properties   4
magnetite   25, 98, 102, 114, 117
malachite   149
manganese   9, 10, 110, 147, 148, 149
   mining   27, 148
manganiferous beds   8, 10
manganite   148, 149
Manor Farm   87, 91

Manstree Cross   46
marine band   5, 13, 15, 17, 18, 21
marine deposits   1, 3, 130, 131, 137,
   142, 153, 154
Mark's Farm   139, 140
marling   146
marlpits   139
Marsdenian   13, 15, 17, 19
   R_2 Zone   19
   R_{2a} Zone   13, 17
   R_{2a1} Zone   15, 18
   R_{2b} Zone   14, 15, 17, 21
   R_{2b1-2} Zone   15
   R_{2c} Zone   17
Marsh Green   86, 87
Marsh Green Hill   87
Marshall Farm   46
Marwood Beds   61
Matford   146
Matford Lane   50
Matford Park Farm   50
maucherite   149
Meadowend   104
megacrysts   112, 116
   layering   115, 116, 171
Melhuish Barton   20
Melton Court   87
Mercia Mudstone Group   29, 33
Merrivale Road   135
metadolerite   26, 117, 118
metalliferous mineralisation   83
metalliferous minerals   146
metamorphism, contact   26, 27
metaquartz-dolerite   26
metaquartzite   77, 87, 89, 90, 91, 92, 95,
   135, 141, 146
metasomatism   26
metatyuyamunite   149
mica   7, 12, 19, 23, 26, 28, 65, 74, 102,
   111, 114
microfloras   34
microgranite   47, 49, 70, 72, 73, 74
Middle Bolealler   85, 135, 142
Middle Heltor Farm   114, 116
Middle Hole   26
millet seed   40, 72
Mincinglake   139
Mincinglake Brook   171
Mincinglake Stream   19
Mincinglake Valley   138
Minehead   87
mines   147
   Anna Maria   147
   Birch   147
   Birch Aller   9
   Birch Allers   147
   Birch Ellers   146, 147
   Bridford   147
   Bridford Barytes   28, 146, 171
   Dunsford   147
   Harehill Plantation   149
   Lawrence   146, 147
   Pound Living   148, 149
   Scanniclift Copse   27, 149
minette   65, 66, 98, 102, 104, 107, 111,
   144, 145, 170
mining   147

miospores   31, 44, 67
Mistleigh Copse   7, 10, 14, 28, 149, 171
modderite   149
mollusc spat   18
molybdomenite   149
monazite   91, 114
Monkerton   50, 52, 58
Monkerton Farm   50
Monkerton Formation   30, 31, 32, 33,
   35, 38, 40, 45, 47, 50, 52, 128, 129,
   153, 155
Monkerton Member   50
montmorillonoid minerals   108
montroseite   149
Mooredge Cottages   52
Moorlake   69
Moorland   62
Mosshayne   56
murchisonite   34, 35, 38, 45, 47, 49, 50,
   54, 56, 68, 72, 73, 74, 75, 111, 112
muscovite   19, 92, 104, 114
Mutterton   85

Nadder Brook   12, 14, 17, 126
Nadderwater   17, 125
Nag's Head   23
Nag's Head Landslip   151
Namurian   3, 5, 13, 16, 17, 18, 31, 33
   faunas   15
nappe emplacement   5, 7
Neadon Farm   9
neodymium and samarium isotope   112
Neopardy   16, 62, 65, 72, 73, 97, 111,
   112, 113
nepheline   102
Nether Exe   56, 136, 137, 142
Netherexe   70, 71
New Barn Cross   148
New Red Sandstone   1, 3, 5, 10, 29, 30,
   31, 32, 33, 35, 38, 59, 68, 104, 121, 123
   unconformity   10
Newbuildings   22, 143
Newcourt Barton   137
Newhall Farm   56
Newton Abbot   137
Newton House   74
Newton Poppleford   94
Newton St Cyres   11, 17, 72, 73, 74, 126,
   129, 133, 134, 138, 142, 144, 146,
   147, 148, 156, 171
Newton St Cyres Breccia   29, 31, 32, 33,
   34, 35, 38, 54, 68, 72, 73, 74, 75, 76,
   98, 111, 112, 139, 144, 148, 153, 156,
   171, 172
Newton Wood   17
Newtown   86, 146
niccolite   149
nickel   79, 149
nitrate   159
nodule   12, 18, 19, 83, 84, 150
   chemistry   79
Norman's Green   133
North Creedy   22
North Wood   21
Northbrook School   40, 44
Northcotts Farm   86
Northridge   11

Northridge Copse 11
Norton Farm 136

Old Dawlish Road 59
Old Heazille Farm 109
Old Wheatley 172
Oldoven Cross 136
Olenekian 33
olivine 96, 102, 103, 104, 107, 108, 110, 118
olivine basalt 63, 103, 105, 107, 108, 109, 171, 172
olivine-dolerite 107, 108, 110
Onaway 58
Orchard Copse 110
Ordovician 87
ores 147, 149
oriented stones 134
orthocone 13, 17, 18, 33
Osbornes Farm 12, 172
ostracod 5, 7, 10
Otter Sandstone 3, 30, 31, 32, 33, 34, 77, 87, 89, 90, 91, 92–94, 95, 128, 134, 139, 141, 142, 144, 153, 156, 157, 158, 159, 160, 171
Otter Valley 137, 159, 160
Otterton Trough 124
overconsolidated 150, 151

Paceycombe Farm 62
Palaeocene 119
palaeocurrent 65, 66, 72, 88, 90, 91, 94
Palaeogene 1, 3, 119, 128
Palaeogene flint gravels 119
palaeomagnetism 4, 96
palaeosol 5, 94
Palaeozoic ridge 90
palynology sample sites 36
palynology samples 32
palynomorphs 3, 15, 22, 31, 45, 68
Paradise Copse 59
parallel lamination 13, 21, 66, 67, 68, 70, 79
parallel-bedding 74
Paris Street 44
Park Farm 62
Park Wood 67
Parsonage House 144
Pathfinder Village 11
patterned ground 134
Peamore House 134, 138
peat 3, 130, 131, 137, 138, 154, 157
Pebble Beds 30
  sedimentary features of 87
pegmatite 116
pegmatitic aplogranite 115
pelycosaurian reptile 32
pencil shales 11
Pendleian 8, 17
  E$_{1a}$ Zone 8, 10, 17, 18
  E$_{1b}$ Zone 17
  E$_{1c}$ Zone 8, 10, 17
peneplanation 34
Penhill Farm 40
Penn Hill 111
Penstone 60

Peradon Farm 86
periglacial 1, 130, 134
  climate 3, 132, 134
Perkin's Village 84
permafrost 134
Permian 1, 3, 17, 22, 29–34, 38, 44, 47, 50, 52, 61, 63, 66, 67, 68, 69, 71, 72, 74, 76, 77, 96, 98, 101, 104, 107, 111, 119, 120, 122, 123, 127, 128, 132, 133, 137, 141, 147, 148
  Lower 113
  palynomorphs 33
Permo-Triassic 1, 3, 120, 124, 132, 133, 134, 139, 142, 146
  boundary 32, 77
Perridge 21, 171
Perridge Tunnel 14, 21
petalite-allanite-pistacite 117
petrogenesis 3, 96
petrography 25, 27, 89
phacopid trilobite pygidium 61
phlogopite 118
photographs 171
photomicrographs 103
phreatophytic plants 34
pillow lavas 25, 27
Pilton Beds 61
Pilton Formation 61
Pin Brook 19, 50
Pinhoe 19, 40, 44, 50, 52, 109, 128, 134, 146, 172
Pinhoe Brickpit 13, 14, 15, 19, 171, 172
Pinhoe Road 44
Pinn Lane 58
pistacite 27, 117
Pitt Hill 73, 112, 171
Pitt Hill, Crediton 156
Pixie Rock 114, 116
plane-bedded 44
plant remains 11, 12, 13, 19, 23, 32, 33, 34, 83, 92
plate subduction 5
playa mudflat facies association 79
Pleistocene 1, 89, 119, 130
Plymtree 86, 87, 133
Pocombe 40, 45, 96, 107, 109, 145
Pocombe Bridge 14
Pocombe Hill 109
Pocombe Stone 144, 145
Pocombe-type lava 110
podzols 141
polianite 148
Polished Stone Values (PSV) 143
pollen 29, 31, 33, 43, 45, 68, 69
pollen analysis 137
Polsloe 135
Polsloe Priory 135
Polsloe Road 135
Poltimore 32, 43, 52, 53, 56, 101, 107, 109
Poltimore Bridge 155
Poltimore Mudstone 30, 31, 51, 52, 153, 154, 155
Pond Wood 46
Pook's Cottages 26
Poole Bere Cottage 26
Poole Farm 70

Poole Grove 26
Pophams Farm 92, 94
popples 145, 146
porphyritic felsite 47
porphyritic granite 47
porphyroblasts 8
porphyry 40, 46, 61, 63, 70, 71, 87, 89, 90, 133, 134
Posbury 59, 63, 96, 107, 108, 145
Posbury Chapel 111
Posbury Clump 63, 108, 145
Posbury Stone 145, 146
Posidonia beds 5, 8, 10, 11, 149
potassium-argon method 40
Pottshayes 62
Pottshayes Farm 62
Pound Lane, Bridford 26
Poundsland 71, 136
prehnite 27, 104
Priorton Hill 22
procolophonid 94
procolophonids 32, 34, 92
prod marks 15, 20
productoid debris 61
pseudomorphs 7, 8, 26
psilomelane 148
public water supply 158, 159, 160
pumice 28, 108
Punch Bowl Tip 68, 72
Punchbowl 139, 160
Pynes Water Works 11, 12, 14, 135
pyrite 12, 117, 147
pyrolusite 104, 148, 149
pyroxene 25, 26, 27, 102, 105, 107, 110, 117, 118
pyrrhotite 120

Quantock Hills 61
quarries 9, 22
  Beggars Roost 88, 89, 90, 91, 92, 94, 139, 142, 171
  Bishop's Court 53, 54, 58, 142, 155
  Blackhill 134, 142, 157
  Blackingstone 25, 112, 114
  Bridford 118
  Bridford Barytes Mine 9, 28, 145, 171, 172
  Budlake 107, 109, 139
  Chilton 22
  Columbjohn Wood 169
  Copse 27
  Crockham 143
  Crossmead 107
  Dowrich 171
  East Henstill 15, 22, 171
  Foxenholes 142
  Haytor 98
  Heavitree 49, 171
  Hillhead 88, 92
  Killerton Park 103, 169
  Knowle 110, 169, 172
  Knowle Hill 63, 102, 103, 104, 169
  Land 143
  Lower Linscombe 171
  Lower Sutton 139
  Matford Park 47
  Meadowend 169

Merrivale   114
Pocombe   42, 103, 106, 109, 169, 171, 172
Posbury   169
Posbury Clump   107, 108
Raddon   66, 103, 107, 108, 169
Rockbeare   142
Rockbeare Hill   91, 139
Ryecroft   25
Scatter Rock   118
School Wood   40, 103, 104, 105, 106, 143, 172
Shillingford Wood   171
Spencecombe   169
Stone Copse   7, 8, 9, 27, 28, 117, 118, 143, 171
Tower Wood   119
Uton   108, 145, 169
Venn Ottery   91, 134, 142, 172
Westcott   114
West Town   104, 110, 169
Whiteball   142
Quarry Gardens   50
Quarrying   143
quartz
  ß-quartz   28
  veins   10, 20, 26
quartz diorite   114
quartz-arenite   40
quartz-biotite   117
quartz-feldspar porphyry   45, 46, 50
quartz-feldspar-biotite porphyry   46
quartz-keratophyre magma   28
quartz-keratophyre vitric-flow tuff   25
quartz-leucodolerite   24, 25
quartz-porphyry   29, 35, 44, 45, 46, 49, 50, 51, 59, 60, 62, 63, 65, 68, 69, 70, 72, 74, 75, 89, 90, 96, 98, 111, 113, 133, 156
quartz-porphyry clast   112
quartz-wacke   19, 20
quartzarenites   89
quartzite   91, 134, 135, 136, 137, 138, 145, 157
Quaternary   1, 3, 10, 42, 130–140, 142, 160

Raddon   65, 66, 70, 107, 108
Raddon Court   66
Raddon Cross   68, 69, 70, 113
Raddon Hill Farm   60, 61, 156, 172
Raddon Hills   1, 59, 60, 61, 132, 172
Raddon lava   66, 108
Raddon Stone   145
radioactive nodules   79, 149
radiocarbon dates   137
radiolaria   5, 8, 9, 28, 117
radiometric age   4, 33, 96, 98, 112
rammelsbergite   149
Ramridge Plantation   21
Ramspit   22, 143
rare earth analyses   34
rare earth and isotopic signatures   75
ratio plot, elements   38
Ratsloe   149
rauisuchians   32, 34, 92, 94
Rb/Sr   34

Reading Beds   119
reclaimed marine deposits   137
Red Cross   108
red deer   130
Red Hill Cross   69
Redhills   18, 19
reduction spots   50, 75, 83, 84, 85, 86
reg type palaeosol   88
regolith   10, 59, 114, 132, 133, 172
remanent magnetisation   98, 120
reptile   32, 34, 92
reptilian footprints   56
residual aeromagnetic anomaly map   99, 100
residual flint gravels   3
residual gravity anomaly profiles   122
resistivity   147
resistivity surveys   87, 129
reticuloceratid   18
Rewe   52, 54, 56, 57, 129, 135, 142
Rewe Cross   75
Rheno-Hercynian tectonic zone   5
Rhode Farm   62, 171
rhodochrosite   148, 149
rhynchosaurs   34, 94
rhyolite   34, 35, 47, 50, 51, 56, 60, 62, 68, 69, 73, 74, 75, 90, 96, 97, 111, 112, 113
rift-valley   34
Rill Farm   83
ripple   63
ripple cross-lamination   65, 72, 79
ripple marks   56
Rivenford Lane   75
River Exe floodplain   135
river floodplain   79, 137
river terrace deposits   66, 74, 94, 130, 133, 134, 141, 142, 153, 155, 157, 158, 160
river terraces   1, 3, 4, 131
rivers   1
  Clyst   1, 56, 133, 137, 142, 157
  Creedy   1, 63, 67, 68, 136, 137, 138, 149
  Culm   1, 85, 135, 136, 137, 142, 157
  Exe   1, 4, 19, 23, 44, 47, 71, 74, 75, 119, 137, 138, 141, 142, 157
  Kenn   104, 133
  Nether Exe   136
  Otter   1, 137, 156, 157
  Teign   1, 20, 28, 117, 119, 136, 137, 147, 157
  Weaver   135, 138
  Yeo   1, 137, 138, 157, 160
Riverside   109
Roach   23
Roach Copse   19, 142
roadstone   143
Rockbeare   86, 87, 133, 134, 135, 142, 171
Rockbeare Hill   87, 92, 94
Rockbeare House   133
Rockclose Plantation   110
Rode Moors   23
Romans   144
root system traces   34
Rougemont   40, 144
Rougemont Castle   109
Rowdon Rock   114, 116

Rowhorne Road   17
Royal Albert Memorial Museum and Art Gallery, Exeter   43
Rudway Barton Exploration Borehole   70, 71
rutile   26, 61, 91
Ruxford Barton   65
Ryecroft Dolerite   25
Ryecroft Sill   8, 24, 25, 26

sabkha-playa lake   3, 34
safflorite   149
Sakmarian   33
saltmarsh   137
Samarium/neodymium   113
samples, chemical analyses of   83
sand and gravel   87, 142
sand-filled pipes   49, 50, 72
Sandford   22, 59, 63, 64, 65, 128, 172
Sandford Road   69
sandstone pipes   87
sandy braidplain   3, 34
sanidine   72, 112
satin spar   84
Scanniclift Copse   8, 25, 26, 27, 28, 128, 149
Scanniclift Copse Spilite   25, 28
Scanniclift Thrust   27, 126
scapolite   116
School Lane   172
School Wood   40, 110, 111
schorl   119
schorlite   118
Scott's Pollard   91
scours   79
Scrawthorn Plantation   147
Scythian   34, 87
sedimentary features   35, 47, 53, 68, 72
seismic activity   76
seismic surveys   137
Senonian   119
sericite   12, 25, 28, 117, 118
Seven Acre Lane   26, 118
Seven Stone Ball   17
shafts   147
Shale Member   141
sheet-flooding   3, 34
sheetflood   44, 48, 51, 54, 63, 67, 68, 72, 74, 75, 7644, 68
Sheldon   126
sheridanite   27
Shermoor Farm   136
Sherwood   133
Sherwood Sandstone Group   29, 30, 31, 33, 77, 78, 87
Shilhay   31, 42, 43, 44
Shillingford Abbot   45, 46, 48, 138
Shillingford Plantation   48
Shillingford St George   45, 46, 50, 156, 171, 172
Shillingford Wood   50, 156
Shippen   10
Shippen Brook   11
Shobrooke   68, 69, 143, 172
Shobrooke Cross   70
Shobrooke House   69
Shobrooke Lake   136

Shobrooke Mill Farm   70
Shuffshayes Farm   135
Shute   70, 74, 75, 113
Shute Cross   75
Shute Sandstone   29, 31, 32, 33, 35, 38, 52, 54, 71, 72, 74, 75, 76, 141, 153, 155, 159
Shutelake   136, 142
Shuttern Brook   148
siderite   12, 19, 147
Sidwell Street   44
Siegenian   87
Silesian   5
silicification   7, 8
sill   7, 24, 26, 27, 143, 149
   albite-dolerite   3, 26, 27, 28
   dolerite   26
sillimanite   90
silver   149
Silverton   23, 29, 31, 38, 59, 60, 61, 62, 65, 66, 68, 70, 71, 107, 108, 136, 142
Silverton Mill   138
Silverton, basalt   71
skarn   98
skutterudite   149
slate   35, 46, 47, 50, 51, 60, 62, 63, 65, 68, 69, 70, 73, 74
slide marks   15
slope stability   151, 153, 157
slopewash   133
slump structure   20
Sm/Nd isotope studies   34
Small Brook   74, 148
smectite   96
soil creep   132, 133
Soil Survey   141
soils   1, 34, 141
   development   132
   Redland   1
sole marks   13, 15, 19, 20, 21, 22, 171
solifluction   3, 130, 132, 133, 157
Somerset   61, 87
Sousson's Wood   112
South Park Copse   11
South Wales   130
South Whimple Farm   56
South Wood   26
Southbrook School   135
Southern area   83
Southfield Farm   91
Sowton   53, 58
Sowton Brook   147
Sowton Lane   58
Spence Combe   68, 108
Spencecombe   63, 107
Spencecombe Lava   171
sphalerite   147, 149
sphene   27, 117, 118
spilite   27
spilitic basalt   8, 25, 27, 28, 143, 149, 171
spinel   117
spirale beds   10
sponge spicules   19
spores   31, 32, 40, 63
Springdale   139, 171, 172
Springdale landfill site   21, 132
springs   159

Sprydoncote   107, 109, 111
Sprydoncote House   109
St Andrew's Road   18
St Cyres Beds   29, 72
St Cyres Breccias   72
St James's Park Halt   155
St Leonard's   135
St Saviour's Way   68
St Saviours Way, Crediton   67
St Swithin's Church   144
St Thomas   135, 136, 155
Stafford Bridge   138
stages, British Quaternary   130
Start Point   90
Station Cross   69
staurolite   61, 90, 91, 93
stems   19
Stephanian   3, 31, 33, 34, 61
Steps Bridge   20, 117, 136
stereograms   125
Stockeydown Farm   73
Stockleigh Pomeroy   134
Stockwell   60, 61, 62, 108, 109
stockwork   7
   ferruginous   7
   manganiferous   7
Stoke Canon   52, 135, 136, 138, 141, 142, 159
Stoke Canon Paper Mill   136
Stoke Hill   19, 139, 160
Stoke Road   13, 17
Stoke Woods   17
Stone   7
Stoneyford   93, 94
Stoneyford Farm   171
Stowford   95
Straight Point Sandstone   30, 77, 79, 84
Straitgate Farm   91
Strete Farm   86, 133
Strete Ralegh   86, 133, 139, 146
Strete Ralegh Farm   86
structure   4, 120
   Devonian and Carboniferous   125
   Permo-Triassic   128
Stumpy Cross   108
Sturridge   143
Sturridge Wood   139
subarkoses   77, 89, 92
submarine fan   3
Sudetic Basin   112
sulphate   159
sulphide   20, 147, 149
Summer's Lane   111, 113
Sweetham   74
syncline   21, 125, 126, 129
   Brampford Speke   129
   Bridford   7, 8, 9, 24, 26
   Christow Common   9, 124
   Doddiscombsleigh   124
   Mistleigh Copse   124
   West Efford   70, 75, 129
taeniate   67, 72
Talaton   86, 90
Tamar Valley   146
Tatarian   32, 33
Tedbridge   23, 59
Tedbridge Copse   23

Tedburn St Mary   10, 11, 12, 14, 125, 141
Teign   149
Teign Chert   5, 7, 8, 9, 10, 11, 14, 17, 18, 116, 117, 128, 141, 145, 146, 147, 149, 151, 153, 171, 172
Teign Valley   20, 24, 28, 125, 141, 143, 146, 147, 149
Teignhead Group   30
Teignmouth Breccias   30
Terley   133
Terraces   135
   Eighth   135
   Fifth   135
   First   136
   Fourth   135, 136, 142
   Second   133, 135, 136
   Seventh   135
   Sixth   135
   Third   135, 136
   undifferentiated   136
terrestrial tetrapod fauna   34
Tertiary   133
tetrahedrite   147
thaw slumping   132
The Danny   172
The Quarries   109
The Quay   45, 46, 51, 144
The Wobbly Wheel   50
thermal metamorphism   26
thermal relaxation   34
thomsonite   27
Thorverton   59, 61, 65, 66, 70, 107, 108, 135, 136, 137, 142, 145, 172
Thorverton Sandstone   29, 31, 32, 33, 35, 38, 65, 66, 68, 70, 71, 98, 102, 107, 108, 109, 128, 141, 153, 155, 158
Threshers   68
thrust   5, 123, 126
Tick Lane   12
tidal channels   137
till   130
tin mineralisation   29
Tinpit Hill   147
titanaugite   25, 27
titanium minerals   91
titanomagnetite   102
Tiverton   61
Tiverton Trough   120
topaz   91
Topsham   1, 53, 58, 77, 135, 136, 137
Topsham Road   49, 172
tor   115, 116
Torbay   31
tourmaline   9, 61, 72, 91, 111, 114, 116, 117
tourmaline breccia   69
tourmaline granite   47
tourmalinised rocks   35, 50, 59, 60, 72, 75, 90
Tourmalinite   35, 49, 60, 68, 69, 73, 74, 87, 89, 95
Tournaisian   5, 7, 22
Tower Wood Gravel   119
Town Barton   12, 65
Town Farm   85
trace fossils   31, 32, 34, 36, 53
trace-element analyses   169

trachytic textures 25
translatent lamination 53
translational slides 140, 151
transmissivity 158, 159
Treasbeare Farm 87, 133
Trebetherick Point 130
tremolite 9, 27, 106, 117, 118
Triassic 1, 3, 29, 31, 33, 34, 77–93
trilete spore 44
trilobite 5, 10, 18
Trobridge House 72
Trusham 143
Trusham Shale 1, 5, 7, 141, 151, 153
tuff 3, 8, 9, 10, 11, 24, 25, 28, 50, 68, 70, 72, 74, 96, 117
turbidites 5, 11, 22, 126
turbiditic sandstone 3, 10, 12, 14, 21, 22, 143
    geochemical studies 10
turbidity currents 13
Turf 137
Turkey Lane 87
Tye Farm 85, 138, 142

Uffculme 88, 91, 92
Ufimian 33
University 138
Up Exe 136, 142
Upham Farm 84
Uphams Plantation 92
Upper Albian 119
Upper Beardon 27
Upper Chalk 119
Upper Greensand 3, 119, 124, 133, 143
Upper Permian 112
Upper Sandstone 30
Upper Sandstones 30
Uppincott 68, 69, 70
Upton Farm 59
Upton Formation 29, 31, 32, 59, 151, 153, 158
Upton Hellions 22
Upton Pyne 74, 148, 149
Upton Pyne Church 172
uranium 79, 83, 113, 149
Uton 63, 65, 107, 108, 145, 159, 171

valley head 132, 133, 134
vanadian mica 149
vanadium 79, 149, 150
Variscan 5, 120, 128, 148
Variscan and post-Variscan events 33
Variscan Orogeny 3, 5, 22, 32, 34, 120, 125, 128, 147
Variscan thrust 123, 128
vein quartz 47, 49, 50, 59, 61, 62, 63, 65, 66, 68, 69, 70, 72, 73, 74, 75, 87, 89, 90, 91, 95, 134, 135, 136, 137, 138, 142
Venn Ottery 89, 90, 142
Venn Ottery Common 85, 91, 141
Venny Tedburn 11, 17
vent breccia 35, 68
ventifact bed 88, 89, 91, 92, 94, 171
ventifacts 88, 89, 92
vertebrate 32, 33, 34, 93
    (reptilian) tracks 34

vertebrate fauna 32
vesicles 102, 104, 108, 109
vesicular basalt 40
vesicular lava 45, 46, 145
Viséan 5, 17
vitrinite 22
volcanic glass 11
volcanic wackes 25
volcanoes 3
vugs 10, 27

wad 8, 149
Wadebridge 147
Warren Gutter Shale 33
Washbeerhayes 62
Washbeerhayes Farm 19
Waste disposal at landfill sites 160
water supply 92, 157–159
Waterleat House 71
weathering 10, 153
Weaver Bridge 135, 142
Webbers 146
Webberton Cross 45, 110, 141
Webberton Wood 40
Webby's Farm 134
Weeke Barton 20, 118
Wellington 142
wells 158, 159, 160
Wessex Basin 124
West Clyst 101, 107, 109
West Clyst Farm 109
West Efford Lane 70
West Hill 87, 142, 171
West Raddon 65, 68, 70, 134
West Raddon Farm 70
West Sandford 22, 63, 64, 65, 67, 68, 102, 104, 108, 133, 171
West Sandford Barton 65, 104
West Town 12, 45, 110, 148
West Town Farm 40, 45, 46, 107, 110, 139, 140
West Town Farm (Ide) 40
West Town, Ide 169
West Wheatley 21
West Wheatley Farm 139
Westacott 114
Westcott 23, 114, 115, 116, 138, 142
Westcott Wood 116
Western Way 143
Westleigh Limestone 8, 19, 59
Westphalian 3, 5, 15, 17, 22, 32, 33, 125
Westwater 11
Westwood 77, 85, 86
Wheal Anna Maria 147
Wheatley 22
Whiddon Farm 46
Whimple 77, 86, 133
Whimple Wood 86
Whipton 40, 44, 46, 50, 51, 128
Whipton Formation 30, 31, 32, 33, 34, 35, 38, 40, 41, 42, 43, 44, 45, 50, 67, 77, 107, 109, 129, 135, 146, 153, 155
Whitestone 1, 11, 12, 17, 125, 132, 139, 140, 171, 172
Whitestone Cross 17
Whitestone Wood 17

whole rock K/Ar method 104
whole-rock analyses 25, 96
wind ripples 54
wind-faceted pebbles 3, 34
wind-rippled sands 79
Windhill Gate 26
Windmill Hill 79, 84, 171, 172
Windout 20
Winscott Barton 136
Winslakefoot 11
Winstode 16
Wishford 102
Wishford Farm 101, 104
Withnoe 34, 90, 113
Wolstonian 130, 137
wood 7
Woodah Farm 8, 9, 25, 28, 128, 147
Woodbury 77, 83, 84, 85, 128, 141, 142, 144, 145, 159, 172
Woodbury Beacon 134
Woodbury Castle 92, 172
Woodbury Church 172
Woodbury Common 89, 91, 92, 129
Woodbury Salterton 84, 85, 172
Woodbury, the 146
Woodlands 9, 11
Woodley Farm 11, 12
Woodrow 135
Woodrow Barton 135
Woodwater Lane 49
Woolcombes Plantation 94, 134
Woolsgrove 62
Worked ground 138, 139
Wyke 72, 74
Wyke Cross 75

X-ray diffraction analysis 96
xenocrysts 104, 105, 106, 110, 111
xenolith 108
xenoliths 104

Yard Downs 172
Yeadonian 17
Yellowford 70, 71, 74, 75, 135, 172
Yellowford Formation 29, 31, 32, 35, 38, 68, 70, 71, 74, 75, 76, 141, 151, 153
Yendacott 75
Yendacott Copse 68, 70, 75
Yendacott Lane 75
Yendacott Manor 70
Yeo 136
Yeo's Farm 46
Yonder Down 71

Zechstein, Late Permian 31
zeolite 27, 96, 104, 107, 108, 110
zinc 147, 149
zinnwaldite 28
zircon 28, 29, 32, 35, 45, 61, 85, 86, 91, 96, 102, 114
zoisite 117, 119, 120, 126, 130, 141, 172

BRITISH GEOLOGICAL SURVEY

Keyworth, Nottingham NG12 5GG
0115 936 3100

Murchison House, West Mains Road, Edinburgh EH9 3LA
0131 667 1000

London Information Office, Natural History Museum
Earth Galleries, Exhibition Road, London SW7 2DE
020 7589 4090

The full range of Survey publications is available through the
Sales Desks at Keyworth and at Murchison House, Edinburgh,
and in the BGS London Information Office in the Natural
History Museum (Earth Galleries). The adjacent bookshop
stocks the more popular books for sale over the counter. Most
BGS books and reports can be bought from The Stationery
Office and through Stationery Office agents and retailers.
Maps are listed in the BGS Map Catalogue, and can be bought
together with books and reports through BGS-approved
stockists and agents as well as direct from BGS.

*The British Geological Survey carries out the geological survey of Great
Britain and Northern Ireland (the latter as an agency service for the
government of Northern Ireland), and of the surrounding continental
shelf, as well as its basic research projects. It also undertakes
programmes of British technical aid in geology in developing countries
as arranged by the Department for International Development and
other agencies.*

*The British Geological Survey is a component body of the Natural
Environment Research Council.*

Published by The Stationery Office and available from:

The Publications Centre
(mail, telephone and fax orders only)
PO Box 276, London SW8 5DT
General enquiries 0171 873 0011
Telephone orders 0171 873 9090
Fax orders 0171 873 8200

The Stationery Office Bookshops
123 Kingsway, London WC2B 6PQ
0171 242 6393 Fax 0171 242 6394
68–69 Bull Street, Birmingham B4 6AD
0121 236 9696 Fax 0121 236 9699
33 Wine Street, Bristol BS1 2BQ
0117 9264306 Fax 0117 9294515
9–21 Princess Street, Manchester M60 8AS
0161 834 7201 Fax 0161 833 0634
16 Arthur Street, Belfast BT1 4GD
01232 238451 Fax 01232 235401
The Stationery Office Oriel Bookshop
The Friary, Cardiff CF1 4AA
01222 395548 Fax 01222 384347
71 Lothian Road, Edinburgh EH3 9AZ
0131 228 4181 Fax 0131 622 7017

The Stationery Office's Accredited Agents
(see Yellow Pages)

and through good booksellers